# The Urban Book Series

The Urban Book Series is a resource for urban studies and geography research worldwide. It provides a unique and innovative resource for the latest developments in the field, nurturing a comprehensive and encompassing publication venue for urban studies, urban geography, planning and regional development.

The series publishes peer-reviewed volumes related to urbanization, sustainability, urban environments, sustainable urbanism, governance, globalization, urban and sustainable development, spatial and area studies, urban management, transport systems, urban infrastructure, urban dynamics, green cities and urban landscapes. It also invites research which documents urbanization processes and urban dynamics on a national, regional and local level, welcoming case studies, as well as comparative and applied research.

The series will appeal to urbanists, geographers, planners, engineers, architects, policy makers, and to all of those interested in a wide-ranging overview of contemporary urban studies and innovations in the field. It accepts monographs, edited volumes and textbooks.

**Indexed by Scopus.**

Xinyu Liu

# Fabricating Nature

## Shenzhen's Urban Parks as Contested Topography

Springer

Xinyu Liu
School of Forestry
Jiangxi Agricultrual University
Nanchang, Jiangxi, China

ISSN 2365-757X ISSN 2365-7588 (electronic)
The Urban Book Series
ISBN 978-981-95-8207-5 ISBN 978-981-95-8208-2 (eBook)
https://doi.org/10.1007/978-981-95-8208-2



This Springer imprint is published by the registered company Springer Nature Singapore Pte Ltd.
The registered company address is: 152 Beach Road, #21-01/04 Gateway East, Singapore 189721, Singapore

**Competing Interests** The author has no competing interests to declare that are relevant to the content of this manuscript.

# About This Book

The division between society and nature in theoretical discourse, as well as the disconnection between landscape history and the history of landscape architecture in historical research, has constrained academia from conducting an in-depth examination of the production of nature in the contemporary urban context. Addressing this issue, this study chooses urban parks in Shenzhen, Guangdong Province, as the research subject, investigating the development history of urban parks from the establishment of Shenzhen City as a reform and opening-up experimental field of mainland China in 1980 to 2020. As an expression of "nature" in the urban environment, how urban parks emerge amidst the large-scale socio-natural transformation in Shenzhen is the main focus of this study. Accordingly, the study seeks to respond to the shortcomings at both the theoretical discourse and historical research levels.

The study establishes a discourse framework through "topography" and uses this as a departing point to examine nature, starting with the experientially visible "physical topography" and then understanding the "topography of life" interwoven in urban social life. Therefore, the study places the development history of urban parks within two intertwined contexts for examination: (1) the political-economic context of Shenzhen since the late 1970s; (2) the geographic-topographic context of Shenzhen within the Pearl River Delta region. This approach diverges from the common art-historical paradigm that revolves around form, style, and other aspects, maintaining a corresponding distance, thereby constituting the significance and innovation of this study. Theoretically, the study attempts to break through the entrenched dichotomy between society and nature in the built environment field and to challenge the logic of the autonomy of urban park forms; empirically, the study endeavors to address the shortcomings in contemporary Chinese urban park research and Shenzhen urban history research.

Based on the aforementioned, the main body of this research will be conducted at two spatial levels: the city scope and the Shenzhen Special Economic Zone (SSEZ) scope. Firstly, the research outlines how the urban landscape of Shenzhen emerges within the political-economic context of globalization and the geographic-topographic context of the Pearl River Delta. On this basis, it observes and explains the basic characteristics and reasons for the spatio-temporal distribution of urban parks within the city scope of Shenzhen. Then, it conducts a detailed study of urban parks built under three typical topographic conditions within the SSEZ. The research method, which intertwines political-economic and geographic-topographic contexts, elevates the study from a simple description of "physical topography" to an in-depth understanding of "the topography of life." Urban parks represent one of the few types of urban spaces that can retain a natural appearance during Shenzhen's rapid urbanization. These parks have transitioned from previously rural collective-owned agricultural lands to spaces provided by local governments for public leisure, urban beautification, and industrial promotion. The study finds that in this process, "nature" has undergone dual transformations in both form and meaning. This transformation of symbolic nature is linked to numerous important issues, including urban marketing, environmental protection, public security management, population resettlement, geopolitics, nation-state dynamics, and global-local relations.

Finally, the research discusses the roles of the state and government in producing nature and the complexities of producing nature within the contemporary urban context. The main conclusions derived from this study are as follows: (1) Urban parks represent a cultural form of both space and nature; (2) There exists a bidirectional shaping between cities and urban parks; (3) Topography can integrate cities and urban parks on a conceptual level.

# Contents

# Chapter 1
# Introduction

## Contents



**Abstract** This chapter begins by examining the binary opposition between society and nature in the built environment and traces its intellectual origins. It then explains why Shenzhen was selected as the case study. Following a critical reconsideration of the concept of urban parks, the author proposes to analyze Shenzhen's urban parks through an integrated political-economic and geographical-topographical framework—an approach that distinguishes this research from traditional art historical paradigms.

**Keywords** Socio-nature · Urban park · Theoretical origin · Research framework

Shenzhen holds extraordinary significance in the trajectory of China's urban construction and development trajectory following the reform and opening up. From its humble beginnings as a small town on the southern border to becoming a crucial


X. Liu, *Fabricating Nature*, The Urban Book Series,
https://doi.org/10.1007/978-981-95-8208-2_1

innovative node in the global urban network today, Shenzhen's urban construction and development undoubtedly represent localized practices within the context of globalization since the 1970s and, to some extent, epitomize the political, economic, and social changes in contemporary Chinese cities. In this transformation, Shenzhen's modern urban landscape rapidly overlays the agricultural landscape of Bao'an County (寶安縣).[1] On the plains, mountains, and coastlines once dominated by agricultural industries, a diverse array of residential areas, skyscrapers, expressways, and urban parks has emerged, interwoven with bustling urban life. To a considerable degree, Shenzhen's urbanization can be interpreted as a large-scale socio-natural evolution.

The urban parks in Shenzhen are not independent of this process, even if they exhibit a natural appearance; instead, they share the same social and historical context with other types of urban landscapes. A detailed examination of the interaction between this context and urban parks may allow us to glimpse the historicity of nature through the urban environment. Consequently, historical research on urban parks will not only focus on the history of their planning, design, and construction but also on the history of the relationship between nature and urban social life.

## 1.1 Research Questions and Rationales

The history of Shenzhen's urban parks is a collection of numerous fragmented or coherent events, figures, expressions, words, and images. It is certainly significant, but the vast amount of information, like an overwhelming sea of smoke, may lead to differing opinions in historical research. Therefore, the selection and organization of historical materials highlight their importance, and the method of selection and organization is evidently related to theory.[2] For historical research, the perspective directly influences the approach, method, and even the conclusions of the study. Thus, it becomes particularly important to clarify the theoretical perspective of the research before delving into the study.

Generally speaking, theory refers to universal discourse, which is a generalization of limited experiences, serving as the framework for our understanding of things and being abstract. Empirical study, on the other hand, is concrete, originating from observations, descriptions, and records of daily life. Our understanding of the world requires theory to establish a basic framework; otherwise, the world would be just a collection of fragmented experiences and impressions. However, at the same time, overly generalized theories can hinder the production of new

[1] Bao'an County was the predecessor of Shenzhen City. In January 1979, the Guangdong Provincial Committee dissolved Bao'an County and renamed it Shenzhen City.

[2] Le Goff, J.雅克·勒高夫. (1988). The New History新史學. In Cai S. Q. (Eds.), *Representing the Past: The Theoretical Perspective of Social History*再現過去:社會史的理論視野. Hangzhou: Zhejiang People's Publishing House杭州:浙江人民出版社. (pp. 92–122).

knowledge. Therefore, specific empirical studies are needed to activate and update these already generalized pieces of knowledge.[3]

Historical research in the field of built environment is greatly influenced by positivist historical scholarship, where literature and historical materials are used to "objectively" create history. Such historical research often leans more toward empirical studies, neglecting in-depth consideration of cognitive frameworks or theoretical perspectives. Social theorist Hans Joas,[4] historian Peter Burke,[5] architectural historian Demetri Porphyrios,[6] garden, and landscape architectural historian Michael Conan,[7] and others all recognize the significant influence of theoretical perspectives on their respective fields.

## *1.1.1 Theoretical Rationales*

### 1.1.1.1 The Gulf Between Society and Nature

The term "nature" maybe one of the most complex words in English, it has had three continuously evolving meanings throughout history: (1) the essential quality and character of something; (2) the inherent force which directs either the world or human beings or both; (3) the material world itself, taken as including or not including human beings.[8] As early as the seventeenth century, Francis Bacon argued that nature could become an object for human control. As a result, nature gradually distanced itself from the realms of mystery, poetry, religion, memory, and identity.[9] Nature was abstracted and defined apart from its social context, aligning with Bacon's political interests as an English aristocrat. The modernist perspective on nature also originated from this separation. In the early eighteenth century, when the third Earl of Shaftesbury laid the aesthetic and philosophical foundation for landscape gardens through his writings, the enclosure movement, driven by the English aristocracy, had been underway for over two centuries and was gradually reaching its climax.[10] In the late eighteenth-century Romantic movement, the term "nature"

[3] Joas, H., & Knöbl, W. (2009). *Social theory: Twenty introductory lectures*. Cambridge University Press. (pp. 1–5).

[4] Joas, H., & Knöbl, W. (2009). *Social theory: Twenty introductory lectures*. Cambridge University Press. (pp. 1–5).

[5] Burke, P. (2005). *History and social theory*. Polity Press. (pp. 16–20).

[6] Porphyrios, D. (1981). *On the Methodology of Architectural History*. London: Academy Editions. (pp. 2).

[7] Conan, M. (1999). Introduction. In Conan, M. (Eds.). *Perspectives on Garden History*. Dumbarton Oaks. (pp. 1–16).

[8] Williams, R. (1985). *Keywords: A Vocabulary of Culture and Society*. New York: Oxford University Press. (pp. 219–224).

[9] Bacon, F. 法蘭西斯. 培根 (2012). *The New Atlantis*新大西島. Beijing: The Commercial Press北京: 商務印書館.

[10] Hunter J. M. (1985). *Land into landscape*, London & New York: George Godwin. (pp. 97–98).

evolved to encompass pleasant connotations such as "goodness" and "innocence." It often refers to "the countryside," "unspoiled places," plants, and creatures other than humans.[11] In the nineteenth century, the urban park movement on both sides of the Atlantic was taking up the natural forms and intellectual threads of landscape gardens.[12] At this time, the term "nature" had conceptually taken on the connotation of excluding the artificial, becoming external to society. This was entirely different from the ancient Greek period, when people viewed nature as a vast living being with which they had emotional connections.[13]

Subsequently, nature became the object of the natural sciences, while society became the object of the social sciences. Robin Collingwood criticized this situation, noting that before the nineteenth century, there was communication between natural science and philosophy. However, this communication disappeared in the nineteenth century, leading to the separation of natural scientists and philosophers into two professional groups lacking mutual understanding.[14] For example, in the early twentieth-century research of the Chicago School of Sociology, Robert Park advocated the use of the "human ecology" approach in sociological research.[15] This was specifically aimed at clarifying the distinction between sociology and natural sciences like animal ecology and plant ecology. Rural areas are considered the origin of nature, while urban areas are seen as the source of society. This dichotomy has been transferred to the organization of academic fields, such as urban studies vs. rural studies, sociology vs. ecology, and so on. As a result, even today, many sociological studies still neglect nature, just as environmental studies continue to overlook urban issues.[16] For a long time, whether in psychology, anthropology, political science, or geography, where nature was considered, the dichotomy between society and nature has been continuously emphasized, intentionally or unintentionally, with nature being perceived as external to society.[17]

Karl Marx broke free from such intellectual threads and insisted on the unity of society and nature. He chastised Ludwig Feuerbach for perceiving nature as eternal, primitive, and purely natural, and held the belief that human labor, intrinsically social, produces, and filters all nature.[18] Alfred Schmidt of the Frankfurt School

[11] Williams, R. (1985). *Keywords: A Vocabulary of Culture and Society*. New York: Oxford University Press. (pp. 220–224).

[12] Crang, M. (2013). *Cultural geography*. New York: Routledge. (pp. 31–36).

[13] Collingwood, R. G. (1960). *The idea of nature*. Oxford University Press. (pp. 3–4).

[14] Collingwood, R. G. (1960). *The idea of nature*. Oxford University Press. (pp. 1–2).

[15] Park, R. E. (1915). The city: Suggestions for the investigation of human behavior in the city environment. *American Journal of Sociology*, 20(5):577–612.

[16] Angelo, H. (2021). *How green became good: Urbanized nature and the making of cities and citizens*. University of Chicago Press. (pp. 8–10).

[17] Smith, N. (2010). *Uneven development: Nature, capital, and the production of space*. University of Georgia Press. (pp. 30–32).

[18] Marx, K., & Engels, F. 卡爾·馬克思, 弗裡德里希·恩格斯 (2001). *Collected Works of Marx and Engels (Vol. 44)* 馬克思恩格斯全集(第44卷). (Compiled and translated by the Editorial Department of Marx-Engels-Lenin-Stalin Works Compilation Bureau of the Central Committee of

extensively studied Marx's concept of nature, arguing that its distinctiveness from various other perspectives lies in Marx's emphasis on the social-historical aspect of nature.[19] More recently, the geographer Neil Smith has pointed out the priority of society over nature, asserting that if nature is not social, it is nothing at all.[20] He is not hesitant to acknowledge the potential anthropocentric tendency in his ideas. Instead, he believes that only by first acknowledging this reality can the problems arising from it be properly addressed.[21] Since entering modern society, particularly after the nineteenth century, nature has become something external to society. The Marxist knowledge tradition, however, has contributed to reintegrating nature into the social context for historical analysis.

In the nineteenth century, from the perspective of Frederick Law Olmsted, nature originated in the countryside and stood in contrast to the city. Therefore, introducing nature into the city could address the social issues of industrial urban areas. He considered nature to be excessively pure, overlooking the potent and intangible social force field of the city. He either failed to perceive or deliberately chose to ignore the ongoing class interactions and conflicts taking place regarding urban parks.[22] Urban parks are one of the few urban spatial types in the high-density modern urban environment that manage to retain a natural appearance. However, ultimately, they are still artificial environments, representing nature with specific meanings. Therefore, adhering to Olmsted's perspective of regarding urban parks as a form of pure nature in opposition to the city, it is evident that times have changed. Especially in the highly globalized present, the production of nature and the production of space intersect and permeate at city, nation-, and global scales.[23] The dichotomy between nature and society, nature and the city, can no longer maintain a significant status in discourse. So, is it possible for the historical study of urban parks to serve as a footnote in bridging the dichotomy between nature and society?

#### 1.1.1.2 The Gulf Between Landscape History and the History of Landscape Architecture

According to the human geographer Denis Cosgrove, the history of landscape architecture is a historical study that explores the interconnectedness of larger landscapes, agricultural productivity, political power, and gardens, it is an examination

---

the Communist Party of China中共中央馬克思恩格斯列寧史達林著作編譯局). Beijing: People's Publishing House北京: 人民出版社. (pp. 88–102).

[19] Schmidt, A. (2014). *The concept of nature in Marx (Vol. 8)*. Verso Books. (pp. 1–3).

[20] Smith, N. (2010). *Uneven development: Nature, capital, and the production of space*. University of Georgia Press. (pp. 08).

[21] Smith, N. (2010). *Uneven development: Nature, capital, and the production of space*. University of Georgia Press. (pp. 38–40; 45–48).

[22] Taylor, D. E. (1999). Central Park as a model for social control: Urban parks, social class and leisure behavior in nineteenth-century America. *Journal of Leisure Research*, 31(4), 420–477.

[23] Smith, N. (2010). *Uneven development: Nature, capital, and the production of space*. University of Georgia Press. (pp. 181–196).

of the historical relationships between gardens and natural systems.[24] Such history of landscape architecture can be considered a landscape history that narrows the spatial scale, and landscape history research has traditionally been the domain where geographers are most involved. The term "landscape" was adopted into the English language as an art term from Dutch around the seventeenth century. Landscape painting specifically pertains to the depiction of inland natural scenery, distinguishing it from portraits and seascapes. By the nineteenth century, "landscape" began to be associated with "gardening" and became related to the design industry. In the nineteenth century, geologists and geographers started employing the term "landscape" to represent a vast stretch of land.[25] In the early twentieth century, Henry Clifford Darby was one of the first to recognize the importance of the interrelationship between geography and history. His 1936 publication, *A Historical Geography of England before A.D. 1800* laid the groundwork for the subsequent emergence of landscape history. In 1955, in a study on the landscape of England, William Hoskins criticized previous geographers for only describing empirical aspects such as terrain and vegetation. He argued that they failed to reveal why and how social activities shaped such landscapes.[26] His research is considered a pioneering work in the field of landscape history.[27] Michael Reed went further to propose that the study of landscape history should focus on human social activities and their relationship with the landscape and environment. He emphasized the need to uncover the reasons behind these processes in research.[28] John B. Jackson did not confine himself to the spatial scale of landscape history. He focused not only on national landscapes but also on rural areas, forests, and even more microscopic elements like parks and buildings. Concerning parks, his attention was directed toward the informal recreational spaces overshadowed by the urban park movement, delving into the underlying aesthetics and power relationships. While he mentioned issues of form and style, they were not central to his overarching purpose.[29]

Similarly, in the historical study of the relationship between humans and nature, garden historians find it relatively challenging to approach the subject from a social and political perspective. This difficulty arises because garden historians have long conducted their work within the framework of art history, focusing on issues such

[24] Cosgrove, D. E. (1998). *Social formation and symbolic landscape*. University of Wisconsin Press. (pp. 142–160).

[25] Wu, J. 吳家驊 (1999). *Landscape Morphology: a comparative study of landscape aesthetics*景觀形態學 景觀美學比較研究. Beijing: China Architecture & Building Press北京: 中國建築工業出版社. (pp. 3–4).

[26] Hoskins, W. G. (1955). *The Making of the English Landscape*. Hodder and Stoughton. (pp. 1–3).

[27] Gao, D. 高岱(2017). William Hoskins and the studies of landscape history威廉·霍斯金斯與景觀史研究. *Academic Research*學術研究, 2017(12), 103–107, 178.

[28] Reed, M. (1990). *The Landscape of Britain: From the Beginnings to 1914*. London: Routledge. (pp. xii–xvi).

[29] Jackson, J. B. (1984). *Discovering the vernacular landscape*. New Haven, Conn: Yale University Press. (pp. 125–130).

as the form, style, design, and construction time of gardens.[30] A compilation method that starts from form rather than problem results in garden history that, at most, presents society and politics as a background, failing to intricately examine the interaction between the social and political context and the gardens. Especially when the subject of study shifts from gardens to landscape architecture, studying urban parks in the manner of fine art becomes more awkward.[31] Approaching the year 2000, Michel Conan raised a crucial inquiry regarding the situation of garden history research: "What is left for the study of gardens when garden design is taken away?" Therefore, he supported the idea of examining the history of gardens via a social anthropological lens.[32] In other words, Conan believes that the study of garden history needs to shift its focus from a perspective that merely provides formal material for design to a discussion of more significant issues. John Dixon Hunt, on the other hand, suggests approaching the study in a "cultural history" manner, paying attention to human interventions in garden sites. He argues that methods from social history and political economic history are equally important for the writing of garden history, alongside existing approaches in art history.[33]

The gulf between landscape history and the history of landscape architecture lies in the following aspects: On the one hand, landscape history excels in studying the interaction between humans and nature on a broad spatial scale, penetrating social and political contexts. While designing and constructing gardens also represent a form of interaction between humans and nature, landscape history appears notably less enthusiastic about this aspect. On the other hand, the history of landscape architecture excels in describing and categorizing the form and style of gardens within a smaller spatial scale. However, it faces challenges in integrating these design aspects into the social and political context for analysis and understanding.

Landscape architecture theorist Marc Trieb identified this problem: landscape designers rarely distinguished the more broadly cultural issues behind specific forms within gardens.[34] This problem arises from the discourse of the history of landscape architecture—if the writing of the history of landscape architecture does not intricately examine the interaction between social, political, and cultural contexts and the forms of gardens, designers evidently cannot discern the differences behind these forms. Although the history of landscape architecture is no longer expected to directly provide material for design, it does not lose its value. Contrarily, following the shift in paradigm, the history of landscape architecture can provide

---

[30] Quest-Ritson, C. (2003). *The English garden: a social history*. David R. Godine Publisher. (pp. 1–3).

[31] Hunt, J. D. (1999). Approaches (New and Old) to Garden History. *Perspectives on Garden Histories. Dumbarton Oaks*. (pp. 77–90).

[32] Conan, M. (1999). From vernacular gardens to a social anthropology of gardening. *Perspectives on Garden Histories*. Dumbarton Oaks. (pp. 181–204).

[33] Hunt, J. D. (2013). *A Cultural History of Gardens in the Modern Age (Vol. 6)*. London: Bloomsbury Publishing. (pp. 1–5).

[34] Trieb, M. (1999). Nature Recalled In Corner, J. (Eds.) *Recovering landscape: Essays in contemporary landscape theory*. Princeton Architectural Press. (pp. 29–44).

theoretical value and significance to design through various perspectives and approaches. This is evidently more crucial. The difficulty described by Trieb arises from discursive deficiencies in the history of landscape architecture, rather than from the designers themselves. James Corner also addressed this issue, proposing to balance the political and physical aspects of design through "plotting" which rooted in hermeneutics.[35] It is evident that only by exploring the origins of landscapes within a historical context rather than solely within the realm of landscape art can the associated ambiguities be clarified.[36]

The study of landscape history can provide excellent theoretical insights into the interaction between society and nature. However, the design and construction of gardens do not align with the main concerns of landscape history research. Theory is non-disciplinary knowledge, whereas the design and construction upon which the discipline of built environment is based are precisely disciplinary and specialized knowledge. This is closely related to the formation of the modern disciplinary system since the nineteenth century.[37] While the importance of theory cannot be overstated, an excessive emphasis on theoretical depth beyond a certain threshold can lead to a disconnection from the discipline of the built environment.[38] Therefore, it is evident that merely adopting the writing methodology of landscape history is not the core objective of the discipline of the built environment.

The spatial form of urban parks, on one hand, originates from the broader spatial context of urban and regional topography, connecting to the landscape history of the city and region. On the other hand, it is influenced by the potent and intangible social force field of the city. So, is it possible to bridge the gap between landscape history and the history of landscape architecture through the historical writing of urban parks? In other words, can the historical writing of urban parks, with the help of the intermediary role of landscape history, connect to the social and political context?

### 1.1.2 Empirical Rationale

Given the world-class and phenomenal pace of development in Shenzhen, it is not an exaggeration to acclaim it as a miracle in the history of global urban development. With the reform and opening up in mainland China in 1978, Bao'an County(寶安縣) in the southern part of Guangdong Province was renamed Shenzhen City

---

[35] Corner, J. (2014). *The landscape imagination: collected essays of James Corner, 1990–2010*. Princeton Architectural Press. (pp. 101–104).

[36] Cosgrove, D. E. (1998). *Social formation and symbolic landscape*. University of Wisconsin Press. (pp. 20).

[37] Moran, J. (2010). *Interdisciplinarity*. Routledge. (pp. 3–16; 74–75).

[38] Zou, H. 鄒暉 (2012). *Fragments and Mirroring: The Twofold Discourse of Comparative Architecture*碎片與比照 比較建築學的雙重話語. Beijing: The Commercial Press北京: 商務印書館. (pp. 33).

in 1979. The following year, 327.5 square kilometers of land in the southern part of Shenzhen City, near Hong Kong, was designated as a special economic zone.

In the landscape of globalization, the restructuring of the global industrial structure in the 1970s has already led to the initial development of the Four Asian Tigers. Although Shenzhen is just situated across the river from Hong Kong, one of the Four Asian Tigers, the economic and social conditions of the two places were incomparable at that time. However, by 2018, Shenzhen's GDP had surpassed Hong Kong, reaching 24,221.98 trillion yuan,[39] and in the same year, Shenzhen, according to the Globalization and World Cities Research Network (GaWC), became an Alpha-level city in mainland China, just after Beijing and Shanghai, becoming an undisputed metropolis in China.[40] This process is also part of Shenzhen's extremely rapid urbanization. In the view of John Friedmann, China's urbanization is a multidimensional social-spatial transformation, and one important dimension is the urbanization of the natural environment.[41]

The Pearl River Delta (PRD), adjacent to the South China Sea, is bordered by the Tianlu Mountains(天露山脈), Yunwu Mountains(雲霧山脈), and Yunkai Mountains(雲開山脈) to the west, the Jiulian Mountains(九連山脈) and Huashi Mountains(滑石山脈) to the north, and the Lianhua Mountains(蓮花山脈) to the east. Internally, it is traversed by four major river systems: the West River(西江), North River(北江), East River(东江), and various deltaic rivers. Consequently, the PRD exhibits the topographical features of "mountains on three sides, facing the sea on one side, convergence of three rivers, and division into eight outlets (三面環山、一面臨海、三江匯合、八口分流)."

Shenzhen, located on the eastern bank of the Pearl River Estuary, is mostly defined by hilly topography within the regional landscape framework. Based on a 1984 survey conducted by the Guangzhou Institute of Geography(廣州地理研究所), the land area of 1890.7 square kilometers can be divided into different types of terrain. Hills make up 44.07% of the total area, while plateaus and plains are almost equal, accounting for 22.35% and 22.12% respectively. Low mountains and terraces make up 4.82% and 5.09% of the land area respectively. In addition, the shoreline spans a distance of 230.0 kilometers.[42] Due to the substantial area of plateaus and plains and the presence of a subtropical monsoon climate, the predominant agricultural activity focus on grain production. Simultaneously, the advantageous

[39] Li, C.H. 李昌鴻 (2019, February 28). *Shenzhen's GDP Surpasses Hong Kong for the First Time, Becoming the Economic Leader of the Greater Bay Area*深圳GDP首次超香港 成大灣區經濟龍頭. Retrieved from http://news.wenweipo.com/2019/02/28/IN1902280011.htm

[40] Globalization and World Cities Research Network (GaWC). (2018, November 13). The World According to GaWC 2018. Retrieved from https://www.lboro.ac.uk/microsites/geography/gawc/world2018t.html

[41] Friedmann, J. (2006). Four theses in the study of China's urbanization. International Journal of Urban and Regional Research, 30(2), (pp. 440–451).

[42] Huang, Z. G. 黃鎮國 (Eds.). (1984). *Shenzhen's Natural Resources and Economic Development*深圳自然資源與經濟開發. Guangzhou: Guangzhou Institute of Geography廣州:廣州地理研究所. (pp. 18–22).

proximity to the sea and the southward relocation of the economic center since the Song Dynasty, Shenzhen has experienced significant growth in its sectors such as fishing, oyster farming, and salt production.[43] In 1980, Shenzhen had a population of 332,900 and an urban built-up area of around three square kilometers, primarily situated in the current Luohu area, and the surrounding area consisted mainly of agricultural grounds. At that time, agriculture continued to play a substantial role in the national economy, accounting for 56.75% of the combined production value of industry and agriculture.[44]

In the context of globalization in the 1980s, to develop the finance and service industries, Hong Kong relocated a large number of industrial factories to the SSEZ in a collaborative model of "front shops and back factories (前店後廠)." These goods were then exported worldwide. This rapidly transformed the urban landscape of this small town on the southern border of China. Firstly, on lands already acquisitioned from farmers by the government and those yet to be acquisitioned, factories and industrial zones emerged. Secondly, corresponding residential areas, theaters, museums, urban parks, and other public service facilities also sprouted from former farmlands, vegetable fields, and fruit orchards. After the mid-1990s, as the land resources in the Luohu area became nearly exhausted, and to continue developing Shenzhen into an international city, the urban center moved westward to Futian. Similarly, Futian Central Business District and its associated public service facilities were constructed on a patch of farmland, fishponds, and villages. During this period, major urban parks were no longer situated in the same topographical conditions as libraries, concert halls, etc., but were instead located on hilly terrain. In the new millennium, under the process of regional integration in the PRD, the Qianhai-Houhai (前海-後海) area near the Pearl River Estuary has become the new urban center. In the 1980s, the Shekou Peninsula shifted from agriculture to industry. However, starting from the late 1990s, the Houhai area, extending from the peninsula to Overseas Chinese Town, continued its transition from industrial activities to high-end industries such as high-tech and modern services. From the perspective of urban landscapes, this process saw the transformation of farmlands, fishponds, and villages into factories and industrial zones. Subsequently, these industrial areas, and even the former sea areas, were filled and transformed into large shopping malls, skyscraper headquarters, and upscale residential areas. During this period, several large parks were built through land reclamation in the former sea areas.

Prior to 1980, Shenzhen possessed merely three urban parks; yet, as of 2019, the city boasted over 1090 urban parks of diverse spatial scales. The urban parks, particularly the larger ones, are constructed on plains, hills, and mountains, or reclaimed land from the sea. In any case, they all undergo a transformation from one form of

---

[43] Bao'an County Local Chronicles Compilation Committee寶安縣地方誌編纂委員會(1997). *Chronicles of Bao'an County*寶安縣誌. Guangzhou: Guangdong People's Publishing House廣州: 廣東人民出版社. (pp. 147–148).

[44] Shenzhen Special Economic Zone Yearbook Editorial Committee深圳經濟特區年鑒編輯委員會 (1986). *Shenzhen Special Economic Zone Yearbook 1985*深圳經濟特區年鑒1985. Shenzhen: Shenzhen Special Zone Yearbook深圳: 深圳特區年鑒. (pp. 581–582).

nature to another. The transformation is driven by the alterations in the landscape of the area, the city, and the region where the parks are located. Landscapes have perpetually been intertwined with social connections and governed by hierarchical systems of power.[45] Meanwhile, the models of describing nature are also provided by social relationships.[46] Therefore, studying the social, economic, and political context behind this landscape change is necessary to understand why the nature represented by urban parks undergoes transformation between different cultural forms. It is only via conducting such research that we can adequately address the regret expressed by sinologist Ezra F. Vogel, who believes that Shenzhen has not received the scholarly attention it deserves.[47]

The story of Shenzhen is not just a historical narrative but also a geographical one. Its story of over forty years is a tale of rapid and large-scale transformation in the relationship between society and nature. It is also a rich experiential ground of history and geography. Given this, is it possible for this experiential ground to provide material for bridging the theoretical gulf between society and nature, landscape history, and the history of landscape architecture?

Hence, the rationale behind selecting Shenzhen as the research subject.

## 1.2 Definitions and Concepts

### *1.2.1 Urban Parks*

According to the definition recognized by the International Federation of Parks and Recreation Administration (IFPRA), urban parks are specific open space areas primarily composed of plants and water features, typically provided for public use.[48] According to the national standards and industry norms of mainland China, a "Public Park" is defined as a publicly accessible open area that is primarily intended for recreational activities. These parks are provided with a wide range of facilities and serve the aims of promoting ecological balance and enhancing aesthetic appeal. Conversely, "Urban Green Space(城市綠地)" refers to a type of land in the city primarily characterized by vegetation and with specific functions and purposes.[49]

---

[45] Cloke, P., Crang, P., & Goodwin, M. (2005). *Introducing Human Geographies (2nd ed.)*. Routledge. (pp 156–167).

[46] Conan, M. (1999). From vernacular gardens to a social anthropology of gardening. *Perspectives on Garden Histories. Dumbarton Oaks*: 181–204.

[47] Vogel, E.F. (2017). Foreword In O'Donnell et al. (Eds.) *Learning from Shenzhen : China's post-Mao experiment from special zone to model city*. Chicago: The University of Chicago Press. (pp. vii–xiv).

[48] Konijnendijk, C. C., Annerstedt, M., Nielsen, A. B., & Maruthaveeran, S. (2013). *Benefits of urban parks. A systematic review. A Report for IFPRA*, Copenhagen & Alnarp. (pp. 1–70).

[49] Ministry of Housing and Urban-Rural Development of the PRC中華人民共和國住房和城鄉建設部, General Administration of Quality Supervision, Inspection and Quarantine of the PRC 中華

A definition is a human-made interpretation that inevitably presupposes and conceals the purposes and judgments of the definer. In fact, both of the above definitions are made from the perspective of formal composition, with the aim of construction utility rather than historical research. In such definitions, urban parks are implicitly treated as value-free objective entities, and their construction is seen merely as an engineering and technical problem. The elements involved are perceived as nothing more than objectively existing entities such as plants, water systems, rocks, and buildings. According to this definition, conducting historical research can easily lead to the loss of the human subjectivity within urban parks, where the influence of social and political contexts on the shaping of urban parks is not adequately reflected. In such cases, the theoretical attempt of this research to bridge the gap between society and nature seems unattainable. As architectural historian Manfredo Tafuri stated, "If it tends to become pure object and, even, pure industrial object, it is clear that one can not even begin to question its historicity,"[50] and the same applies to urban parks.

Urban parks in the Western context originated from the nineteenth century. During that period, the glamour of modernity emanating from capitalist industrial cities gradually faded, and urban issues such as hygiene, morality, and crime became a nightmare. Parks were considered as an antidote in addressing these problems,[51] Countries such as the United Kingdom, Germany, and the United States have actively engaged in the urban park movement.[52] Furthermore, the key figures driving the urban park movement, whether it be Hirschfeld's advocacy of "German courtyards" and "national parks,"[53] or Olmsted's emphasis on national identity in urban park design and wilderness conservation,[54] all reflect a nationalist undertone in their ideologies. In the Chinese context, the term "park(公園) "was initially

人民共和國國家品質監督檢驗檢疫總局. (2016). *Park Design Standards公園設計規範: GB51192–2016*. Beijing: China Architecture & Building Press北京: 中國建築工業出版社; Ministry of Housing and Urban-Rural Development of the PRC中華人民共和國住房和城鄉建設部. (2017). *Basic Terminology Standard for Landscape Architecture風景園林基本術語標準: CJJ/T91–2017*. Beijing: China Architecture & Building Press北京: 中國建築工業出版社.

[50] Tafuri, M. (1980). *Theories and history of architecture*. London; New York: Granada. (pp. 41).

[51] Crompton, J. L. (2013). The health rationale for urban parks in the nineteenth century in the USA. *World Leisure Journal*, 55(4), 333–346; Jones, D. (2015). *Crime, protest, community, and police in nineteenth-century Britain*. Routledge. (pp. 1–32).

[52] Birnbaum, C. A. & Karson, R. S. (2000). *Pioneers of American landscape design*. New York: McGraw Hill. (pp. 61–63; 96–99; 107–109; 277–281); Shirahata, Y.白幡洋三郎. (2014). *A History of Modern Urban Parks: Origins of Westernization* 近代都市公園史:歐化的源流. Beijing: New Star Press 北京: 新星出版社. (pp. 16–49).

[53] Shirahata, Y.白幡洋三郎. (2014). *A History of Modern Urban Parks: Origins of Westernization* 近代都市公園史:歐化的源流. Beijing: New Star Press 北京: 新星出版社. (pp. 124).

[54] Duempelmann, S. (2014). Herrington S. Plotting Time in Landscape Architecture. *Studies in the History of Gardens & Designed Landscapes*, (34):1–14; Liu, X., & Wang, X. 劉新宇,王曉俊 (2021). A Study of F. L. Olmsted and His Designing Ideas from the Perspective of American History in 19th Century 19世紀美國史視角下的奧姆斯特德及其設計思想. *Chinese Landscape Architecture*中國園林. 37(02), 106–111.

mentioned in the "North History(北史)" to describe official gardens.[55] Although China has a tradition of public gardens in ancient times,[56] the concept of modern urban parks, as influenced by ideas from Kang Youwei (康有為, 1858–1927), Liang Qichao (梁啟超,1873–1929), and other bourgeoisie in the late Qing Dynasty who traveled to the West, began to emerge. In their contemplation, urban parks were no longer seen as public gardens dominated by colonizers or mere tourist attractions. Instead, they became intricately linked to the construction of modern Chinese cities and the cultivation of modern civilization. These ideas were propagated by various Chinese newspapers and publications around 1900, emphasizing their roles in hygiene and civic morality.[57]

In both Eastern and Western contexts, modern urban parks, in their modern sense, have been led by the upper-middle-class bourgeoisie within the framework of modern nation-states. As expressed by Japanese historian Shirahata Yozaburo, parks can be seen not so much as meeting the demands of urban residents as an urban apparatus desired by the ruling class.[58] Additionally, the purpose behind the construction of parks by these upper-class segments is to educate the public and transmit their own values, such as civic consciousness and patriotic sentiments, to the middle and lower classes through the urban parks.[59] It can be seen that urban parks are not merely engineering and technical issues, but rather natural installations shaped by the social and historical context of the city. Therefore, the definition of urban parks should not simply overlay their public attributes based on their physical structure but should include the social and historical context that shapes them within their definition. Thus, this research comprehends urban parks as both a cultural form of nature and a cultural form of space. It encompasses not only the physical aspects of urban parks but also the associated images, words, and discourses. As a study within the field of built environment, this research primarily emphasizes the physical representation of urban parks.

In such a definition, publicity alone is insufficient for green spaces in urban areas to be classified as urban parks. A more important premise is that these green spaces need to be incorporated into macro-level political and economic plans. In other words, urban parks are, on the one hand, urban green spaces open to all citizens and

---

[55] Chen, Z.陳植 (1930). *On Urbanism and Parks*都市與公園論. Beijing: The Commercial Press北京: 商務印書館. (pp. 1–2).

[56] Mao, H. S.毛華松 (2014). *Formation Analysis of Chinese Ancient Parks—Analysis of Urban Parks Development in Song Dynasty*論中國古代公園的形成——兼論宋代城市公園發展. Chinese Landscape Architecture中國園林, 30(01), (pp. 116–121).

[57] Lin, Z.林峥 (2022). *Park Beijing: Cultural Production and Literary Imagination*公園北京 文化生產與文學想像. Beijing: Peking University Press北京: 北京大學出版社. (pp. 7–9); Lin, Z. 林峥 (2019). *"Homeland Nostalgia" of the "Worldly Man": On the Park Narratives in Kang Youwei's Overseas Travelogues.* "世界人"的鄉愁——論康有為海外遊記中的公園書寫. Collection of Studies on Modern Chinese Literature中國現代文學研究叢刊, (01), 151–166.

[58] Shirahata, Y.白幡洋三郎 (2014). *A History of Modern Urban Parks: Origins of Westernization*近代都市公園史:歐化的源流. Beijing: New Star Press北京: 新星出版社. (pp. 3–6).

[59] Lin, Z.林峥 (2022). *Park Beijing: Cultural Production and Literary Imagination*公園北京 文化生產與文學想像. Beijing: Peking University Press北京: 北京大學出版社. (pp. 4–6).

possessing a public nature; on the other hand, they are a component of macro-level goals of the state or local government. Therefore, urban parks represent a shared solution between bottom-up demand and top-down governance practices. The motivations of the latter are typically concealed within generalized discourses.

It is also necessary to clarify that the aforementioned nature and space are not a pair of separate concepts. According to Neil Smith, the production of space is a logical result of the production of nature.[60] Smith's concept of "nature" includes various interpretations. The study's topic, termed as "nature" by Anglo Hilary, refers to the "everyday signifiers of nature," including urban parks, community gardens, rivers, mountains, street trees, tiny animals, and other similar elements.[61] The term "cultural form" denotes that space, nature, and related concepts are all generated by human social activities. This aims to dispel the notion that they are fixed and universally unchanging. In the field of archaeology, Chris Gosden advocates focusing on the "cultural biography" of "objects," examining the processes of their production, exchange, and consumption in specific social contexts.[62] In the field of the built environment, Hsia Chu-joe(夏鑄九) advocates examining different spatial discourses and physical spatial forms as "cultural forms of space" to understand their manifestations in political, economic, and cultural processes.[63] Hunt's elaboration on the concept of "cultural history" broadly refers to extensive human interventions in gardens, encompassing social, economic, political, and anthropological dimensions.[64] Although the expressions vary, culture can be understood as the intertwining and interaction between human social activities and objects. These elements are inseparable and defy individual delineation and emphasizing human subjectivity.

### 1.2.2 Socio-Nature

The term "socio-nature" arises from a critique of the dichotomy between society and nature.[65] In the *A Dictionary of Geography* by Oxford, socio-nature is defined as "an ecosystem that has been socially created," wherein "social relations are inherently ecological and that ecological relations are inherently social." In the *International Encyclopedia of Geography*, the concept of socio-nature is described

---

[60] Smith, N. (2010). *Uneven development: Nature, capital, and the production of space*. University of Georgia Press. (pp. 092–094).

[61] Angelo, H. (2021). *How green became good: Urbanized nature and the making of cities and citizens*. University of Chicago Press. (pp. 3).

[62] Gosden, C., & Marshall, Y. (1999). The cultural biography of objects. *World archaeology*, 31(2), 169–178.

[63] Hsia, C. J. 夏鑄九 (Ed. & Trans.). (1989). *Readings in Social Theories and the Cultural Form of Spaces*空間的文化形式和社會理論讀本. Taipei: Ming Wen Book Co., Ltd. 臺北:明文書局.

[64] Hunt, J. D. (2013). A Cultural History of Gardens in the Modern Age (Vol. 6). London: Bloomsbury Publishing. (pp. 1–5).

[65] Mayhew, Susan (2015). *A Dictionary of Geography*. Oxford University Press.

as "a concept that is used to argue that society and nature are inseparable and should not be analyzed in abstraction from each other." Its roots can be traced to two main sources: (1) the Marxist historical materialism tradition emphasizing temporality and process and (2) the post-structuralist tradition emphasizing hybridity.[66]

On the historiographical level, dividing "society-nature" into two terms for discussion is not impossible. In the early twentieth century, when the Annales School established the field of social history, its representative figure Marc Bloch had already recognized the ambiguity of the term "social." However, he still believed that precisely because of the term's ambiguity, it was created according to the purpose of history, there is actually no such thing as economic and social history, only history as a whole, and the entire history is social history. Thus, the Annales School advocated for a "total history" approach that broke down the barriers and distinctions between disciplines.[67] Eric Hobsbawm, heavily influenced by the Annales School, also believed that social history could not be a specialized discipline, going as far as to say that it was no longer necessary to define. Of course, he also distinguished three forms of social history:[68] (1) the history of the poor or lower classes; (2) the history of lifestyles, customs, and everyday life; (3) economic history. The attention to "nature" on the historiographical level can also be traced back to the efforts of the Annales School in the first half of the twentieth century to expand the field of history.[69] However, its development and flourishing are primarily evident in the rise of environmental history in the late twentieth century.[70] In the field of environmental history, nature is not merely seen as an objective existence or a subject of scientific research; rather, the interplay between human social activities and nature becomes the focus of environmental historians. At this point, "society" returns to the discussion of nature.

The dichotomy between society and nature is a product of Enlightenment modernity. Sociologist Bruno Latour argues that Enlightenment modernity presupposes a "progressive" future through such divisions. However, over time, humanity encounters not separateness but rather a blending, with more and more elements being incorporated into human civilization. Therefore, there is a need to study how natural and artificial entities, as well as political and social factors, mutually translate with

[66] Bear, Christopher (2017), "Socio-Nature," in Richardson, Douglas; Castree, Noel; Goodchild, Michael F.; Kobayashi, Audrey (eds.), *International Encyclopedia of Geography: People, the Earth, Environment and Technology*, Oxford, UK: John Wiley & Sons, Ltd. (pp. 1–5).

[67] Le Goff, J.雅克·勒高夫. (1988). The New History新史學. In Cai S. Q. (Eds.), Representing the Past: The Theoretical Perspective of Social History再現過去:社會史的理論視野. Hangzhou: Zhejiang People's Publishing House杭州:浙江人民出版社.

[68] Hobsbawm, E. J. From social history to the history of society從社會史到社會的歷史. In Cai S. Q. (Eds.), Representing the Past: The Theoretical Perspective of Social History再現過去:社會史的理論視野. Hangzhou: Zhejiang People's Publishing House杭州:浙江人民出版社.

[69] Febvre, L. (2013). A geographical introduction to history. Routledge.

[70] Hughes, J. D. (2016). *What is environmental history?*. John Wiley & Sons. (pp. 50–53).

each other.[71] Geographer Kenneth Olwig opposes viewing geography as "ahistoric" and proposes the "society/nature problematic," arguing that the cognitive separation of society and nature in geography is a retreat.[72] David Harvey has also criticized René Descartes for theoretically considering "society" as a closed system with a relationship to another closed "biosphere." He believes that such a mechanical separation has led to the contemporary ecological problems at the level of thought, as in practice, it is challenging to distinguish the boundaries between "society" and "nature".[73] This entrenched dichotomy may indeed influence some readers' understanding of this research. Perhaps, in their view, the author should list separately the social and natural factors involved in the development process of urban park construction, establish a matrix or index system, and then study how these factors interact with each other to bridge the gap between society and nature. To such a suggestion, I could respond by borrowing from Smith's words: "Two things can only interact or reflect each other if they are defined in the first place as separate."[74] The birth of the term "society-nature" criticizes precisely this kind of separate definition. In other words, society is inherent in nature, and nature is also inherent in society.

In the field of built environment, attention to the socio-nature often focuses on its relationship with designed landscapes. As Ian L. McHarg viewing ecology through a natural science paradigm, Anne W. Spirn similarly characterized his leadership in landscape architecture at the University of Pennsylvania as "ahistorical."[75] Moreover, she also found that another significant figure in landscape architecture, Olmsted, was also not adept at considering social and political processes, leading to his works often being subverted and transformed. Works in the field of landscape architecture are often not recognized as human-made constructions due to their natural appearance, hindering the understanding of conflicting environmental values.[76] The American urban park movement in the nineteenth century centered around figures like Olmsted, has also been criticized by Anita Berrizbeitia. She contends that in these urban parks, everyday urban life and aesthetic practices are disconnected. However, in her study of the Amsterdam Bos, she identified a distinction: the productive processes of the park are closely linked to the industrialization processes of

---

[71] Zhang, K.張亢 (2020). *Reconstructing Nature and Politics: On Bruno Latour's Political Ecology*重構自然與政治:論拉圖爾的政治生態學. Journal of Dialectics of Nature自然辯證法通訊, 42(01), 19–26.

[72] Olwig, K. R. (1980). Historical geography and the society/nature "problematic": the perspective of JF Schouw, GP Marsh and E. Reclus. *Journal of Historical Geography*, 6(1), 29–45.

[73] Harvey, D. (2016). *The ways of the world*. Oxford: Oxford University Press. (pp. 136–176).

[74] Smith, N. (2010). *Uneven development: Nature, capital, and the production of space*. University of Georgia Press. (pp. 106–107).

[75] Spirn, A. W. (2000). Ian McHarg, landscape architecture, and environmentalism: Ideas and methods in context. *Environmentalism in landscape architecture. Dumbarton Oaks*, 22, 97–114.

[76] Spirn, A. W. (1996). Constructing nature: the legacy of Frederick Law Olmsted. In Cronon, W. (Eds.). *Uncommon ground: Rethinking the human place in nature*. WW Norton & Company. (pp. 91–113).

the surrounding city. In other words, urban parks manifest social and political structures through material production patterns—not isolating the park from capital but making it an entity of production within the city.[77] Thus, in the historiographical and theoretical currents of research since the late twentieth century, the study of urban parks cannot avoid the concept of "society." It is precisely the ambiguity of the term "society" that allows for the possibility of breaking down disciplinary barriers in the historical study of urban parks and for the reclamation of the significance of "nature" within the urban context at the historiographical level, contributing to new knowledge.

It can be observed that whether in geography, urban studies, or the built environment, exploring nature divorced from its social context has come under scrutiny. The dichotomy between society and nature, delineated by Enlightenment modernity, has nearly exhausted its explanatory power in contemporary knowledge production and is no longer able to maintain its discursive status.

### *1.2.3 Topography*

In *Webster's Dictionary* and *WordNet Dictionary*, "Topography" refers to: (1) the description of a particular place, town, manor, parish, or tract of land; especially, the exact and scientific delineation and description in minute detail of any place or region; (2) the configuration of a surface and the relations among its man-made and natural features; (3) topography—precise detailed study of the surface features of a region.[78]

"Topography" originates from the Greek words "topos" and "graphia," which signify "place" and "writing," respectively.[79] Thus, the original meaning of "topography" refers to the writing, editing, or representation of a place, with human subjectivity inherent in its meaning. However, the Enlightenment, especially the scientific and professional specialization that followed in the nineteenth century, led to the detachment of Western modern geography and surveying from the relationship between topography and human phenomena such as diseases, livelihoods, and politics. Similarly, in Eastern literatures such as *The Art of War* (孫子兵法), *Huainan Zi* (淮南子), and *The Craft of Gardens* (園冶), the humanistic

[77] Berrizbeita, A. (1999). The Amsterdam Bos: the modern public park and the construction of collective experience. In Corner J. (Eds.). *Recovering Landscape: Essays in Contemporary Landscape Architecture*. (pp. 186–203).

[78] Chen, J. P. 陳潔萍 (2011). *Place-Writing: The Study on Topography in the Expanded Field in Contemporary Architecture, Urbanism, and Landscape Design*書寫場地:當代建築、城市、景觀設計中擴展領域的地形學研究. Nanj ing: Southeast University Press南京:東南大學出版社. (pp. 11–13).

[79] M'Closkey, K., & VanDerSys, K. (2023). Topography as the Groundwork for Landscape Design—Interview With Karen M'Closkey and Keith VanDerSys. *Landscape Architecture Frontiers*, 11(4).

interpretations of topography have also gradually faded, reducing topography to mere physical terrain, inert matter devoid of value, and abstract contours and contour lines.[80] In this sense, "topography" and "society-nature" are conceptually very close. However, David Leatherbarrow highlights two key distinctions between topography and nature. (1) Topography is not just a material entity; (2) the manifestation of topography is inseparable from human nature or artifacts. However, the "nature" he refers to here is the commonly understood nature, not "society-nature." He also expresses agreement with some understandings of nature that can help clarify topography. Therefore, to a large extent, understanding the concept of "society-nature" can help better understand the concept of "topography."

In this study, topography encompasses more than just the visible and tangible physical topography, like plains and mountains, in the physical world. Instead, it encompasses a broader concept that includes physical topography, surface vegetation, geological structures, and their integration with social life. This understanding of topography is, in fact, an extension of the "society-nature" theoretical concept. In the narrow sense, topography should be considered alongside vegetation, water systems, and buildings as elements of landscape architecture, which refers to "physical topography." However, in a broader sense, topography encompasses these elements as well as the social life interwoven among them, thus constituting "topography of life."

Why does this study use the concept of "topography" as a starting point rather than directly addressing "nature"?

Firstly, the meaning of "nature" is overly abstract and complex, whereas topography is more concrete and closely aligned with the built environment. Although topography is not entirely synonymous with nature, it provides a means to approach, describe, and study nature within the context of the built environment.

Secondly, and more importantly, in historical studies within the built environment field, understanding of natural changes often carries a linear evolutionary hue: from the first nature (wilderness/primitive nature) to the second nature (agriculture/practical nature), and then to the third nature (gardens/aesthetic nature).[81] This classification, guided by the study of garden history, emphasizes human activities' impact on nature but also implies a clear distinction between the second and third natures. However, in contemporary natural production, productive and practical nature does not completely oppose aesthetic and artistic nature, nor does the latter replace the former. Topography is well-suited for this situation: it studies nature that is intertwined and blended within social life, rather than the environment categorized into hierarchical and evolutionary stages. From a topographical perspective, understanding nature transitions from hierarchical to non-hierarchical, from clear delineation to fuzzy boundaries, thereby better capturing the authentic and complex relationship between humans and nature.[82]

[80] Liu, D. Y. 劉東洋(2011). Urban notes VII: A view of topography from my rear window城市筆記之七:從我家後窗望見的地形學. *The Architect*建築師, 2011(03), 105–111.

[81] Hunt, J.D. (2000). Greater Perfections: The Practice of Garden History. London: Thames & Hudson. (pp. 32–75).

[82] Chen, J. P. 陳潔萍 (2011). *Place-Writing: The Study on Topography in the Expanded Field in Contemporary Architecture, Urbanism, and Landscape Design*書寫場地:當代建築、城市、景

## 1.2.4 Shenzhen

### 1.2.4.1 Space Range

1. Scope of Shenzhen City

In January 1979, the Guangdong Provincial Party Committee of the Communist Party of China(中共廣東省委) issued the *Decision on the Establishment of Shenzhen and Zhuhai Cities*(關於設立深圳市和珠海市的決定), revoking Bao'an County and establishing Shenzhen City. The total land area of the city remained at 2020 square kilometers as recorded in the original *Chronicles of Bao'an County* (寶安縣誌) until before 1994.[83] However, due to ongoing land reclamation, territorial adjustments, and other factors, the calculated results for the area have not been consistent over different periods since then. The spatial scope of Shenzhen City referred to in this study corresponds to the total land area calculated in the official urban land use and change survey in 2011 (1997.47 square kilometers[84]), as indicated by the red line in Fig. 1.1.

2. Scope of Shenzhen Special Economic Zone (SSEZ)

In May 1981, the Shenzhen Municipal Party Committee (深圳市委) submitted a report to the Guangdong Provincial Committee titled *Request for Instructions on the Scope and Management of the Shenzhen Special Economic Zone* (關於深圳經濟特區範圍和管理的請示報告). The boundary of the SSEZ was established as "commencing at Beizaijiao(背仔角) in the eastern part of the Dapeng Corner(大鵬角), stretching in a southwesterly direction towards Shekou…proceeding in a northerly direction along the Wutongshan mountain (梧桐山) and Yangtaishan mountain(羊臺山) ranges…, including a total area of 327.5 square kilometers," as indicated by the region situated to the south of the purple line in Fig. 1.1. In October 1981, Shenzhen resumed jurisdiction over Bao'an County, covering the area outside the SSEZ. Between June 1982 and December 1984, an 84.6-kilometer-long administrative boundary, known as the "Second Line Customs," was established to separate the special economic zone from Bao'an County.[85]

In May 2010, the State Council granted approval for the *Request for Extending the Scope of the Shenzhen Special Economic Zone*(關於延伸深圳經濟特區範圍的

---

觀設計中擴展領域的地形學研究. Nanj ing: Southeast University Press南京:東南大學出版社. (pp. 23–24).

[83] Shenzhen Urban Planning and Land Development Research Center深圳市規劃國土發展研究中心 (2019). *Land Resources of Shenzhen City*深圳市土地資源. Beijing: Science Press北京:科學出版社. (pp. 64–65).

[84] Shenzhen Urban Planning and Land Development Research Center深圳市規劃國土發展研究中心 (2019). *Land Resources of Shenzhen City*深圳市土地資源. Beijing: Science Press北京:科學出版社. (pp. 60).

[85] Shenzhen Museum深圳博物館(2019). *Historical Changes in the Administration Boundary (Second Line) of Shenzhen Special Economic Zone*深圳經濟特區管理線(二線)歷史變遷研究. Shenzhen: Haitian Publishing House深圳:海天出版社. (pp. 33; 40–47).

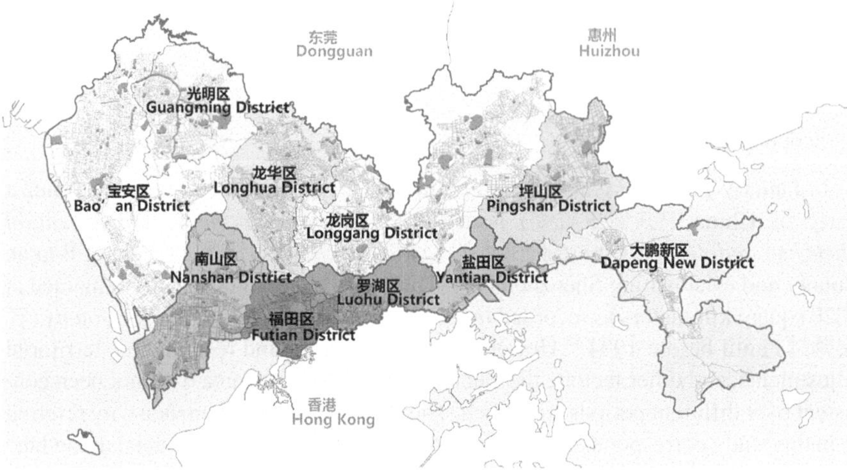


**Fig. 1.1** Spatial scope of Shenzhen city and the special economic zone

請示), so consenting to the expansion of the SSEZ to encompass the entire city. This represented a pivotal milestone in commencing the comprehensive advancement both within and outside the Second Line. Nevertheless, for the sake of convenience in expression, this study still refers to the spatial area south of the Second Line, covering 327.5 square kilometers, as the special economic zone area, and designates the spatial area of 1997.27 square kilometers as the city scope to avoid confusion. Additionally, the boundaries of administrative districts in Shenzhen have varied in terms of scope and names during different historical periods.[86] In this study, the administrative districts mentioned are based on the 10 administrative districts formed after the official establishment of Guangming District in February 2018. Complex historical changes in administrative divisions before this date are not considered to avoid causing mis-interpretation.

Thus, this study is conducted on two levels in terms of spatial scope: the first level focuses on the historical and geographical context, morphological changes, and historical interpretation of the spatiotemporal distribution of urban parks within the city scope; the second level selects various topographical conditions within the special economic zone to explore case studies of urban parks constructed on different topographic conditions.

[86]Cui, X. S.崔孝松 (2016). An Explanatory Study on the Historical Evolution of Administrative Divisions in Shenzhen深圳行政區劃歷史沿革考釋. *Practice and Theory of SEZS*特區實踐與理論, (04), 71–74.

#### 1.2.4.2 Time Range

In August 1980, the 15th meeting of the Standing Committee of the Fifth National People's Congress approved the *Regulations on the Guangdong Special Economic Zone*(廣東省經濟特區條例) proposed by the State Council, endorsing the establishment of the special economic zone in Shenzhen. This laid the foundation for a new institutional framework for the construction of the special economic zone cities, including urban parks. In the same year, the *Program for Urban Development of Special Economic Zone of Shenzhen Municipality* (深圳市經濟特區城市發展綱要) *(1980)* marked the earliest instance of systematic consideration for the planning of the urban park system within the special economic zone. In May 2018, under the supervision of the Shenzhen Municipal Urban Management Bureau(深圳市城市管理局) and led by the Shenzhen Park Management Center(深圳市公園管理中心), a comprehensive survey of basic information on urban parks within the city scope of Shenzhen was initiated. The final results were compiled in November 2019. The data and basic information at the municipal level on which this study relies are based on this survey. Simultaneously, this marks the government's first systematic investigation into the fundamental information of parks citywide, holding significant symbolic importance. Additionally, two significant planning documents pertaining to the urban spatial development and industrial growth of Shenzhen, namely The *Plan for the Coordinated Development of the PRD metropolitan area* (珠江三角洲城鎮群協調發展規劃) *(2004–2020)* and the *Urban Master Plan of Shenzhen Municipality* (深圳市城市總體規劃) *(2010–2020)* both set 2020 as the planning horizon.

Therefore, this study covers the time range from 1980 to 2020.

## 1.3 Objective, Significance, and Innovation

### *1.3.1 Research Objectives*

#### 1.3.1.1 The Historicity of Urban Parks in the Topography of Shenzhen

Over the past 40 years, urban parks built in Shenzhen are scattering throughout the city, with their construction and opening period extending to the present day. These numerous and vast urban parks, whether built on flat coastal plains, undulating mountainous hills, or reclaimed land, had different land use patterns before being developed as parks. They also vary significantly in terms of topographical aspects, including area, slope, and elevation. The topographical conditions of their chosen locations may seem to have formed unintentionally in countless site selections. In reality, it is neither entirely natural nor arbitrary but rather inherent in the urbanization process of Shenzhen. The historical context of the site selection of these urban parks is, on the one hand, related to the overall topographical conditions within the city scope and, on the other hand, associated with the social and historical context of Shenzhen's urbanization process.

Therefore, exploring the relationship between the temporal-spatial distribution of urban parks and the topographical conditions under the corresponding social and historical context is a necessary step to bridge the gap between nature and society.

#### 1.3.1.2 The Evolution of Ownership and Morphology During Urban Park Construction

Shenzhen is characterized by a multi-centered, cluster-based urban spatial structure, which results in variations in the timing of development and construction as well as the formation of urban landscapes in different urban areas. This provides a favorable intermediate scale for research. Over different periods, the transformation of spatial-industrial relationships in urban areas has led them to evolve gradually from agricultural landscapes to urban landscapes, with urban parks emerging amidst such transitions. Understanding the evolution of urban district landscapes lays the groundwork for a better comprehension of the changes in land use for urban parks. Whether transforming from plains, hills, or coastal areas into urban parks, it is an evolution in form. However, the morphological transformation of urban space must be based on changes in land ownership. Before the establishment of Shenzhen as a city, its industrial structure was primarily dominated by agriculture, and most of its land was collectively owned by rural villages. After the implementation of a series of urban policies, these collectively owned lands were nationalized. Simultaneously, the relationships between people and land attached to the land underwent changes, marking a transformation of nature across different cultural forms. The newly built urban parks, however, do not manifest in a spatial form devoid of any previous traces, as they are ultimately built upon the former agricultural fields, mountains, forests, fishponds, or coastal areas. Through planning and design, these topography and vegetation condition transform into subsequent urban parks, showcasing their meaning in the urban context.

Therefore, the study will explore the changes in ownership and form of land used for urban park construction, thereby illustrating how the nature of different cultural forms undergo transformations within specific social and historical context.

#### 1.3.1.3 The Conflicts and Connections Between Urban Policies, Nature, and Everyday Life

The construction of urban parks undoubtedly receives explicit support from urban policies. In general discourse, the construction of urban parks is considered self-evidently reasonable. However, conflicts or gaps often exist between top-down urban policies and bottom-up everyday life. Individuals who are compelled to abandon their land and change their lifestyle due to the construction of urban parks often become "the Anonymous"[87] figures in historical research, remaining unnoticed. For example, temporary residents who once lived in Meilin Park and Lianhuashan Hill, as well as displaced oyster farmers along the coast of Shenzhen Bay. Furthermore, these urban policies directly shape the urban parks as new cultural forms of nature, such as lychee forests in Lizhi Park used for marketing, and the introduction of color-foliage and flowering plants in various mountain parks. They are all nature that represent in everyday urban life.

Therefore, exploring and uncovering the conflicts and connections between urban policies and everyday life in the process of urban park construction is also a necessary step in examining urbanization from the perspective of socio-natural transformations.

#### 1.3.1.4 The Relationship Between Urban Parks and Globalization

From a domestic perspective, Shenzhen is considered the experimental field for mainland China's reform and opening-up policies. However, from a global standpoint, Shenzhen serves as a nodal city that leads mainland Chinese cities into the global city network system. Understanding the urban development and construction of Shenzhen cannot be approached separately at any particular level. Whether it's constructing the Luohu Center to attract investment from Hong Kong, building the Futian Central Business District to create an "international city," or constructing the Qianhai-Houhai Center for regional integration in the PRD and Greater Bay Area, all represent efforts by Shenzhen to integrate into globalization and lead its industries into the world market. In fact, the construction of urban parks in Shenzhen is closely related to globalization in aspects of national gate image, foreign trade, urban beautification, the construction of a world-famous flower city, the transformation of the Shenzhen-Hong Kong border defense line, and the construction of international communities.

Currently, the global perspective in historiography has evolved from theoretical exploration to the study of various specific themes, transcending the boundaries of nation-states to understand history, which has become a new trend in historiography.[88] For this research, exploring the history of urban parks within the social and historical context of Shenzhen's integration into globalization facilitates a more macroscopic and profound comprehension of urban parks.

[87] Zhu, J. F.朱劍飛 (2021). What is "Everyday Life"? 什麼是"日常生活"? Prospects and Planning in Architecture建築學的遠望與策劃. *Times Architecture*時代建築, 2021(05). (pp. 14–17).

[88] Yu, P. 於沛 (2007). *Globalization and global history*全球化和全球史. Beijing: Social Sciences Academic Press北京:社會科學出版社. (pp. 1–10).

## 1.3.2 *Theoretical Significance and Innovation*

### 1.3.2.1 Attempting to Break the Binary Discourse in the Field of Built Environment

As mentioned earlier, since the Enlightenment, the binary framework of society and nature has shaped the modern academic system and knowledge production. This results in the study of natural elements in urban areas, primarily emphasizing their physical or biochemical mechanisms while overlooking the political-economic logic behind them. Due to the assumption that a pure environment does not exist, these technical studies examining natural elements in urban settings in terms of natural sciences often face a crisis of meaning. Within this dualistic framework, on one hand, the understanding of urban ecological problems is normalized, neutralized, and depoliticized.[89] On the other hand, and closely related to this study, it leads to a concentration of research on urban parks in directions such as the heat island effect, biodiversity, and behavioral research. Especially in the Chinese context, the social and historical aspects of nature in urban park research are almost overlooked. Just as environmental historian Donald Worster criticizes that "there is little history in the study of nature, and there is little nature in the study of history."[90]

In the field of historical research on urban parks, the phenomenon criticized by Worster presents itself in two aspects. Firstly, historical studies of urban parks often fail to adeptly respond to the societal and historical contexts that shape them. At most, these studies present such contexts as background information without thoroughly examining the interactive relationships between urban parks and their social and historical contexts. The second aspect is that even when some studies manage to respond, they are primarily based on artificial elements within the parks, such as buildings, structures, squares, and other man-made features. In these studies, nature is still treated merely as decoration or a container that does not actively participate in historical processes.

In the process of shaping the modern knowledge production system and disciplinary groups, disciplines such as architecture, urban planning, and landscape architecture, collectively referred to as the built environment disciplines, have been defined as applied disciplines, historically lacking theoretical speculation to examine current issues.[91] Today, the relationship between architecture and politics is increasingly widely discussed.[92] However, the relationship between nature and

---

[89] Robbins, P. (2020). *Political ecology: A critical introduction*. West Sussex: John Wiley & Sons, (pp. 8–21).

[90] Worster, D. (1984). History as natural history: an essay on theory and method. *The Pacific Historical Review*, 1–19.

[91] Hsia, C. J. 夏鑄九 (Ed. & Trans.). (1989). *Readings in Social Theories and the Cultural Form of Spaces*空間的文化形式和社會理論讀本. Taipei: Ming Wen Book Co., Ltd. 臺北:明文書局. (pp. 1–4).

[92] Hsia, C. J. 夏铸九 (2020). *Representation of Space: Rupture and repair*空间再现: 断裂与修复. Shanghai: Tongji University Press上海: 同济大学出版社；Zhu, J. F. 朱剑飞 (2018). *Forms and*

politics, more precisely, the relationship between nature and politics in urban settings, is seldom addressed in the built environment disciplines. Breaking through the dichotomy between society and nature may allow the built environment disciplines to "return to nature"—not only in the sense of returning to nature as a material entity but also returning to a vibrant nature that is embedded in everyday life in conceptual and social dimensions.

This study will start from the natural topography of the city, discussing how urban landscapes and urban parks are produced in the context of the interaction between topography and politics. Based on this premise, in Chaps. 4, 5, and 6, the study will investigate urban parks located on plains, hills, and reclaimed land, respectively. The aim is to explore how the natural elements of urban parks are intertwined with the social and historical context, attempting to rectify the shortcomings in historical research that have persisted under the long-standing dualistic framework.

#### 1.3.2.2 Attempting to Break Through the Autonomy Discourse

In Western architectural discourse since the advent of modernism, two fundamentally different value judgment systems are concealed: "heteronomy" and "autonomy." The former believes that architectural forms primarily arise based on external factors, with the core value lying in creating a connection between architecture and external factors. The latter contends that the acquisition of architectural forms should be independent, possessing a rigorous internal logical structure.[93]

In Western classical art, formal principles occupy a significant discourse position. Consequently, describing urban parks as artistic works indeed holds dominant discourse power in driving practice. For instance, Olmsted once referred to himself as an artist, thereby increasing his leverage in the planning and design game.[94] However, as the discourse system of classical art disintegrates and moves toward contemporary art,[95] describing the planning and design of urban parks as practices in the sense of Western classical art today is unlikely to retain dominant discourse power. This is due to the diminishing influence of the bourgeoisie of the nineteenth century in urban park planning and design, and their discourse system regarding art and urban parks has also undergone corresponding changes. Therefore, for the plan-

*politics: An Approach to thinking in architecture.*形式与政治:建筑研究的一种方法. Shanghai: Tongji University Press上海: 同济大学出版社.

[93]Yuan, S. N., & He, B. J. 苑思楠, 何蓓潔 (2020). The Transformation of Architectural Theories and Education after Modernism in between Autonomy and Heteronomy: Speculation and Orientation現代主義之後自律與他律價值演變中的建築理論與教學——思辨與定位. *The Architect*建築師, 2020(02), 40–47.

[94]Olmsted, F. L., Schuyler, D., Kaliss, G., & Schlossberg, J. (2015). *The Papers of Frederick Law Olmsted (Volume III)*. Johns Hopkins University Press. (pp. 304).

[95]Zhu, Q.S.朱青生(2000). *Nobody is an artist, and nobody is not an artist*沒有人是藝術家，也沒有人不是藝術家. Beijing: Commercial Press北京:商務印書館. (pp. 1–15).

ning and construction of urban parks, relying on the introverted logic of Western classical art theory may not be very applicable. Today, the intricate interplay between globalization and nation-state significantly influences local urban construction and development. Therefore, turning toward understanding external factors may perhaps regain dominant discourse power in practice.

Architectural historian Leonardo Benevolo argues that society takes precedence in architecture. He suggests that the emergence of modernist architecture is not due to technical or artistic reasons but rather to the new social, economic, and political relationships brought about by the Industrial Revolution. He also criticizes the approach of modern architectural history writing that focuses on form and art as starting points.[96] Similarly, geographer Neil Smith has highlighted the precedence of society over nature. He emphasizes that this priority is not something externally imposed by the "social" but is inherent in its own attributes.[97] Clearly, in these concepts, the construction of space and nature is not based on a logic of autonomy. The nature and space in the experiential world are, in fact, a system of representations determined by social practices and historical processes that define meaning.[98,99]

This study focuses on the social and historical context shaping urban parks, rather than the physical forms of urban parks themselves. Therefore, the orientation of the research path leans more toward "heteronomy" than "autonomy." This also constitutes an important aspect of the theoretical significance of this study, namely, attempting to break through the discoursive logic hidden in most urban park historical research: the form of urban parks is subject to some inevitable, specific, and independent formal rules beyond politics and economics. Most of the time, this autonomy, as understood by Colin Rowe,[100] is based on classical formal principles. In short, uncritically naturalizing the formal principles of classical gardens, considering them as constant rules that urban park historical research and design practices should unquestionably follow or inherit, is precisely the general discourse that this study seeks to break through. After all, historical research is not meant to perpetuate traditional culture but to question existing discourses.[101]

---

96 Tournikiotis, P. (2001). *The historiography of modern architecture*. Massachusetts: The MIT Press. (pp. 93–99;109).

97 Smith, N. (2010). *Uneven development: Nature, capital, and the production of space*. University of Georgia Press. (pp. 47–48;87–91).

98 Cosgrove, D. E. (1998). *Social formation and symbolic landscape*. University of Wisconsin Press. (pp. 1–12).

99 Castells, M. (1983). *The city and the grassroots: A cross-cultural theory of urban social movements*. London: E. Arnold. (pp. 303).

100 Colin Rowe, Robert Slutzky. 柯林·罗, 罗伯特·斯拉茨基 (2007). Transparency (Trans. by Jin, Q.Y. & Wang, Y.J. 金秋野 王又佳). *Beijing: China Architecture & Building Press*北京: 中國建築工業出版社. (pp. 23–55); Yuan, S. N., & He, B. J. 苑思楠, 何蓓潔 (2020). The Transformation of Architectural Theories and Education after Modernism in between Autonomy and Heteronomy: Speculation and Orientation現代主義之後自律與他律價值演變中的建築理論與教學——思辨與定位. *The Architect*建築師, 2020(02), 40–47.

101 Luo, X. 羅新 (2020). Rebels Who Refuse to Conform: Criticism, Doubt, and Imagination有所不為的反叛者 批判、懷疑與想像力. *Shanghai: Shanghai Joint Publishing Company*上海: 上海

### 1.3.3 *Empirical Significance and Innovation*

#### 1.3.3.1 Attempting to Fill the Gaps in Contemporary Urban Park Research in China

In 1969, architectural historian John Mass criticized the American architectural history field for focusing solely on classical architecture, leaving the abundant construction of contemporary industrial buildings as a realm "feared to tread." In his view, this involved various methodological issues within Western architectural history research traditions.[102] Mass's criticism, although occurring over 50 years ago in the United States, is still relevant and deserving of attention in today's Chinese context.

Over the last four decades of reform and opening up, China's urban development has experienced significant and rapid progress, with urban parks sprouting up across the country like mushrooms after rain. Nevertheless, a substantial proportion of current research has been dedicated to investigating traditional gardens with a long history. Research on Chinese urban parks tend to concentrate on the modern parks built between 1840 and 1949, covering topics such as nationalism, colonialism, civic public life, and Sino-Western cultural integration. There is limited research on park construction between 1949 and 1978, and almost no research on contemporary parks after the reform and opening up.

Research on parks before the reform and opening-up period, and even traditional gardens, certainly holds significance. However, this does not imply that urban parks constructed after the reform and opening up should be left in a state of research neglect. In fact, these contemporary urban parks are evidently more lively, more worthy of contemplation, and also more likely to be integrated with theories. The existence of this research gap may be attributed to several reasons: (1) Scholars such as Tong Jun (童寯,1900–1983), Liu Dunzhen (劉敦楨,1897–1968), Zhou Weiquan (周維權, 1927–2007), Zhang Jiaji (張家驥,1932–2013), Yang Hongxun (楊鴻勳,1931–2016), and others have already explored methods and frameworks in their study of traditional gardens. Historians choosing previously unexplored topics following these paths may face smaller risks in terms of methodology. (2) Traditional gardens, as private properties reflecting the interests and aspirations of ancient literati, have rich cultural connotations. Studies on them mainly follow the methodology of art history. However, the user and spatial production processes of traditional gardens are significantly different from contemporary urban parks. This makes existing methodology less applicable to the study of modern urban parks, posing evident risks if applied indiscriminately. (3) Historians generally believe that there should be a considerable temporal distance from the research object before it can be studied. This is in contrast to the perspective of Marc Bloch, who recognized that

三聯書店. (pp. 007–011).

[102] Maass, J. (1969). Where architectural historians fear to tread. *Journal of the Society of Architectural Historians*, 28(1), 3–8.

history is "nothing other than inquiry" and that it can be momentary convulsions or the most lasting developments.[103]

Given the unique significance of Shenzhen in contemporary China, this study focuses on the urban parks of Shenzhen, aiming to address the lack of discourse on contemporary urban parks in historical research.

#### 1.3.3.2 Attempting to Rectify the Deficiencies in Shenzhen's Urban History

After the year 2000, research on the urban history of Shenzhen[104] and urban construction history of Shenzhen[105] has gradually emerged from different perspectives. However, in these studies, references to urban park construction are only briefly mentioned. Apart from retrospective articles and project introductions[106] published by planners, designers, and project managers who have personally participated in urban park construction, there is almost no scholarly or analytical research. This is partly due to the aforementioned lack of research methods, and on the other hand, it is also related to the rapid construction in Shenzhen, which makes it difficult to archive, organize, and publicly disclose historical documents in a timely manner. Nevertheless, it needs to be acknowledged that an urban park is the best way to study urban history from a natural perspective.

In fact, research activities on the urban history and urban construction history of Shenzhen are more or less closely related to various levels of government,

---

[103] Bloch, M. (1992). *The historian's craft*. Manchester University Press. (pp. 17–18).

[104] O'Donnell, M. A., Wong, W., & Bach, J. (Eds.). (2019). *Learning from Shenzhen: China's post-Mao experiment from special zone to model city*. University of Chicago Press; Du, J. (2020). *The Shenzhen experiment: The story of China's instant city*. Harvard University Press.

[105] Zhang, Y. Q. 張勇強 (2004). *Research on Self-Organization of Urban Space Development: A Case Study of Shenzhen*城市空間發展自組織研究——深圳為例 [Doctoral dissertation, Southeast University東南大學].

Chen, Y. X. 陳一新 (2015). *Shenzhen Futian CBD: A Thirty-year History of Planning and Construction*深圳福田中心區(CBD)城市規劃建設三十年歷史研究. Nanjing: Southeast University Press南京:東南大學出版社; Shenzhen Municipal Planning and Land Resources Commission深圳市規劃和國土委員會 (2010). *Urban Planning Practice in Shenzhen's Fifteen Years of Reform and opening-up: 1980-1995*深圳改革開放十五年的城市規劃實踐: 1980-1995年. Shenzhen: Haitian Publishing House深圳: 海天出版社.

[106] He, F. 何昉 (2010). Shenzhen Landscape Architecture: From the Inheritance of Landscape Architecture Mechanism of China to the Creation of Regional Style深圳園林:從繼承中國風景園林體系到地域風格的創建. *Landscape Architecture*風景園林, 2010(05), 70–72; Meng, Z. Z. 孟兆禎 (2020). Chinese Traditional Garden Flourishing in Shenzhen: In Honor of the 40th Anniversary of Shenzhen Special Economic Zone大鵬風舞中國傳統園林——紀念深圳特區成立40周年. *Landscape Architecture*風景園林, 27(10), 10–12; Li, Z. S. 李錚生 (1994). Planning of Shenzhen's Green Space System深圳市綠地系統規劃. *City Planning Review*城市規劃, 1994(01), 53–55; Shenzhen Urban Management and Comprehensive Law Enforcement Bureau深圳市城市管理和綜合執法局. (2022). *Stories of Shenzhen Parks*深圳公園的故事. Publishing House of Shenzhen Press Group深圳:深圳報業集團出版社.

inevitably influencing the research methods, perspectives, viewpoints, and conclusions. The urban construction of Shenzhen has long been described as an ideal city built on a "blank slate,"[107] overlooking the rich pre-reform history of this land, avoiding discussions on many complex social relationships, neglecting the shaping impact of the original topography on the urban form. Portrayal of this kind isolates Shenzhen's urban construction in the historical context that only after the reform and opening-up era. Specifically, how the land used in the early urban development of the SSEZ transitioned from rural collective land to state-owned land, how this transformation triggered the urbanization of the land and its physical topography changes, and how this transition concealed the transformations of human are crucial questions. If the urban construction of Shenzhen over the past forty years is a contemporary "myth" radiating a divine light, avoiding these questions makes it difficult for historical research to penetrate the shadows behind the divine light and the individuals beneath those shadows. In reality, it misses the opportunity to examine urbanization from the perspective of socio-nature transformations. Urban parks, as a type of urban space that preserves natural features to the maximum extent, provide the best opportunity to initiate such scrutiny. As Edward W. Said stated, historical research should involve: "scouring alternative sources, exhuming buried documents, reviving forgotten (or abandoned) histories."[108] This aligns with the reminder from urban scholar Du Juan that historical research about Shenzhen should remember: "its humble beginnings of mere rice fields."[109]

This study, on the one hand, has unearthed and supplemented a large amount of historical materials for the historical research of urban parks in Shenzhen. On the other hand, in the selection of these historical materials, it tends to lean on the socio-nature relationship, providing corresponding perspectives to observe the interaction between micro-level individuals under urbanization and urban policies.

## 1.4 Structure of the Book

Based on the research questions, definitions, and objectives outlined above, the core content of this thesis, comprising Chaps. 3, 4, 5, and 6, will progress on two levels of detail: The first level (Chap. 3) will focus on the political-economic and geographic-topographic contexts of the temporal-spatial distribution of urban parks within the city scope. The second level (Chaps. 4, 5, and 6) will conduct more detailed case studies of urban parks constructed under different topographic

[107] People's Daily Commentary Department人民日報評論部 (2020, October 26). *Embarking on Splendor on a Blank Sheet of Paper*在一張白紙上演繹精彩. People's Daily人民日報(005); Ma, X., Zhang, L., & Bai, X. J.馬瑢,張淩,白景雪 (2009, December 20). *A Blank Sheet of Paper Lets Dreamers Freely Express Themselves* 一張白紙任理想者揮灑筆墨. Shenzhen Special Zone Daily深圳特區報(A03).

[108] Said, E. W. (2012). *Representations of the Intellectual*. New York: Vintage Books. (pp. xviii).

[109] Du, J. (2008). Don't underestimate the rice fields. *Urban Transformation*, 218–225.

conditions within the special economic zone. As shown in Fig. 1.2, the specific approach is outlined as follows.

Chapter 3 will provide a comprehensive understanding of the construction and development status of urban parks within the city scope, divided into three parts: Firstly, it will present the landscape history of Shenzhen, examining how the geographic-topographic features of Shenzhen intertwine with the political-economic context of globalization to produce the urban landscape of Shenzhen; Secondly, it will divide the period from 1980 to 2020 into four stages, describing the temporal-spatial distribution and evolution characteristics of urban parks within the city scope. Finally, it will interpret these distribution and evolution characteristics within the history of Shenzhen's urban development. These three parts correspond to the three sections of Chap. 3.

Chapters 4, 5, and 6 will respectively focus on the Shagnbu-Luohu Park Group built on the plain, the Meilin-Gangxia Park Group built on the mountainous and hilly topography, and the Shenzhen Bay Park Group built on reclaimed land. These chapters will examine how different topographies are integrated into the social and historical context of urban park construction from the intertwined perspectives of

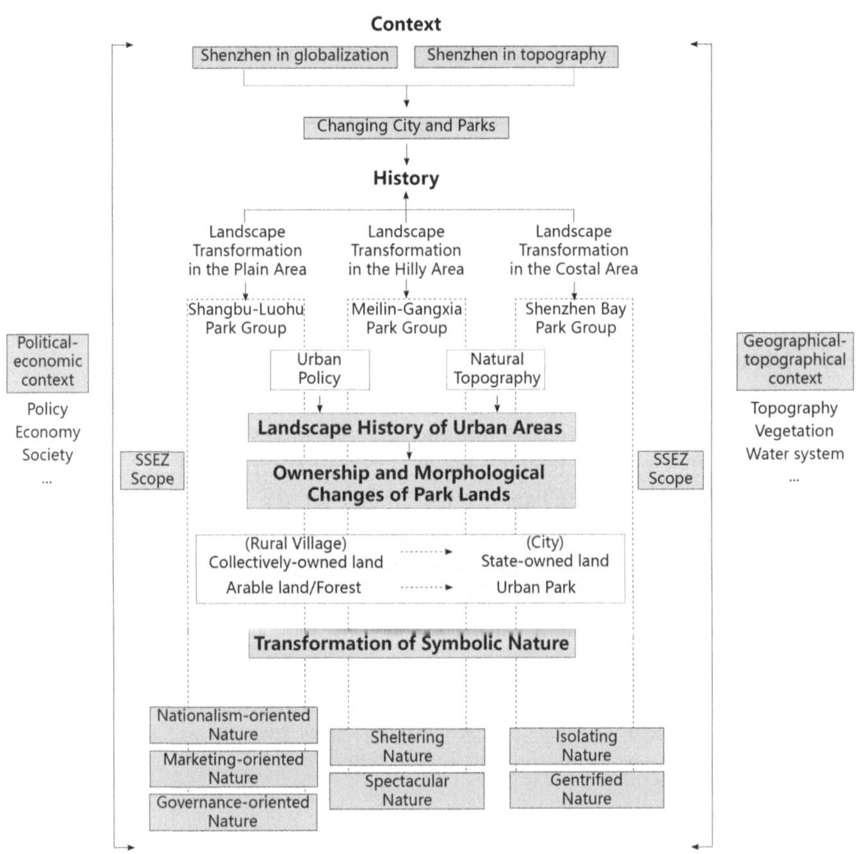


**Fig. 1.2** The research framework

political-economic and geographic-topographic contexts, and how they have contributed to the urban development within specific contexts. As shown in Fig. 1.3, the westward shift of urban centers within the SSEZ over the past 40 years coincides with the opening time of three park groups. This is not merely a coincidence, but rather a reflection of both geographical and historical factors. Geographically, the plains, mountainous, and hilly areas, and reclaimed land represent the three typical topographical characteristics upon which urban construction within the SSEZ is based. Historically, the construction of the Luohu Center, which began in the 1980s, featured urban parks primarily located on plains; the construction of the Futian CBD, initiated in the mid-1990s, featured urban parks mainly situated in mountainous and hilly areas; and the Qianhai-Houhai Center, which started around 2010, featured urban parks primarily built on reclaimed land. In such a historical-geographical or socio-natural context, urban parks within the SSEZ are interwoven into its urban life. From a positivist perspective, topography simply represents the objective spatial form of the earth's surface. But understanding it in historical writing should not stop there. As David Leatherbarrow understands, the study of topography would "lead nowhere" if it meant only an interest in the physical aspects of the land. In his opinion, "cultural connotation" and "life practice" ought to be the focus of topographic research.[110]

While this study may provide materials or inspiration for writing a systematic and comprehensive history of urban park planning, construction, and design in Shenzhen, this kind of history is not the primary focus of this research. The intention of this study is to use Shenzhen as a case study to explore the relationship between urban parks and urban social life, thus addressing the theoretical dilemma of the disconnect between society and nature, as well as between landscape history and the history of landscape architecture within the built environment domain. Strategically, the research will begin by examining the "physical topography" and then attempt to articulate the concept of "topography of life," specifically selecting

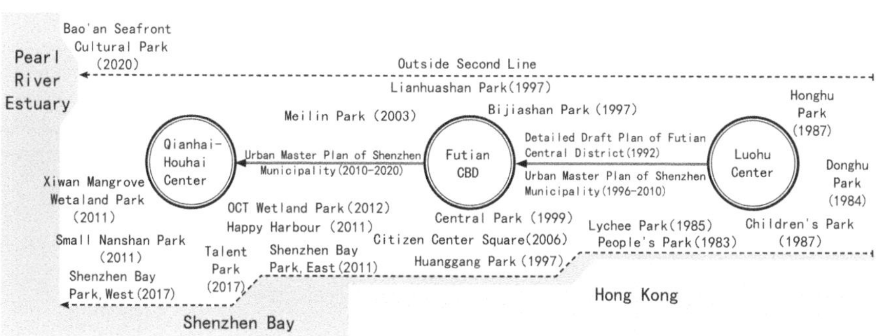


**Fig. 1.3** Diagram of the distribution of large urban parks inside SSEZ

[110] Leatherbarrow, D. Is Landscape Architecture? In Doherty, G., & Waldheim, C. (Eds.). (2015). *Is landscape...?: Essays on the identity of landscape*. Routledge. (pp. 327–338).

urban parks situated on three typical physical topographies within the economic special zone: plains, hilly areas, and reclaimed land. The goal is to explain how these parks intertwine within specific social and historical contexts at different times and in different ways. We chose these three physical topographies as case studies to provide insights into the transformations of nature within the Shenzhen city scope, thereby addressing the original intent of the research.

Based on such a perspective, Chaps. 4, 5, and 6 will follow a parallel framework, each consisting of three parts: The first part will delve into the landscape history of the urban area where the park group is located, elucidating the social and historical context of the emergence of the urban landscape starting from the changes in natural terrain in the area. This context is also the context in which urban parks emerge; The second part will discuss the ownership and morphological changes of the land used for these urban parks, which are the prerequisites for all changes to occur; The third part will select 2–3 aspects of the construction and development process of urban parks to elaborate on the transformation of symbolic nature, i.e., the transition of nature between different cultural forms. Additionally, it will analyze in detail how urban parks built under different topographic conditions represent vibrant local practices within the context of globalization.

# Chapter 2
# Conceptual Background

## Contents



**Abstract** This chapter primarily outlines the conceptual background of the study. Through a review of research spanning three areas—human-land relations, urban studies, and the history of urban parks—the author points out that historical studies of urban parks have often overlooked the critical issue of the historicity of natural elements.

**Keywords** Human-land relations · Urban studies · History of urban parks · Conceptual background

In Hunter's view, urban parks historically, metaphorically, and symbolically exist between nature and the city.[1] According to this understanding, the historical study of urban parks requires both an understanding of nature and an understanding of the

---

[1] Hunt, J. D. (2014). *A world of gardens*. Reaktion Books. (pp. 241–256).



X. Liu, *Fabricating Nature*, The Urban Book Series,
https://doi.org/10.1007/978-981-95-8208-2_2

city. The following is divided into (1) research on human–land relationships; (2) urban studies; (3) historical research on urban parks, for a comprehensive review, and the literature is organized roughly in chronological order.

## 2.1 Human–Land Relationship

### *2.1.1 Human–Land Relationship in Human Geography*

Since the 1970s, two complementary yet distinct traditions of geographical research have gained widespread influence:[2] the first is the Marxist tradition (Marxist geography), which focuses on how nature is transformed into various human uses under different social conditions; the second is the cultural geography tradition, which concentrates on the changes in the concept of nature, exploring the meaning of nature to different societies and how societies represent nature in various forms, such as text or images.

Marxist geography seeks to analyze human–land relationship issues within a broader social and political context. As early as the mid-nineteenth century, Marx's critique of capitalism was actually based on the dialectical relationship between nature and society. Noel Castree points out the geographical significance of Marxism, arguing that the commodity economy not only changes the physical structure of the natural world but also transforms the relationship between humans and land. In the capitalist mode of production, landscapes, bodies, or cellular structures can all be transformed into commodities.[3] Marx emphasized the historical unity of nature and society, arguing that labor integrates nature into the commodities of society and all that nature is the product of human labor.[4] This inspired many geographers during the social and economic crises of the 1960s and 1970s. David Harvey criticized the positivism that emerged in the 1950s in geography, as well as the reliance on positivist thinking in social theory, physics, engineering, and many other disciplines. He advocated for the use of dialectical methods to deepen the understanding of social–ecological processes, arguing that changes in space,

---

[2] Whatmore, S. (2007). Culture-Nature. In P. Cloke et al. (Eds.), *An Introduction to Human Geography*. (Trans. by Wang Zhihong王志弘 et al.). Taipei: Chuliu Book Co., LTD. (pp. 006–016).

[3] Castree, N. (1995). *The nature of produced nature: materiality and knowledge construction in Marxism*. Antipode: a radical journal of geography, 27 (1), 12–48.

[4] Marx, K., & Engels, F. 卡爾·馬克思, 弗裡德里希·恩格斯 (2001). *Marx-Engels Collected Works (Volume 44)* 馬克思恩格斯全集(第44卷). (Trans. by Central Compilation and Translation Bureau of Marxist, Leninist, Stalinist Works of the Central Committee of the Communist Party of China 中共中央馬克思恩格斯列寧史達林著作編譯局). Beijing: People's Publishing House北京: 人民出版社. (pp. 88–102).

environment, and place are not neutral but related to human practices of dominance.[5] Neil Smith also criticized positivism, viewing nature, space, and uneven development as intertwined with capital, necessitating a Marxist relational perspective to explain the natural and spatial basis of landscapes. Edward Soja recognized that space is not merely a physical environment but also a product of social relations and power. In summary, within the tradition of Marxist geography, the natural world cannot be conceptually or materially separated from human society; the politics of nature is, in essence, the politics of society.

Cultural geography focuses on the representation of human–land relationships in different cultures and the relationship between different forms of representation and views of nature. This tradition values the concept of nature, considering the real natural entity to have the same significance as nature represented in texts and paintings. Therefore, Cosgrove believes that "a landscape is more palpable but no more real, nor less imaginary, than a landscape painting or poem…,"[6] viewing landscapes as one of the ways people observe nature within culture. He has also studied the relationship between cultural products like nature poetry, landscape paintings, and industrial capitalism.[7] J.B. Jackson focused on contemporary American landscapes, moving beyond his earlier emphasis on the visual to advocate for understanding landscapes as environments of human activity rather than abstract spaces. This led him away from formalism and style toward vernacular landscapes.[8] Mike Crang observed that within the context of nation-states, the state would establish national images through specific symbols, and he has also elaborated on how power and meaning are conferred on human–land relationships.[9] Wendy Darby, on the other hand, investigates the aesthetic discourse of British nationalism by tracing the literary and artistic origins of landscapes. Additionally, she analyzes the origins of the idea of London-centered home counties being shaped into quintessential zones by British literature.[10] In summary, cultural geography recognizes the intricate connections between real and represented nature, and it is particularly concerned with the hidden meanings behind different forms of representation.

Around the turn of the new millennium, a new materialist geography emerged. For instance, Sarah Whatmore has reservations about both of the aforementioned

[5] Harvey, D.哈威 (2010). *Justice, Nature and the Geography of Difference*正義、自然和差異地理學. (Trans. by Hu D. P. 胡大平). Shanghai: Shanghai People's Publishing House上海: 上海人民出版社. (pp. 7; 52).

[6] Cosgrove, D., & Daniels, S. (Eds.). (1988). *The iconography of landscape: essays on the symbolic representation, design and use of past environments (Vol. 9)*. Cambridge University Press. (pp. 1–2).

[7] Cosgrove, D. E. (1998). *Social formation and Symbolic Landscape*. Univ of Wisconsin Press. (pp. 223–253).

[8] Jackson, J. B. (1984). *Discovering the vernacular landscape*. Yale University Press. (pp. 1–4).

[9] Crang, M. (2003). *Cultural Geography*. (Trans. by Yang Shuhua et al.). Nanjing: Nanjing University Press. (pp. 6; 18–34).

[10] Darby, W. J. (2020). *Landscape and identity: Geographies of nation and class in England*. Routledge. (pp. 1–8).

traditions: the Marxist tradition overemphasizes the ability of capital to produce nature, neglecting the active role of natural entities and processes in shaping human–environment relationships; and the cultural geography tradition fixates on human imagination, overlooking the diverse forms of life. Therefore, Whatmore emphasizes the need to animate geographical landscapes.[11] Among Western scholars, Peter Jackson, Alan Latham, and Derek McCormack share similar views.[12,13]

### 2.1.2 Human–Land Relationship in Environmental History

Compared to human geography, which also regards human–land relationships as a crucial part of its research, urban environmental historiography has an academic lineage of just over 30 years. However, its two predecessors, urban history and environmental history, have earlier origins, each becoming relatively independent branches of history during the 1960s and 1970s.[14]

Shane Ewen views urban history as the historical study of cities placed within broad economic, social, political, cultural, and spatial systems, further situating them within wider regional, national, international, and even transnational networks for examination.[15] Urban history truly emerged as an important field of historiography, mainly in the 1960s, with Jim H. Dyos being a notable figure in its development.[16] At this time, burgeoning leftist movements spurred historical research to move away from previous political dominations, starting to focus on the marginalized groups within cities and contesting the previously elite tradition of historical writing. Issues such as gender, race, and class inequalities, as well as environmental crises, received scholarly attention,[17] exemplified by works like "Poverty and Progress: Social Mobility in a Nineteenth-Century City."[18] Hence, this also laid the groundwork for the rise of environmental history studies in Europe and America. Donald Hughes considers environmental history a discipline that understands

---

[11] Whatmore, S. (2007). Culture-Nature 文化-自然. In P. Cloke et al. (Eds.), *An Introduction to Human Geography*人文地理概論. (Trans. by Wang Zhihong王志弘 et al.). Taipei: Chuliu Book Co., LTD.巨流圖書股份有限公司 (pp. 006–016).

[12] Jackson, P. (2000). *Rematerializing social and cultural geography*. Social & Cultural Geography, 1(1), 9–14.

[13] Latham, A., & McCormack, D. P. (2004). Moving cities: rethinking the materialities of urban geographies. *Progress in human geography*, 28(6), 701–724.

[14] Xiao, X. D. 肖曉丹 (2018). *A Historiography of European Urban Environmental History*歐洲城市環境史學研究. Chengdu: Sichuan University Press成都:四川大學出版社. (pp. 26).

[15] Ewen, S. *What is Urban History*. Cambridge: Polity Press, 2016. (pp. 1).

[16] Fraser, D., & Sutcliffe, A. (1983). *The Pursuit of urban history*. E. Arnold. (pp. ix–x).

[17] Hou, S. (2013). *Cities Without Borders: From American Urban History to Urban Environmental History*. Journal of Renmin University of China, 27(03), 20–29.

[18] Thernstrom, S. (1964). *Poverty and progress: Social mobility in a nineteenth century city*. Harvard University Press.

human behavior and thought throughout history by studying the changing relationships between humans and nature across different eras.[19] In the 1960s and 1970s, environmental history was marked by significant works such as Rachel Carson's Silent Spring, which criticized the human misuse of chemical substances that not only harmed wildlife but also directly threatened human survival.[20] Another key work is Roderick Nash's "Wilderness and the American Mind," which describes the process of Americans adjusting their views on wilderness, suggesting that the shift from hostility toward wilderness to its admiration reflected changes in American environmental thought.[21] By the mid-1980s, scholars like Donald Worster and Richard White had relatively systematically reviewed the research and theoretical methods in American environmental history.[22]

Before the 1990s, urban history research in Europe and America seldom viewed cities as manifestations of human-land relationships. In 1992, William Cronon's publication of "Nature's Metropolis" broke through this limitation by dissolving the traditional oppositions found in environmental history works between city and countryside, society and nature. Grounded in the flow and commodification of key materials like timber and corn, it revealed the close connection between the metropolis of Chicago and its vast rural hinterland.[23] Following Cronon, in 1994, Christine M. Rosen and Joel Tarr co-edited a "Special Issue on Environment and City" for the Journal of Urban History, in which their extensive research elaborated on the interactions between cities and natural environments, arguing that urban historians needed to enhance, rather than disperse, societal understanding of environmental change.[24] This special issue also included research by other scholars on the topic, collectively pushing urban history to begin considering nature at a theoretical level, marking the formation of urban environmental history as a study of human–land relationships within the urban context.

After the turn of the millennium, research in urban environmental history gradually shifted from theoretical discussions to a broader range of empirical studies, with its subjects becoming more diversified, covering garbage, urban parks, energy, water resources, pollution, natural disasters, biodiversity, and more. For example, Martin

---

[19] Hughes, J. D. (2016). *What is environmental history?*. John Wiley & Sons. (pp. 1–7).

[20] Carson, R. (2002). *Silent spring*. Houghton Mifflin.

[21] Nash, R. F. (2014). *Wilderness and the American mind*. Yale University Press.

[22] White, R. (1985). American environmental history: the development of a new historical field. *Pacific Historical Review*, 54(3), 297–335; Worster, D. (1984). History as natural history: an essay on theory and method. *Pacific Historical Review*, 53(1), 1–19.

[23] Cronon, W. (1992). *Nature's metropolis: Chicago and the Great West*. W.W. Norton & Company. (pp. xv–22).

[24] Rosen, C. M., & Tarr, J. A. (1994). The importance of an urban perspective in environmental history. *Journal of Urban History*, 20(3), 299–310.

V. Melosi has studied themes such as garbage and energy in cities,[25,26] viewing urban environmental history not as a separate field but as a perspective or way of thinking about cities from the angle of human–land relationships.[27] Matthew Gandy, taking New York as a case study, argued that the emergence of urban nature involves not only transformations in capital but also changes in the role of the state, the rise of the culture of the natural metropolis, and the interweaving of various aspects of urban life, with urban parks becoming new commodities under the interplay of real estate and landscape aesthetics.[28] William Solecki and Cynthia Rosenzweig researched the biodiversity of the New York metropolitan area, suggesting that biodiversity is not a natural outcome but a product of ecological and social interactions, connecting local, regional, and global scales.[29] David Blackbourn conducted research on the transformation of various types of water systems in Germany, describing how Germans built cities, constructed dams, and ultimately established modern Germany, while also reflecting on the consequences of conquering nature.[30] Paul Robbins traced the relationship between suburban green spaces and chemical pollution in the United States, suggesting that the lawns in front of suburban homes are actually reconstructions of nature by chemical products.[31] Inn Douglas's research on urban environmental history observed the impact of globalization on cities, discussing urban parks and green spaces and using London as an example to show the historical evolution of green spaces.[32] Globalization is also a theme in the works of Lisa Benton-Short and John Rennie-Short, which cover a wide range of urban environmental aspects such as water, air, waste, disasters, and inequality. They observed that natural disasters are natural processes manifested through socio-political and economic arrangements, suggesting that cities are constructed by both natural and social environments.[33] Aside from research involving disasters, most studies are written based on the logic of "nature as a victim," including Cronon's "Nature's Metropolis." This unidirectional action of humans on nature clearly deviates from the original intention of

---

[25] Melosi, M. V. (2004). *Garbage in the Cities: Refuse reform and the Environment*. University of Pittsburgh Press.

[26] Melosi, M. V. (2000). *Effluent America: Cities, industry, energy, and the environment*. University of Pittsburgh Press.

[27] Bao, M. H.包茂宏 (2004). Martin Melosi and his Study of the U.S. Urban Environmental History 馬丁·麥樂西與美國城市環境史研究. *Journal of Chinese Historical Geography*中國歷史地理論叢, (04): 114–126+160.

[28] Gandy, M. (2003). *Concrete and clay: reworking nature in New York City*. MIT Press. (pp. 5).

[29] Solecki, W. D., & Rosenzweig, C. (2004). Biodiversity, biosphere reserves, and the Big Apple: a study of the New York Metropolitan Region. *Annals of the New York Academy of Sciences*, 1023(1), 105–124.

[30] Blackbourn, D. (2007). *The conquest of nature: Water, landscape and the making of modern Germany*. Random House. (pp. 1–20).

[31] Robbins, P. (2012). *Lawn people: How grasses, weeds, and chemicals make us who we are*. Temple University Press.

[32] Douglas, I. (2013). *Cities: an environmental history*. Bloomsbury Publishing. (pp. 1–22; 350–398).

[33] Benton-Short, L., & Short, J. R. (2013). *Cities and nature*. Routledge. (pp. 1–22).

urban environmental history to study the interactive relationship between cities and the natural environment. Ted Steinberg's publication, "Gotham Unbound: The Ecological History of Greater New York," goes beyond Cronon by clearly presenting the cooperation and conflict between New York, a city on an estuary, and nature.[34] Moreover, in these studies, urban nature is not discussed in isolation but is debated at the interface of society and nature.

The start of urban environmental history research by Chinese scholars was relatively late, beginning around the year 2000 with historical studies on ancient urban environments. Examples include Cheng Suiying's research on Kaifeng during the Tang and Song dynasties,[35] Zhu Shiguang's study on the relationship between the Guanzhong area and the Han dynasty's Chang'an city,[36] and Yu Xixian's research on the changes in Beijing's natural environment.[37] Additionally, research on urbanization in China has seldom involved the social and political processes of nature, with recent studies beginning to emerge around modern cities. For instance, Cao Mu conducted research on the modern functional changes of Tianjin's Qiangzi River, involving issues of environmental rights, water resource distribution, commercial economy, territorial sovereignty, etc., to explore the interaction between nature and cities.[38] Li Yingchun and Meng Xinyu examined the spatial changes of the riverside grasslands along the Bund in modern Shanghai and their social drivers.[39] Wang Yu and Shi Qiujie studied Shenzhen, presenting the phased evolution of the city-water relationship and the changing power structures behind it.[40]

[34] Steinberg, T. (2014). *Gotham Unbound: The Ecological History of Greater New York*. Simon and Schuster. (pp. xv–40).

[35] Cheng, S. Y. 程遂營 (2002). *Research on the Ecological Environment of Kaifeng during the Tang and Song Dynasties*唐宋開封生態環境研究. China Social Sciences Press中國社會科學出版社.

[36] Zhu, S. G. 朱士光 (2000). The Relationship between the Ecological Environment Characteristics of the Guanzhong Region in the Western Han Dynasty and the Capital City Chang'an西漢關中地區生態環境特徵與都城長安相互影響之關係. *Journal of Shaanxi Normal University: Philosophy and Social Sciences Edition*陝西師範大學學報:哲學社會科學版, 29(3), 8.

[37] Yu, X. X. 於希賢 (1995). Preliminary Study on the Historical Natural Environment Changes in Beijing北京市歷史自然環境變遷的初步研究. *Journal of Chinese Historical Geography*中國歷史地理論叢, (01), 45–58.

[38] Cao, M. 曹牧 (2022). Resources, Environment, and Rights: Modern Pollution Transformation and Impact of Wangzi River in Tianjin資源、環境與權益——天津牆子河的近代排汙轉型與影響. In Gao, B., & Hu, H. 高波, 胡恒 (Eds.), *How Much Space Is There in "Space"? Revisiting Multidimensional History*"空間"還有多少空間?重放多維度歷史. Beijing: China Social Sciences Press北京:中國社會科學出版社.

[39] Li, Y. C., & Meng, X. Y. 李穎春,孟心宇 (2023). Detailed Study on the Spatial Changes of the Bund Riverside Grassland in Modern Shanghai近代上海外灘濱江草地空間變遷詳考. In Ma, X. Q., Tsukada, T., & Tang, H. R. 馬學強,塚田孝,唐惠榮 (Eds.), *Proceedings of the International Academic Symposium on Spatial Morphology and Historical Cultural Evolution of Waterfront Cities*濱水城市空間形態與歷史文化演進國際學術研討會論文集. Beijing: Commercial Press北京: 商務印書館. (pp. 62–82).

[40] Wang, Y., Man, C. X., Xu, A. L., et al. 王雨,滿彩霞,徐瑷琳等 (2022). Spatial Transformation and Environmental Governance Innovation in China from the Perspective of Urban Political

### *2.1.3 Trend and Enlightenment*

From the above discussion, it is clear that the human–land relationship is a crucial component of both human geography and environmental history, representing a commonality between the two disciplines. In fact, at least in Europe, human geography serves as one of the most significant theoretical sources for environmental history.[41]

Entering the twentieth century, the understanding of human–land relationships in geography gradually freed itself from environmental determinism. In the 1950s, humanistic geography, grounded in phenomenology, criticized positivism-based quantitative geography, and both were later critiqued by Marxist scholars, with David Harvey emerging as one of the most important figures of that period. Concurrently, cultural geography, which emphasized studying society through representations of nature and was represented by scholars such as Cosgrove, Jackson, and Cronon, also made its mark. After the new millennium, scholars like Whatmore argued that previous human geography overly emphasized human agency without adequately considering the active role played by natural entities, suggesting a "return to materialism." Indeed, academic history demonstrates that the understanding of human–land relationships has evolved over time, and the concept of "returning to materialism" may offer theoretical insights into exploring the agency of natural entities like urban parks in promoting urban development. After all, when discussing urban parks within the context of built environment studies, their materiality is an aspect that cannot be overlooked.

Urban environmental history, originating in the 1990s, emerged from the union of urban history and environmental history research. From this point, cities began to be seen as manifestations of human–land relationships within the study of environmental history. Subsequent efforts by scholars like Rosen, Tarr, and Melosi brought a theoretical focus on nature within urban history. After the new millennium, a wide range of empirical studies began to emerge, covering various aspects of the relationship between cities and nature, including urban parks, though nature was often described within the logic of being a victim. Steinberg's research stepped out of this logic, presenting a bidirectional interaction between cities and nature. Meanwhile, the study of urban environmental history in China, which began to take off in the new millennium, has also started to pay attention to the social and political processes of nature within cities in recent years. Urban environmental history and urban history research are interconnected, emphasizing the analysis of changes in natural elements within cities against a broad backdrop of political and economic contexts. This perspective provides theoretical justification for analyzing the historical context of urban parks in cities like Shenzhen.

Ecology: An Analysis Based on the Dynamic Evolution of the Urban-Water Relationship in Shenzhen城市政治生態學視角下的空間轉型與中國環境治理創新——基於深圳城——水關係動態演變的分析. *Progress in Geography*地理科學進展, 41(09), 1755–1769.

[41] Xiao, X. D. 肖曉丹 (2018). *A Historiography of European Urban Environmental History*歐洲城市環境史學研究. Chengdu: Sichuan University Press成都:四川大學出版社. (pp. 219).

## 2.2 Urban Studies

### 2.2.1 *Studies on Global Cities*

In the 1980s, the concept of global or world cities propelled urban studies into new heights. In 1982, John Friedmann and Goetz Wolf proposed the hypothesis of "world city," suggesting that the global economy was no longer imperial but rather a series of controlled markets and production units by transnational capital organizations, and world cities were the material manifestation of this control.[42] Subsequently, Saskia Sassen contributed to the theoretical construction of this hypothesis with the concept of "global city,"[43] as exemplified in works such as "The Global City: New York, London, Tokyo" and "Cities in a World Economy."[44] These works addressed macro-issues such as capital flows, trade, and finance, as well as issues of inequality and ecology. Scholars like Aljun Appadurai[45] and David Harvey[46] have also conducted urban studies within the context of globalization.

Furthermore, in the process of forming global cities, new urban spatial forms gradually emerged. Manuel Castells conceptualized mega-cities as new spatial forms in the networked information age. With the aid of information technology, corporate headquarters and production units could spatially separate and better access land, labor, and innovation capabilities. He also highlighted the relationship between transnational urban elites and the flow spaces of arts, culture, and leisure.[47] Allen Scott similarly acknowledged the relationship between the emerging cognitive–cultural economy and what he termed "global city-regions." He observed that urban spaces detached from production gradually took on functions such as high-tech research and development, financial services, upscale retail, entertainment, and culture. These spaces concurrently became "learning regions," constructing an "innovative milieu."[48] The importance of high-tech parks was noted by both Castells and Peter Hall. Their collaborative work, " Technopoles of the world: The making of twenty-first century industrial complexes," analyzed the relationship between globalization, the information revolution, high-tech parks, and their political foundations.[49]

---

[42] Friedmann, J., & Wolff, G. (1982). World city formation: an agenda for research and action. *International journal of urban and regional research*, 6(3), 309–344.

[43] Sassen, S. (1991). *The global city: New York, London, Tokyo (2nd ed.)*. Princeton University Press. (pp. 1).

[44] Sassen, S. (2018). *Cities in a world economy*. Sage Publications. (pp. 235–272).

[45] Appadurai, A. (2000). Grassroots globalization and the research imagination. *Public Culture*, 12(1): 1–19.

[46] Harvey, D. (2016). *The ways of the world*. Profile Books. (pp. 136–176).

[47] Castells, M. (2011). *The Rise of the Network Society: The Information Age: Economy, Society, and Culture Volume I (2nd ed., Vol. 12)*. Wiley. (pp. 434–452).

[48] Scott, A. J. (2012). *A world in emergence: cities and regions in the 21st century*. Edward Elgar Publishing (pp. 64–78).

[49] Castells, M. & Hall P. (2014). *Technopoles of the world: The making of 21st century industrial complexes*. Routledge (pp. 1–6).

Finally, the relationship between global cities and nation-states is also a focal point. As early as Friedman's hypothesis, he observed that transnational capital sought to minimize state intervention in the flow of financial capital, information, and goods.[50] Subsequently, Sassen noticed the denationalization of corporate elites,[51] Castells observed how networks formed by multinational corporations transcended national boundaries and identities,[52] and Appadurai advocated for post-national world thinking and research methods.[53] These scholars, to varying degrees, argued that nation-state power structures were gradually diminishing in the wave of globalization. However, Neil Brenner disagrees with this viewpoint. He emphasizes the role of the state in spatial restructuring and criticizes scholars like Sassen for diminishing the role of the state in cities. Brenner argues that urban spaces and territorial organizations, such as state institutions, are mutually constitutive of globalization.[54] Similarly, Shatkin Gavin criticizes researchers of global cities for overlooking cultural, geographical, and institutional differences among different cities. He suggests shifting focus from models of global or world cities to a more substantive examination of global and local actors in specific contexts.[55]

### 2.2.2 Studies on Chinese Cities

Since the 1960s, the viewpoints of Max Weber[56] and John K. Fairbank[57] have been criticized for their apparent "Western-centric" bias within the academic environment of that time. Paul Wheatley acknowledged Weber's notion of the isomorphic relationship between royal lineage and state government but criticized Weber for emphasizing the economic functions of ancient Chinese cities while neglecting their

[50] Friedmann, J., & Wolff, G. (1982). World city formation: an agenda for research and action. *International journal of urban and regional research*, 6(3), 309–344.

[51] Sassen, S. (2018). *Cities in a world economy*. Sage Publications. (pp. 235–272).

[52] Castells, M. (2011). *The Rise of the Network Society: The Information Age: Economy, Society, and Culture Volume I (2nd ed., Vol. 12)*. Wiley. (pp. 206–209).

[53] Appadurai, A. (1996). *Modernity at large: cultural dimensions of globalization*. University of Minnesota Press. (pp. 158–177).

[54] Brenner N. 尼爾·博任納 (2019). *City, Territory Planet:Essays in Critical Urban Theory*城市、地域、星球 批判城市理論 (Trans. by Li Z.G.李志剛 et al.). Beijing: Commercial Press北京:商務印書館. (pp. 41–42).

[55] Shatkin, G. (2007). Global cities of the South: Emerging perspectives on growth and inequality. *Cities*, 24(1), 1–15.

[56] Weber, M. (2008). *Confucianism and Taoism*儒教與道教 (Tran. by Hong, T. F. 洪天富). Nanjing: Jiangsu People's Publishing House南京:江蘇人民出版社. (pp. 3–37; 91–114).

[57] Teng, S., & Fairbank, J. K. (1982). *China's response to the West: a documentary survey, 1839–1923*. Harvard University Press.

role as ceremonial centers.[58] Edward W. Said's "Orientalism"[59] and Paul A. Cohen's "Discovering History in China"[60] are two significant works that critique Western-centric perspectives. While they do not primarily focus on cities, both argue against viewing Chinese issues through the lens of presupposed Western rationality. After the 1960s, G. William Skinner became one of the most important scholars in Western urban studies. He divided the study of modern Chinese cities into nine regions based on geographical conditions. Skinner's advocacy for a research approach grounded in the specific context of China influenced many Western scholars who subsequently studied Chinese cities,[61] such as William T. Rowe's research on modern Hankou[62] and Philip C. C. Huang's study of rural areas in the Yangtze River Delta.[63]

Since the late 1970s, China's reform and opening-up policy have propelled the country into a new phase of globalization, prompting overseas scholars to delve deeper into the urban conditions of China. Studies by Richard Kirkby,[64] Terry Cannon and Alan Jenkins, and others have observed urbanization during China's planned economic era and its unique historical, political, and economic conditions.[65] In the new millennium, Lu Duanfang approached urban studies during the planned economic era by examining urban form. Lu argued that Chinese work units (單位) were not only social institutions but also workplaces, distinct from both capitalist and Soviet cities, emerging as a unique result between capital accumulation and labor reproduction in the context of socialist Third World countries.[66] The research by these overseas scholars has early on addressed the shortcomings in the study of urbanization in China during the planned economic era and, to some extent, recognized the distinctiveness of China's urban issues compared to the West.

Compared to research on Chinese cities during the planned economic era, the urban conditions in China after the reform and opening-up period have attracted more attention from overseas scholars. John Logan focuses on the globalization, migration, and market-oriented reforms involved in the transition of Chinese cities

---

[58] Wheatley, P. (1971). *The pivot of the four quarters: a preliminary enquiry into the origins and character of the ancient Chinese city.* Edinburgh University Press.

[59] Said, E. W. (2003). *Orientalism*. Penguin.

[60] Cohen, P. A. (2010). *Discovering history in China: American historical writing on the recent Chinese past*. Columbia University Press.

[61] Skinner, G. W. (1977). *The City in late imperial China*. Stanford University Press.

[62] Rowe, W. T. (1992). *Hankow: Conflict and community in a Chinese city, 1796–1895*. Stanford University Press.

[63] Huang, P. C. (1990). *The peasant family and rural development in the Yangzi Delta, 1350–1988*. Stanford University Press.

[64] Kirkby, R. J. R. (1985). *Urbanisation in China: town and country in a developing economy, 1949–2000 AD*. Croom Helm.

[65] Cannon, T., & Jenkins, A. (Eds.) (2002). *The geography of contemporary China: the impact of Deng Xiaoping's decade*. Routledge.

[66] Lu, D. (2006). *Remaking Chinese urban form: modernity, scarcity and space, 1949–2005*. Routledge.

from a planned economy to a market economy.[67] Terry McGee, on the other hand, emphasizes spatial issues, suggesting that globalization is a significant driving force behind the spatial restructuring of Chinese cities.[68] Additionally, he pays attention to the transformation between rural land and state-owned land. During the same period, Friedman proposed that understanding China's urban issues requires considering its historical uniqueness and the relationship between urban and rural areas. He outlined seven dimensions for studying urbanization in China, including the urbanization of the natural environment, which is closely related to this study.[69] Laurence Ma and others acknowledge that the party-state is the ultimate decision-maker in China's urban economic transformation and capital accumulation. This differs from the Western model that China has rejected. Therefore, studying Chinese cities should emphasize a political-economic perspective.[70]

Chinese scholars' research on Chinese cities initially began in the first half of the twentieth century, initiated by sociologists and geographers.[71] While these early studies held value, they rarely treated urbanization as a social process. During the planned economy era from the 1950s to the late 1970s, urbanization was considered a phenomenon unique to capitalist countries, and there were hardly any Chinese scholars studying urbanization in China.[72] In the late 1970s, Wu Youren's publication of "On the Issue of Socialist Urbanization in China (關於中國社會主義城市化問題)" initiated the study of urbanization in China.[73] Subsequently, the "small town thesis (小城鎮論)," represented by Fei Xiaotong, became a relatively mainstream viewpoint.[74] In the 1990s, Chinese scholars began discussing issues related to Chinese cities and urbanization, with Gu Chaolin and Cui Gonghao being

[67] Logan, J. R. (2002). Three challenges for the Chinese city: globalization, migration, and market reform. The new Chinese city: Globalization and market reform. (pp. 3–21).

[68] McGee, T., Lin, G. C., Wang, M., Marton, A., & Wu, J. (2007). *China's urban space: development under market socialism (Vol. 29)*. Routledge.

[69] Friedmann, J. (2006). Four theses in the study of China's urbanization. *International journal of urban and regional research*, 30(2), 440–451.

[70] Ma, L. J. (2002). Urban transformation in China, 1949–2000: A review and research agenda. *Environment and planning A*, 34(9), 1545–1569.

[71] Wu, J. C.吳景超 (2008). *The Fourth Types of Nations*第四種國家的出路. Beijing: Commercial Press北京:商務印書館；Xu, S. L.許仕廉 (2022). *Population Issues in China*中國人口問題. Beijing: Commercial Press北京:商務印書館; Wang, Y. Y. 王益厓 (1935). A Study on Urban Geography of Wuxi無錫都市地理之研究. *Acta Geographica Sinica*地理學報, 1935(03), 23–63; Shen, R. S., & Sun, M. X. 沈汝生,孫敏賢 (1947). A Study on Urban Geography of Chengdu成都都市地理之研究. *Acta Geographica Sinica*地理學報, 1947(Z1), 14–30; Zhao, S. Q., & Bai, Y. Z. 趙松喬,白秀玲 (1950). Preliminary Study on Urban Geography of Nanjing南京都市地理初步研究. *Acta Geographica Sinica*地理學報, 1950(02), 39–71.

[72] Cui, G. H. 崔功豪 (1989). Review and Prospect of Research on China's Urbanization in Last Decade近十年來中國城市化研究的進展. *Areal Research and Development*地域研究與開發, (01), 1–5+62.

[73] Wu, Y. R. 吳友仁 (1979). On the Issue of Socialist Urbanization in China關於我國社會主義城市化問題. *City Planning Review*城市規劃, (05), 13–25.

[74] Fei, X. T. 費孝通 (1986). *On Small Towns and Others*論小城鎮及其他. Tianjin: Tianjin People's Publishing House天津: 天津人民出版社.

particularly representative figures. Gu Chaolin observed that urbanization and industrialization in China were both initiated by the government, with urban development and rural urbanization occurring simultaneously. Cui Gonghao acknowledged the existence of a dual urbanization pattern in China and analyzed the bottom-up urbanization centered around rural township enterprises. He also explored the relationships and mechanisms among the government, farmers, and capital under state policies.[75] Furthermore, scholars such as Yao Shimou, Xu Guodi, and Zhang Yuxing during this period focused on the gradual rise of Chinese metropolitan areas and related spatial forms and industrial issues.[76]

Around the year 2000, many scholars began to lean toward studying Chinese cities within the context of globalization, while the role of information technology in the formation of metropolitan area began to receive attention. Gu Chaolin noticed the shift from industrial to information societies internationally and from a planned to a market economy domestically. He believed it was essential to understand the development of Chinese cities from a global perspective. He also observed the gradual formation of Chinese metropolitan areas, noting that international cities like Shenzhen and Hong Kong would enter the global city network.[77] Yeung Yue-man analyzed the rise of the PRD metropolitan area from a global perspective, considering factors such as global industrial division and the flow of transnational capital.[78] Zhang Jingxiang analyzed the rise of various metropolitan areas and regional organizations worldwide, along with the changes in power and control mechanisms behind them. Although the focus of Zhang's article was not specifically on Chinese cities, it provided a global perspective for the study of Chinese metropolitan areas.[79] Zhu Yingming, while not extensively discussing globalization, elucidated the emergence of networked metropolitan areas in China based on information technology and their relationship with communication technology, service industries, commodity consumption, and upstream and downstream companies in industrial

75 Cui, G. H., & Ma, R. C. 崔功豪,馬潤潮 (1999). Urbanization from Bellow in China: Its Development and Mechanisms中國自下而上城市化的發展及其機制. *Acta Geographica Sinica*地理學報, 1999(02), 12–21.

76 Yao, S. M. 姚士謀 (1992). Characteristics, Types, and Spatial Distribution of Urban Agglomerations in China我國城市群的特徵、類型與空間佈局. *Urban Problems*城市問題, 1992(01), 10–15+66. Xu, G. D.徐國弟 (1994). Conceptualization of Development Strategy Positioning for the Beijing-Tianjin-Hebei Urban Agglomeration京津冀城市群體發展戰略定位構想. *Urban Development Research*城市發展研究, 1994(02), 18–22. Zhang, Y. X. 張宇星 (1995). Spatial Fractal Characteristics of Urban and Urban Agglomeration Morphology城市和城市群形態的空間分形特性. *New Architecture*新建築, 1995(03), 42–46.

77 Gu, C. L., 顧朝林et al. (1999). *Economic Globalization and Urban Development in China: A Study on the Development Strategy of Chinese Cities across the Century*經濟全球化與中國城市發展——跨世紀中國城市發展戰略研究. Beijing: Commercial Press北京: 商務印書館.

78 Yang, R. W. 楊汝萬 (2004). *Asia-Pacific Cities in the Context of Globalization*全球化背景下的亞太城市. Beijing: Science Press北京: 科學出版社出版.

79 Zhang, J. X. 張京祥 (2004). *Transformation of Regional Development and Planning in the Globalization Background*全球化背景中的區域發展與規劃轉變. Urban Planning International 國外城市規劃, 2004(03), 1–4.

chains.[80] At that time, globalization was almost unquestionably seen as inherently correct. Zhang Tingwei critically pointed out that this phase of globalization represented a new stage in capitalist development, leading to certain contradictions between global capital and local cities.[81] He also noted the relationship between neoliberalism, urban management, and urban competitiveness within the theoretical context of urban planning in China.[82] Compared to the 1990s, when research mainly focused on the basic characteristics, macro strategies, and spatial forms of metropolitan areas, the studies during this period had a broader global perspective. They linked global issues with Chinese issues and deeply examined the interconnections between technology, industry, and space.

In the past decade, more specific studies have emerged focusing on various metropolitan areas such as the PRD, Yangtze River Delta, and Beijing-Tianjin-Hebei region. For example, Luo Xiaolong conducted research on the development and evolution of the relationship between Shenzhen and Hong Kong within the framework of economic and institutional integration.[83] Zhang Xianchun and others studied the relationship between regional governance frameworks and land development rights, using the example of the Shenzhen-Shantou Special Cooperation Zone.[84] Wu Qungang and Yang Kaizhong researched the relationship among public policies, industries, and population dynamics in the development of the Beijing-Tianjin-Hebei region.[85] In addition to institutional-level research, the development of new districts has also been explored within the framework of regional integration. Liu Yan studied the development of the Binhai New District under the Beijing-Tianjin-Hebei framework.[86] Chen Hongsheng and others studied the Qianhai Cooperation Zone in Shenzhen and the Hengqin Cooperation Zone in Zhuhai within the context

[80] Zhu, Y. M. 朱英明 (2001). Trends in the Regional Development of Urban Agglomerations in China我國城市群區域聯繫發展趨勢. *Urban Problems*城市問題, 2001(06), 22–24.

[81] Zhang, T. W. 張庭偉 (2003). The Misunderstood Globalization and the Misled Selling Place in China對全球化的誤解以及經營城市的誤區. *City Planning Review*城市規劃, 2003(08), 6–14.

[82] Zhang, T. W. 張庭偉 (2004). Neo-liberalism, City Management, City Governance and City Competitiveness in China新自由主義·城市經營·城市管治·城市競爭力. *City Planning Review*城市規劃, 2004(05), 43–50.

[83] Luo, X. L., & Shen, J. F. 羅小龍,沈建法 (2010). From "Front Shop, Back Factory" to Hong Kong-Shenzhen Metropolis: The Changing Relationship between Hong Kong and Shenzhen in the Past Three Decades從"前店後廠"到港深都會:三十年港深關係之演變. *Economic Geography*經濟地理, 30(05), 711–715.

[84] Zhang, X. C., Luan, X. F., Yue, W. Z., et al. 張衛春,欒曄帆,岳文澤等 (2023). Spatial Restructuring of Land Development Right under the Regional Governance Framework區域治理框架下的土地發展權空間重構研究. *City Planning Review*城市規劃, 47(08), 21–28.

[85] Wu, Q. G., & Yang, K. Z. 吳群剛,楊開忠 (2010). Thinking on the Integrated Development of Beijing-Tianjin-Hebei Region關於京津冀區域一體化發展的思考. *Urban Problems*城市問題, 2010(01), 11–16.

[86] Liu, Y., & Liu, L. D. 劉焱,劉利東 (2010). Research on Tianjin Binhai New Area under Beijing-Tianjin-Hebei Region Frame京津冀區域框架下的濱海新區發展研究. *Urban Development Studies*城市發展研究, 17(05), 18–21.

of the Guangdong-Hong Kong-Macao Greater Bay Area.[87] Liu Yongjing and others researched the development mechanisms, spatial organization, and governance models of cross-border new areas in China.[88] Furthermore, the construction of infrastructure in the context of regional integration has attracted significant academic attention. Zhou Yunyuan, Liang Chengwen, Deng Huanbin, Luo Zixin, and others have focused on issues related to the construction of transportation infrastructure under the framework of regional integration.[89] Dai Bin and Huang Huang have looked into regional tourism integration in the Beijing-Tianjin-Hebei region.[90] Chi Wenxiu and Lin Guangsi, Wang Fuyuan and Wang Kaiyong have focused on issues related to greenways and ecological recreational spaces in the PRD metropolitan area under the backdrop of regional integration.[91]

Research on urban development in Shenzhen can be broadly categorized into two main types: (1) studies focusing on its urbanization process and urban policies; and (2) studies examining the urban space and urban nature.

Gu Chaolin et al. conducted a macro-analysis of the development of Shenzhen's new industrial space at the turn of the millennium, focusing on the impact of economic globalization. They provided directional recommendations for the

[87] Chen, H. S., Li, Z. G., Xiao, Y., et al. 陳宏勝,李志剛,肖揚等 (2022). Research on the Development of New Cross-Border Districts from the Perspective of Regional Function: A Comparison between Qianhai and Hengqin Cooperation Zones基於區域功能視角的邊界新區發展研究——對深圳前海合作區與珠海橫琴合作區的比較. *City Planning Review*城市規劃, 46(04), 21–29.

[88] Liu, Y. J., Luo, X. L., Tian, D., et al. 劉永敬,羅小龍,田冬等 (2014). Preliminary Study on China's Cross-border New District in Forming Mechanisms, Spatial Organization, and Governance Mode中國跨界新區的形成機制、空間組織和管治模式初探. *Economic Geography*經濟地理, 34(12), 41–47.

[89] Zhou, Y. Y. 周運源 (2010). Several Thoughts on the Construction of Nan Station New City in the Regional Economic Integration Development of the Pearl River Delta南站新城建設在珠三角區域經濟一體化發展中的若干思考. *Urban Insight*城市觀察, (S1), 122–125; Liang, C. W., Liu, W., & Zhang, Y. Y. 梁成文,劉威,張玉一 (2010). Rapid Transit Planning of Shenyang under the Background of Regional Integration區域一體化背景下瀋陽市快速軌道交通線網規劃. *City Planning Revies*城市規劃, 34(03), 71–75; Deng, H. B., Zhang, Z. H., & Deng, Y. Z. 鄧煥彬,張智慧,鄧遠征 (2011). Study on Countermeasures of Regional Transportation Integration in the Pearl River Delta Region珠三角區域交通一體化對策研究. *Transpo World*交通世界(運輸.車輛), 2011(11), 128–130; Luo, Z. X., Zhou, S. H., & He, J. M. 羅子昕,周素紅,何嘉明 (2019). Cross-border Social Networks: A Case Study of Guang-Fo Metro, China區域一體化下城市居民跨邊界社會網路構建——以廣佛地鐵為例. *Urban Development Studies*城市發展研究, 26(06), 70–78.

[90] Dai, B., & Huang, H. 戴斌,黃璜 (2016). Theories and Strategic for the Tourism Integration: The Case Study of the Beijing-Tianjin-Hebei區域旅遊一體化的理論建構與戰略設計——以京津冀為例. *Human Geography*人文地理, 31(03), 128–135.

[91] Chi, W. X., & Lin, G. S. 茌文秀,林廣思 (2021). Study on the Safeguard Mechanism of Large-scale Landscape Planning Projects——A Case Study of the Greenway Network in Pearl River Delta大尺度景觀規劃專案的實施保障機制研究——以珠三角綠道網為例. *Chinese Landscape Architecture*中國園林, 37(09), 25–30; Wang, F. Y., & Wang, K. Y. 王甫園,王開泳 (2019). Connectivity and Distribution Pattern of Regional Greenways and Ecological Recreation Spaces in the Pearl River Delta Urban Agglomeration珠江三角洲城市群區域綠道與生態遊憩空間的連接度與分佈模式. *Progress in Geography*地理科學進展, 38(03), 428–440.

industrial–spatial development of Shenzhen, including vigorously developing high-tech industrial clusters and developing industrial clusters in the eastern region.[92] Luo Qing et al. systematically reviewed the development and changes in various industries in Shenzhen, including agriculture, industry, high-tech industries, logistics, finance, and commerce. They also conducted problem analysis and provided policy recommendations.[93] Yeung Yue-man et al. (2009) retrospectively evaluated the performance of the five special economic zones established in China in the 1980s and argued that despite various uncertainties, the establishment of Shenzhen's special economic zones was successful, even though global economic turmoil often affected China.[94] They analyzed population structure, technological innovation capabilities, culture, housing, and manufacturing industries, and believed that a headquarters economy strategy would help enhance Shenzhen's internationalization level. Among overseas scholars, Carolyn Cartier (2018) analyzed the political-economic context of the concept of "Special Zone," which played a significant role in driving urbanization in China, and its evolving significance. She argued that China's urban development is shifting from reliance on special zones to focusing on urban agglomerations.[95] Mary Ann O'Donnell proposed "learning from Shenzhen" and viewed Shenzhen as a new form of China's participation in economic globalization under the leadership of the Communist Party of China. Discussions on cadre appointments, university construction, and other aspects elaborated this viewpoint.[96]

In research on urban space and urban nature, Meng Jianmin et al. divided the evolution of Shenzhen's urban spatial form into four stages, combining it with industrial development. Chen Keshi et al., on the other hand, categorized the evolution and development of Shenzhen's cluster-based spatial structure into five different periods. They also identified the problems faced by the cluster-based spatial structure in the early 2010s and proposed corresponding strategies.[97] Aside from these, studies on Shenzhen's overall urban spatial form mostly appear in the form of

---

[92] Gu, C.L., Chen, G., Huang, C.Y., et al. 顧朝林,陳果,黃朝永等 (2001). On the development of new industrial space in Shenzhen: The result of economic globalization, industrial restructuring, and relocation論深圳新工業空間開拓——經濟全球化、產業結構重建與轉移的結果. Economic Geography.經濟地理, (03), 261–265.

[93] Luo, Q.H.羅清和 (2007). Economic development and industrial growth: A study on the industrial growth and development trends i n Shenzhen經濟發展與產業成長：深圳產業成長與發展趨勢研究. Shanghai: Shanghai Sanlian Bookstore上海: 上海三聯書店.

[94] Yeung, Y., Lee, J., & Kee, G. (2009). China's Special Economic Zones at 30. *Eurasian Geography and Economics*, 50(2), 222–240.

[95] Cartier C. (2018). From 'Special Zones' to Cities and City-regions in China In Developmentalist Cities? Interrogating Urban Developmentalism in East Asia. Bril.

[96] O'Donnell, M.A., Wong, W. and Bach, J. (2017). Introduction: Experiments, Exceptions, and Extensions In O'Donnell et al. (Eds.) Learning from Shenzhen: China's post-Mao experiment from special zone to model city. Chicago: The University of Chicago Press.

[97] Chen, K.S., Yang, R., & Liu, B.B. 陳可石,楊瑞,劉冰冰 (2013). Study on the evolution and development of Shenzhen's cluster spatial structure深圳組團式空間結構演變與發展研究. Urban Development Studies城市發展研究, 20(11), 22–26.

doctoral dissertations. For instance, Luo Pei revealed the dynamics and patterns of Shenzhen's urban form evolution after analyzing indicators such as population, GDP, and built-up area.[98] Wang Fuhai's research, similar to Luo Pei's methodology, starts with empirical forms and ultimately deduces corresponding dynamic mechanisms.[99] While this research approach may not emphasize theoretical contributions, it clearly focuses on controlling urban spatial form and its functionality, demonstrating strong practicality. However, there are also studies that explore Shenzhen's urban spatial form from corresponding theoretical perspectives. For example, Zhang Yongqiang utilized self-organization theory to interpret relevant phenomena, suggesting that Shenzhen's rapid development is the result of the combination of spatial self-organization and hetero-organization.[100] Luan Feng, adopting a structuralist perspective, conducted a comparative study on the evolution of urban spatial forms in Shenzhen and Xiamen, arguing that these causal mechanisms involve structural factors and agents.[101] The interaction between the social context of Shenzhen's urbanization and its natural environment, which theoretically relates to this research, has received increasing attention in recent years. For instance, Shi Peijun et al. studied landscape urbanization in Shenzhen;[102] Qu Xiao et al. researched fish communities under the background of Shenzhen's urbanization;[103] Wang Yu et al. examined the relationship between Shenzhen's urban–water relationship from a political ecology perspective;[104] Fan Yating and Xiao Jing et al. focused on the study of Shenzhen's waterfront spaces;[105] Zhang Liping, Jiang Yu, and Gao Wenqi, among others,

[98] Luo, P. 羅佩(2007). Research on the evolution of urban form in Shenzhen深圳城市形態演進研究 (PhD thesis). Sun Yat-sen University中山大學.

[99] Wang, F.H.王富海(2003). Research on the evolution of urban space in Shenzhen深圳城市空間演進研究 (PhD thesis). Peking University北京大學.

[100] Zhang, Y.Q. 張勇強(2003). Study on the self-organization of urban spatial development: The case of Shenzhen城市空間發展自組織研究——深圳為例 (PhD thesis). Southeast University東南大學.

[101] Luan, F. 欒峰(2004). Study on the causal mechanisms of rapid urban spatial form evolution since the reform and opening-up改革開放以來快速城市空間形態演變的成因機制研究 (PHD thesis). Tongji University同濟大學.

[102] Shi, P., et al. 史培軍 等 (2012). Landscape urbanization and ecological infrastructure construction: Shenzhen case景觀城市化與生態基礎設施建設:以深圳為例. Beijing: Science Press北京:科學出版社.

[103] Qu, X., Liu, H., Yang, M., et al. 屈霄,劉晗,陽敏,等 (2022). Characteristics of fish communities and driving factors under urbanization in typical river basin in Shenzhen City, China城鎮化背景下深圳典型流域魚類群落特徵及驅動因數. *Acta Ecologica Sinica*生態學報, 42(24), 10029–10040.

[104] Wang, Y., Man, C., Xu, A., et al. 王雨,滿彩霞,徐璦琳,等(2022). Spatial transformation and Chinese environmental governance innovation from an urban political ecology perspective: An analysis of the Shenzhen's evolving waterscape城市政治生態學視角下的空間轉型與中國環境治理創新——基於深圳城——水關係動態演變的分析. *Progress in Geography*地理科學進展, 41(09), 1755–1769.

[105] Fan, Y.T., Cui, Q.Y., & Xiao, J. 范雅婷,崔秦毓,肖靖 (2021). Shenzhen Bay Waterfront in the Agenda of Leap-forward Urbanism: Historical and Reflective Studies on Its Formation and

conducted research on water quality, water environment, and river habitats under the background of urbanization.[106] These studies have begun to address the relationship between urbanization and natural elements within the city, primarily focusing on water environments and waterfront spaces, while other forms of natural elements receive less attention.

### 2.2.3 Trend and Enlightenment

During the 1960s and 1970s, the global shift toward neoliberalism in political and economic practices led to widespread discussion and acceptance of theories on global cities since the early 1980s. Overall, research on global cities shifted from macroeconomic issues such as international capital flows, trade, and finance to more specific topics in various disciplines such as ecology, social stratification, and spatial studies. Concerning spatial issues, attention turned to the relationships between metropolitan areas, corporate headquarters, high-tech parks, and information technology. Regarding the relationship between global cities and nation-states, relevant theories exhibited a trend from de-emphasizing the role of the nation-state to a renewed recognition of its importance. The developmental history of Shenzhen over the past forty years not only represents a testing ground for China's reform and opening-up policies but also signifies a continual endeavor to integrate into the global city network. Therefore, understanding the evolution of global city theories facilitates an examination of Shenzhen from a global perspective.

Foreign scholars' research on Chinese cities and related issues is marked by a critical examination of biases toward figures like Weber and Fairbank, signaling a shift towardstudies advocating for entering the specific context of Chinese cities, led by scholars such as Skinner and Rowe. From the mid-1980s to the turn of the millennium, foreign scholars paid more attention to urbanization during China's planned economic era. In the new millennium, foreign scholars represented by Logan, McGee, and Ma focused on post-reform and opening-up Chinese cities,

---

Publicity躍進式城市化視角下的深圳灣濱水帶沿革及其公共性演變研究. *Chinese & Overseas Architecture*中外建築, (11), 26–32; Fan, Y., Wu, K., Cheng, J.K., et al. 范雅婷,吳昆,程家輝,等(2023). Study on the ªNature-Society° Mutual Feedback Relationship in the Evolution of Urban Waterfront Space: a Case Study of Shenzhen城市濱水空間演化中的“自然-社會”互饋關係研究——以深圳為例. *Design Community*住區, (05), 100–107.

[106] Zhang, L.P. 张利泙(2023). Long-term Characteristics of Drinking Water Quality in Rapidly Urbanization Area: A case study in Shenzhen reservoir (1991—2018) 快速城市化地区饮用水源水质长期变化特征分析——以深圳水库(1991—2018年)为例. *Guangdong Water Resources and Hydropower*广东水利水电, (08), 31–35; Jiang, Y., Zhang, Q.Y., Li, F.D., et al. 蒋钰,张秋英,李发东,等 (2023). Impact of rapid urbanization on Shenzhen water environment快速城市化对深圳水环境的影响分析. *Pearl River*人民珠江, 44(08), 8–17, 24; Gao, W., Lu, Y., Qu, X., et al. 高雯琪,陆颖,屈霄,等(2021). River habitat assessment under urbanization: A case study in Shenzhen 城镇化背景下河流生境评价——以深圳市为例. *Acta Ecologica Sinica*生态学报, 41(22), 8783–8793.

emphasizing to varying degrees the uniqueness of Chinese urbanization and its relationship with globalization. Chinese scholars began researching urban issues in the first half of the twentieth century but did not gain widespread influence before the reform and opening-up period. Afterward, discussions by scholars like Wu Youren and Fei Xiaotong on urban issues gained widespread attention. In the 1990s, Chinese scholars such as Gu Chaolin and Cui Gonghao recognized the close relationship between urban and rural issues in China, with the rise of metropolitan areas also receiving initial attention. Around the turn of the millennium, the impact of globalization on Chinese urban development and the rise of metropolitan areas garnered closer attention, with emerging topics including information technology, global industrial division of labor, cross-border capital flows, services, and consumer goods. In the past decade, specific studies involving major metropolitan areas, such as governance of new urban areas, regional public policies, regional tourism, and ecological recreational spaces, have received close attention. Indeed, in those studies on Shenzhen's urban development, there has been a predominant focus on urban space and its construction and developmental patterns. Studies often delve into the evolution of urban spatial structures, industrial development, and policy implications. Conversely, research on urban nature, particularly in recent years, has gained momentum. However, this focus has primarily been on water environments, with other types of natural elements within the city receiving relatively less attention, especially from a historical perspective. Overall, research on Chinese cities has shifted its focus from viewing cities individually to viewing them through the lens of metropolitan areas, from urban studies to urban-rural studies, and from macro issues to specific topics, all of which provide insights for the study of Shenzhen.

## 2.3 Historical Research on Urban Parks

### *2.3.1 Researches by Chinese Scholar*

Regarding the origins of urban parks in China, the prevailing view, as advocated by scholars such as Li Deying, Chen Yunxi, and Lei Yi, posits that Chinese urban parks are an import from abroad and emerged during China's modern era, specifically between 1848 and 1949.[107] Wang Jintao, Luo Huali, Wu Renshu, Liu Qiqi, and Wang Xin, have conducted research on ancient Chinese public gardens, suggesting

[107] Li, D. Y.李德英 (2003). Social Conflicts in Parks: A Case Study of Parks in Modern Chengdu 公園裡的社會衝突——以近代成都城市公園為例. *Historical Review*史林, 2003(01), 1–11+123; Chen, Y. Q. 陳蘊茜 (2005). On The Conflict between Colonialism and Nationalism in Everyday Life: From a perspective of the Parks in Modern China日常生活中殖民主義與民族主義的衝突——以中國近代公園為中心的考察. *Journal of Nanjing University (Philosophy, Humanities and Social Sciences)* 南京大學學報(哲學.人文科學.社會科學版), 2005(05); Lei, Y.雷頤 (2008). Parks: Past and Present公園古今事. *Yanhuang Chunqiu*炎黃春秋, 2008(04), 65–68.

that many ancient gardens already possessed public characteristics.[108] Among them, Mao Huasong's viewpoint stands out; he argues that the notion that Chinese parks emerged in the nineteenth century is inconsistent with the evolution of Chinese urban civilization and directly proposes that Chinese urban parks originated in the Song Dynasty.[109] However, it is inappropriate to conclude that ancient China had what we now refer to as "parks," simply because ancient texts used the term "park (公園)" to refer to green spaces with public characteristics in cities. In short, publicity is not a sufficient condition for green spaces to become parks; more importantly, the broader and more institutionalized economic and political forces behind park construction are crucial. As Lin Zheng understands, urban parks were established in the context of nineteenth-century modern nation-states by the middle and upper-class bourgeoisie with the aim of enlightening the public and fostering patriotic sentiments.[110] Another part of these parks was established in the context of colonialism by political and business figures of the colonial powers in their colonies or semi-colonies.[111] These parks were the precursors and prototypes of today's so-called "parks," embodying modern significance. Therefore, Peng Changxin's viewpoint is worth agreeing with, that is, understanding Chinese modern parks within the context of nineteenth-century global park construction activities.[112]

Before the early 2000s, there were only a few preliminary explorations in historical research on urban parks. Concerning domestic parks, Li Min explored the achievements and development of urban park construction since the founding of the

[108] Wang, J. T. 王勁韜 (2011). On the Public Nature of Chinese Ancient Gardens and Its Impact on Urban Life: Taking Gardens of Song Dynasty as an Example中國古代園林的公共性特徵及其對城市生活的影響——以宋代園林為例. *Chinese Landscape Architecture*中國園林, 27(05), 68–72; Luo, H. L. 羅華莉 (2012). *The Research on the Narrative of Ancient Chinese Public Garden*中國古代公共園林故事性研究(Master's thesis, Beijing Forestry University北京林業大學); Wu, R. S. 巫仁恕 (2010). From Sightseeing to Tourism: Changes in Suzhou's Tourism Activities and Spaces from the 16th to the Early 20th Centur y從遊觀到旅遊:16至20世紀初蘇州旅遊活動與空間的變遷. In R. S. Wu, B. Kang, & M. L. Lin巫仁恕 康豹 林美莉 (Eds.), *Viewing Modernity in China from Urban Perspective*從城市看中國的現代性. *Taipei: Institute of Modern History, Academia Sinica*臺北:中央研究院近代史研究所; Liu, Q. Q., & Wang, X. 劉琪琪, 王欣 (2018). Research on the Development of Moonlake Public Garden in Ancient City of Ningbo寧波月湖古代城市公共園林流變研究. *Landscape Architecture*風景園林, 2018(01).

[109] Mao, H. S.毛華松 (2014). Formation Analysis of Chinese Ancient Parks——Analysis of Urban Parks Development in Song Dynasty論中國古代公園的形成——兼論宋代城市公園發展. *Chinese Landscape Architecture*中國園林, (01), 116–121.

[110] Lin, Z.林崢 (2022). *Park Beijing*公園北京. Beijing: Peking University Press北京: 北京大學出版社. (pp. 4–6).

[111] Peng, C. X., & Wang, Y. T. 彭長歆,王黜婷 (2021). Landscaping outside the City Wall: Canton's Gardens in Late Qing Dynasty and the Moder Transformation of Lingnan Gardens城外造景:清末廣州景園營造與嶺南園林的近代轉型. *Chinese Landscape Architecture*中國園林, 37(11), 127–132；Yu, Y., & Zhang, Q. Y. 余洋,張琦瑀 (2021). Product of Cultural Change: Analysis of Evolution Process and Mechanism of Modern Parks in Northeast China文化變遷的產物:東北近代公園演化進程與機制解析. *Landscape Architecture*風景園林, 28(06), 32–37.

[112] Peng, C. X. 彭長歆 (2014). The Beginning of Public Park in China: The American and English Gardens of Thirteen Factories of Canton中國近代公園之始——廣州十三行美國花園和英國花園. *Chinese Landscape Architecture*中國園林, 30(05), 108–114.

People's Republic of China and proposed strategies for building modern urban parks with Chinese characteristics.[113] Wang Juyuan reviewed the basic course of urban greening and park construction since the founding of the People's Republic of China, aiming to summarize experiences, lessons, and prospects for the new century.[114] Similarly, Liu Shanghua has studied the general course of urban park construction and management since the founding of the People's Republic of China.[115] Apart from these relatively systematic studies, there was also Zhang Jingsan's research on modern parks in Shanghai[116] and Hu Qifang's study on modern parks in Jiangsu Province.[117] Regarding foreign parks, Xu Hao and Zhang Junhua investigated the historical origins and evolution of Japanese urban parks, discussing their significance as references for green space planning in China.[118] Wu Renwei introduced foreign park movements, park systems, ecological movements, and anticipated trends in the development of urban green spaces in the twenty-first century.[119] During this period of historical research, many researchers had backgrounds in design, and their focus was not on delineating the social and historical context of urban parks but rather on supplementing planning and design practices. This was a basic characteristic of the early stage of historical research on urban parks.

After this period, urban park historical research in the field of humanities and social sciences gradually broke away from the basic characteristics mentioned above, with these studies first emerging in the mid to late 1990s and becoming more prominent after 2000. The earliest examples include Xiong Yuezhi's research on Shanghai's Zhang Garden (張園)[120] and Shi Mingzheng's study on the

---

[113] Li, M.李敏 (1987). Chinese Modern Parks: Development and Evaluation中國現代公園——發展與評價. *Beijing: Beijing Science and Technology Press*北京:北京科學技術出版社出版.

[114] Wang, J. Y.汪菊淵 (1992). Retrospect and Prospect of Urban Greening and Landscape Construction in China我國城市綠化、園林建設的回顧與展望. *Chinese Landscape Architecture*中國園林, 1992(01), 17–25.

[115] Liu, S. H. 柳尚華 (Eds.) (1999). Contemporary 50 Years of Chinese Landscape Architecture中國風景園林當代50年. *Beijing: China Architecture & Building Press*北京:中國建築工業出版社.

[116] Zhang, Z. E.張志恩 (1996). Historical Evolution and Historical Evaluation of Lu Xun Park魯迅公園的歷史沿革與歷史評價. *Chinese Landscape Architecture*中國園林, 1996(04), 22–24.

[117] Hu, Q. F. 胡其舫 (1999). Overview, Significance, and Influence of Modern Parks in Jiangsu Province江蘇近代公園概貌及其意義和影響. *Chinese Landscape Architecture*中國園林, 1999(06), 68–71.

[118] Xu, H. 許浩 (2002). A Study on History of Japanese Modern Urban Green Space and System對日本近代城市公園綠地歷史發展的探討. *Chinese Landscape Architecture*中國園林, (03), 62–65; Zhang, J. H. 章俊華 (2001). City Green Belt Area in Japan (A)日本城市綠地空間(上). *Chinese Landscape Architecture*中國園林, (05), 38–42.

[119] Wu, R. W.吳人韋 (1998). Development History of Urban Green Spaces in Foreign Cities國外城市綠地的發展歷程. *City Planning Review*城市規劃, (06), 39–43.

[120] Xiong, Y. Z. 熊月之 (1996). A Study of Zhang Garden: A Public Space in Late Qing Dynasty Shanghai張園晚清上海一個公共空間研究. *Memories and Archives*檔案與史學, 1996(06), 31–42；Xiong, Y. Z. 熊月之 (1998). Opening of Private Gardens and Expansion of Public Spaces in Late Qing Shanghai晚清上海私園開放與公共空間的拓展. *Academic Monthly*學術月刊, 1998(08), 73–81.

transformation of Beijing's imperial gardens into urban parks,[121] both attempting to depict the changes in urban society and space through parks. Post-2000, Li Deying used modern Chengdu parks as an example, viewing parks as new public spaces and carriers of social conflicts between different classes and groups.[122] Dai Haibin studied Beijing Central Park and its integration with various aspects of modern social life, such as entertainment, education, commerce, and culture.[123] Chen Yunqian conducted an in-depth study on the relationship between modern Chinese urban parks and nationalism, colonialism, and entertainment spaces.[124] Xiong Yuezhi also researched modern parks in Shanghai, involving social interactions, ethnic characteristics, and visitor analysis.[125] Wang Min studied the process of modern Shanghai parks transitioning from exclusion of Chinese people to openness against the backdrop of the decline of the British Empire and the rise of Chinese nationalism in the 1920s.[126] Zhu Nan studied the changes in Chinese understanding of parks from the mid-nineteenth century to the early twentieth century and argued that parks have cultural enlightenment and political indoctrination functions.[127] Zheng Qiongxian and Liu Yangling studied how modern Shanghai parks constrained the authoritarian power of the late Qing dynasty and the early Republican government.[128] These studies differ significantly from the research in the field of landscape architecture before

---

[121] Shi, M. (1998). From imperial gardens to public parks: The transformation of urban space in early twentieth-century Beijing. *Modern China*, 24(3), 219–254.

[122] Li, D. Y. 李德英 (2003). Social Conflicts in Parks: A Case Study of Parks in Modern Chengdu 公園裡的社會衝突——以近代成都城市公園為例. *Historical Review*史林, 2003(01), 1–11+123 ; Li, D. Y. 李德英 (2000). Urban Public Spaces and Social Life: A Case Study of Modern Urban Parks城市公共空間與社會生活——以近代城市公園為例. *Urban History Research*城市史研究, 2000(Z2), 127–153.

[123] Dai, H. B. 戴海斌 (2005). The Central Park and Beijing Society in Early Republic of China中央公園與民初北京社會. *Social Sciences of Beijing*北京社會科學, (02), 45–53+121.

[124] Chen, Y. Q. 陳蘊茜 (2005). On The Conflict between Colonialism and Nationalism in Everyday Life: From a perspective of the Parks in Modern China日常生活中殖民主義與民族主義的衝突——以中國近代公園為中心的考察. *Journal of Nanjing University (Philosophy, Humanities and Social Sciences)* 南京大學學報(哲學.人文科學.社會科學版), (05) ; Chen, Y. Q. 陳蘊茜 (2006). Reconstruction of Space and the Worship towards Sun Yat-sen: An Investigation with Yat-sen Park as its Center in Republic of China空間重組與孫中山崇拜——以民國時期中山公園為中心的考察. *Historical Review*史林, 2006(01), 1–18+123 ; Chen, Y. Q. 陳蘊茜 (2004). Changes in Tourism and Entertainment Spaces from the Late Qing Dynasty to the Republic of China: An Examination Centered on Parks論清末民國旅遊娛樂空間的變化——以公園為中心的考察. *Historical Review*史林, 2004(05), 93–100+124.

[125] Xiong, Y. Z.熊月之 (2013). The Parks and Daily Life in Modern Shanghai近代上海公園與社會生活. *Journal of Social Sciences*社會科學, 2013(05), 129–139.

[126] Wang, M.王敏 (2011). Open of Shanghai's Parks in Modern Times論近代上海公園開放. *Historical Review*史林, 2011(01), 1–13+188.

[127] Zhu, N.朱楠 (2017). Observations Overseas in Late Qing Dynasty and the Introduction of Park Concepts晚清海外見聞與公園觀念的引進. *Cultural Studies*文化研究, (01), 274–286.

[128] Zheng, Q. X., & Liu, Y. L. 鄭瓊現,劉鴦淩 (2013). The Restrictive Function of Parks: A Case Study of the Power Struggle between Modern Shanghai Parks and State Power公園的限權功能——以近代上海公園與國家權力的博弈為例. *Social Science Front*社會科學戰線, 2013(11), 166–170.

the early 2000s in several aspects: (1) Researchers mostly come from the fields of history or cultural studies rather than having a design background. (2) The research is not aimed at supplementing planning and design practices but rather focuses on analyzing the relationship between urban parks and urban social life, showing a stronger historical scholarly professionalism. However, these studies all emphasize urban parks as a new type of modern urban space and correlate them with corresponding social contexts. Their natural attributes are almost neglected, leading to little substantial difference between urban parks and urban squares. In short, in terms of theoretical framework, the relationship between nature and history, nature and society are fragmented. Although Dai Yifeng emphasizes the importance of theoretical perspectives in the historical study of urban parks,[129] this issue is not addressed from the standpoint of a unified theory of nature-society dialectics.

Following the emergence of historical research on urban parks in the humanities and social sciences around 2010, there was a significant increase in historical research on urban parks in the built environment field. Researchers began to focus on the social and historical context of urban parks, primarily those built in modern times. For instance, Zhou Xiangpin and Chen Zhehua studied the design history of parks in Shanghai, while Wang Dandan researched the opening process of public parks in Beijing during the early Republic of China period.[130] Wang Zhiyong and Sun Xinwang, among others, studied the park system outlined in the Republican-era "Capital Plan (首都計畫)".[131] Zhu Zhenjun and Wang Xin researched the construction of self-built public parks in modern Wuxi,[132] while Zhao Jijun examined the historical changes of Wuchang's Shouyi Park.[133] Li Baoyong and Gu Xinren focused on the planning and construction of green spaces in parks in Nanchang during the Republic of China period.[134] Yu Yang and Zhang Qiyu researched Northeastern modern parks and their heritage preservation

[129] Dai, Y. F.戴一峰 (2011). A Multianalysis on the Space of City in Modern China—Focus on Park of City多元視角與多重解讀:中國近代城市公共空間——以近代城市公園為中心. *Journal of Social Sciences*社會科學, (06), 134–141.

[130] Wang, D. D.王丹丹(2012). Study of the Opening Process of Public Gardens in Beijing in the early Republic of China (1914–1929) 民國初期(1914–1929年)北京公共園林開放初探. *Landscape Architecture*風景園林, 2012(06), 101–103.

[131] Wang, Z. Y., Li, C., & Sun, X. W. 王智勇,李純,孫新旺 (2017). Interpretation, Evaluation, and Enlightenment of the Park System Planning in Nanjing Capital Plan民國南京《首都計畫》中公園系統規劃的解讀、評價及啟示. *Chinese Landscape Architecture*中國園林, 33(11), 81–86.

[132] Zhu, Z. J., Yan, C. Y., Wang, X., et al. 朱震峻,嚴晨怡,王欣等 (2017). On the Historical Status of Wuxi Modern Gardens論無錫近代園林的歷史地位. *Chinese Landscape Architecture*中國園林, 33(10), 47–51.

[133] Zhao, J. J. 趙紀軍 (2011). A Historical Study on the Evolution of Shouyi Park, Wuchang武昌首義公園歷史變遷研究. *Chinese Landscape Architecture*中國園林, 27(09), 70–73.

[134] Li, B. Y., Wu, C. S., Liu, M., et al. 李寶勇,吳春生,劉牧等 (2020). Research and Enlightenment on Planning and Construction of Modern Nanchang Park and Green Space in the Republic of China民國南昌公園綠地規劃建設研究及啟示(1926–1949). *Chinese Landscape Architecture*中國園林, 36(08), 101–106.

and utilization.[135] Duan Jianqiang and Zhang Hua studied the "city-park interaction" mechanism in the urbanization process of modern Shanghai.[136] Wang Na and Deng Ge researched the spatial changes in Shenyang Zhongshan Park.[137] Mao Huasong studied the process of modern urban landscape parks and their contributions to public space supply, conservation of natural and cultural resources, and new planning ideas.[138] Zhang Yichi studied the functional changes of urban parks in Shanghai's leased territory at the turn of the nineteenth and twentieth centuries in the context of changes in the city's population and the formation of an international port trade network, attempting to delve into the specific context of urban parks in China.[139] In addition to research on modern urban parks, there were also studies on urban parks built after the founding of the People's Republic of China. For example, Zhao Jijun conducted a detailed study on urban greening and park construction during the planned economic era, elucidating the influence of macro social backgrounds such as the Soviet experience and the Cultural Revolution on urban greening and parks.[140] Liu Yishi compared three parks in Beijing under the socialist background with the typical model of Soviet cultural recreation parks,[141] and also examined the utilization strategy of historical and cultural resources in Taoranting, arguing that Taoranting Park was a place for national mobilization, mass education, and forging "new people."[142] Except for Zhang Yichi and Liu Yishi, these studies almost all aimed to promote traditional culture and supplement current practices. Around 2020, there emerged some historical research on foreign parks, which can

[135] Yu, Y., & Zhang, Q. Y. 余洋,張琦瑀 (2021). Product of Cultural Change: Analysis of Evolution Process and Mechanism of Modern Parks in Northeast China文化變遷的產物:東北近代公園演化進程與機制解析. *Landscape Architecture*風景園林, 28(06), 32–37.

[136] Duan, J. Q., & Zhang, H. 段建強,張樺 (2021). Interactions between Urban Space and Landscape: Research on Transformation Mechanism of Urban Public Parks in Shanghai's Urbanization from Early Modern to Nowadays城園互動:上海近現代城市化中公共園林轉型機制研究. *Landscape Architecture*風景園林, 28(06), 38–43.

[137] Wang, N., Deng, G., & Zhao, W. T.王娜,鄧舸,趙偉韜 (2019). Research on the Historical Transition of Space Utilization of Zhongshan Park in Shenyang City瀋陽中山公園空間利用的歷史變遷研究. *Chinese Landscape Architecture*中國園林, 35(02), 139–144.

[138] Mao, H. S., Wang, X. C., & Wu, Y. H. X. 毛華松,王雪純,吳映華夏(2021). Park-oriented Development of Modern Urban Landscape: Cognition, Construction, and Achievements近代城市山水風景公園化:認知、營建與事功. *Landscape Architecture*風景園林, 28(06), 25–31.

[139] Zhang, Y. C. 張亦弛 (2020). Transforming Urban Parks in the International Settlement of Shanghai at the turn of the 19th and 20th Centuries十九、二十世紀之交的上海公共租界城市公園功能變遷研究. *Chinese Landscape Architecture*中國園林, 36(06), 125–129.

[140] Zhao, J.J. 趙紀軍 (2009). 60 years of landscape policies and development in China (2): Soviet experience新中國園林政策與建設60年回眸(二)蘇聯經驗 *Landscape Architecture*風景園林, (02), 98–102; Zhao, J. J. 趙紀軍 (2009). 60 years of landscape policies and development in China (4): Garden revolution新中國園林政策與建設60年回眸(四) 園林革命. *Landscape Architecture*風景園林, (05), 75–79.

[141] Liu, Y.S. 劉亦師 (2017). Leisure and Recreation Under Socialism: A Preliminary Study on New Urban Parks in Beijing in the Early 1950s社會主義城市休閒空間——建國初北京新公園建設探論. *New Architecture*新建築, 2017(06), 35–41.

[142] Liu, Y.S. 劉亦師 (2018). "People's Fairyland": A Study on Taoranting Park"人民的樂園"——陶然亭公園營建記 (1952–1966). *Archicreation*建築創作, 2018(05), 162–169.

be divided into two categories: the first category studies urban parks built in the nineteenth century, represented by Liu Xinyu and Wang Xiaojun's research on the Boston park system,[143] Zhang Yichi and Wang Junzhe's research on Parisian urban parks,[144] and Fan Simin and Xu Xiaoming's research on the Chicago lakefront park system.[145] The second category studies the relationship between the rise of neoliberalism or urban entrepreneurialism and urban parks since the 1960s, represented by Zhang Tianjie and Cheng Bingqian's research on parks in Singapore,[146] and Li Lingyue and Wang Lan's research on the High Line Park.[147]

After 2020, scholars began to pay attention to the history of urban parks built in China after the reform and opening up. Although the number of such studies is relatively small, they still represent certain significance. For example, Luo Yuchen and Lin Guansi studied the urban parks built in Guangzhou during the 1980s and 1990s, suggesting that they exhibit a blend of Chinese and Western gardening cultures, emphasizing aesthetics and ecology. They boldly integrated Western styles while inheriting traditional garden styles.[148] Liu Fangxin and Zhao Jijun studied urban green space construction in China under the "ecological concept" during the early period of reform and opening up,[149] as well as the cultural output in the creation of Chinese gardens overseas in the 1980s.[150] Both of these studies, to varying degrees, still aim to provide historical insights for contemporary practices. After 2020, new research has emerged that connects the social context through the natural attributes

---

[143] Liu, X., Xue, Q., & Wang, X. 劉新宇,薛求理,王曉俊 (2022). Morphological Transformation of Boston Park System during the 19th Century and Its Reasons十九世紀波士頓公園體系的形態演變歷程及其動因分析. *Chinese Landscape Architecture*中國園林, 38(09), 134–139.

[144] Zhang, Y., & Wang, J. 張亦弛,王珺哲(2019). Jean-Charles Adolphe Alphand and the Promenades in the 19th Century Paris阿爾方與19世紀巴黎城市公園建設. *Landscape Architecture*風景園林, 26(11), 108–112.

[145] Fan, S., Liu, Z., & Xu, X. 范思敏,劉志成,許曉明 (2020). Development and Evolution of the Chicago's Lakefront Park System芝加哥湖濱公園系統發展演變歷程. *Landscape Architecture*風景園林, 27(02), 66–71.

[146] Zhang, T.J., & Cheng, B.L.張天潔,程秉鈐 (2021). City Cultural Capital and Local Rebuilding: Analysis of Singapore Park Construction and Renewal Process城市文化資本與地方重塑:新加坡公園建設與更新歷程解析 (1968–2020). *Landscape Architecture*風景園林, 28(06), 56–63.

[147] Li, L., Li, W., & Wang, L. 李凌月,李雯,王蘭 (2021). Urban Entrepreneurialism and Green Renewal of Industrial Heritage: Mechanisms and Impacts — Implications and Reflections on Highline Park都市企業主義視角下工業遺產綠色更新路徑及其影響——廢棄鐵路蛻變高線公園. *Landscape Architecture*風景園林, 28(01), 87–92.

[148] Luo, Y.C., Xiao, L., Lv, B.W., & Lin, G.S. 羅雨晨,蕭蕾,呂博文,林廣思 (2023). Historical Facts and Characteristics of the Construction of Guangzhou Parks in the 1980s and 1990s1980~1990年代廣州公園的建設史實與特點. *South Architecture*南方建築, (06), 42–51.

[149] Liu, F., Zhao, J., & Han, Y. 劉方馨,趙紀軍,韓依紋 (2021). Research on the Characteristics of Urban Green Space and Practice in China from the Perspective of "Ecological View" in the Initial Stage of Economic Reform and opening-up改革開放初期"生態觀"視角下中國城市綠地發展與實踐特點研究. *Landscape Architecture Academic Journal*園林, 38(12), 68–74.

[150] Liu, F.X., & Zhao, J.L.劉方馨,趙紀軍 (2021). Cultural Export and Representation in the Making of Overseas Chinese Gardens in the 1980s20世紀80年代境外中國園營造中的文化輸出與再現. *Chinese & Overseas Architecture*中外建築, (01), 86–90.

of urban parks, reflecting urban parks as a cultural form of nature. For instance, Liu Yishi researched the concept of planting urban street trees in modern China and its relationship with popularizing knowledge, disciplining citizens, and nationalism.[151] He argues that street trees reflect the process of nation construction in modern China. Liu Xinyu and Xue Qiuli studied the parks and green space construction in the economic special zone of Shenzhen in the 1980s, suggesting a close relationship between urban park construction at that time and attracting foreign investment and national rejuvenation.[152] Shen Shanshan and Bao Zhiyi traced the history of green space construction in Hangzhou's Zhongshan Park since the Republic of China era, suggesting that the process of plant landscape changes not only serves beautification functions but also reflects its role as a political indoctrination medium and carrier of historical culture.[153] On this note, the research of Peng Haoxin is also noteworthy. He analyzed the phenomenon of the relocation of parks and green spaces in Taipei City to areas outside of flood defenses from a political economic perspective. Peng argues that this phenomenon roots in the inability of cities to sustain capital reproduction by developing parks within flood defenses. Thus, it is a shared solution between urban growth alliances and anti-growth alliances.[154]

The writing on the history of urban parks in Shenzhen mainly consists of retrospective articles and project introductions by planners and project managers who have been involved in park construction planning and management. On the occasion of the 30th and 40th anniversaries of the establishment of the SSEZ, Mr. Meng Zhaozhen recalled the contributions of the design team from Beijing Forestry University to Shenzhen's urban greening construction, emphasizing the application of traditional garden design concepts.[155] He Fang divided Shenzhen's landscape and garden construction into three stages, highlighting its contemporary characteristics and significance. Especially on the occasion of the 40th anniversary, he focused more on the practical aspects of the relationship between landscape gardens and the city,

---

[151] Liu, Y.S.劉亦師 (2016). A Preliminary Study on the Rise and Popularization of Roadside Tree Planting in the Republic China民國時期行道樹種植觀念之形成與普及探述. *Landscape Architecture*風景園林, (04), 80–87.

[152] Liu, X.Y., Wang, X.J., & Xue, Q.L.劉新宇,王曉俊,薛求理 (2022). The Practice and Representation of Urban Parks in Shenzhen Special Economic Zone in the 1980s 20世紀80年代深圳經濟特區城市公園的實踐與表現. *Landscape Architecture Frontiers*景觀設計學, 10(04), 70–83.

[153] Shen, S., Hu, W., Yang, F., et al. 沈姍姍, 胡汪涵,楊凡等 (2022). Historical Changes of Plant Landscape in Zhongshan Park of Hangzhou杭州中山公園植物景觀歷史變遷探析(1912–2021). *Chinese Landscape Architecture*中國園林, 38(09), 99–104.

[154] Peng, H.X. 彭皓炘(2010). *Ambiguous Parks——A Political Economy Analysis of the Riverside Park Development in Taipei (Master's thesis, Graduate Institute of Building and Planning*曖昧的公園─臺北市公園綠地向堤外移轉的政治經濟學分析, College of Engineering, National Taiwan University臺灣大學建築與城鄉研究所.

[155] Meng, Z.Z. 孟兆禎(2020). Chinese traditional garden flourishing in Shenzhen: In honor of the 40th anniversary of the Shenzhen Special Economic Zone大鵬風舞中國傳統園林——紀念深圳特區成立40周年. *Landscape Architecture*風景園林, 27(10), 10–12; Meng, Z.Z. 孟兆禎 (2010). The United Endeavour for the Construction Programs in Shenzhen In Honor of the 30th Anniversary of Shenzhen Special Economic Zone眾志成城的深圳建設 紀念深圳特區30周年. *Landscape Architecture*風景園林, 2010(05), 56–57.

but all emphasized the application of traditional garden design concepts.[156] In addition, Liu Jiaqi, Yang Laili, Zhang Xinsi, Zhu Weihua, and others also reviewed and summarized the development of Shenzhen's urban parks and greening projects.[157] Other studies mainly focus on specific projects, such as Li Zhisheng's planning of Shenzhen's urban green space system,[158] Fu Keqin's planning of Shenzhen Garden Expo Park,[159] Zhang Xinsi's construction of People's Park,[160] He Fang and Ye Feng's overall planning of Lotus Hill Park,[161] David Thompson et al.'s planning and design of OCT Happy Coast and Beihai Wetland Park,[162] Zhu Jie's planning and design of Talent Park, Chen Tao et al.'s construction of ecological landscape forests,[163] and so on. Most of these retrospectives and introductions are written by planners and project managers and do not emphasize the exploration of research objects from a historical perspective, so they can hardly be considered strictly historical research. We can observe that the academic and analytical aspects of the historical research on urban parks in Shenzhen still have significant room for improvement and potential.

---

[156] He, F. 何昉 (2010). Shenzhen Landscape Architecture: From the Inheritance of Landscape Architecture Mechanism of China to the Creation of Regional Style. For the Celebration of the 30th Anniversary of the Founding of Shenzhen Special Economic Zone深圳園林:從繼承中國風景園林體系到地域風格的創建　寫在深圳經濟特區成立30周年之際. *Landscape Architecture*風景園林, (05), 70–72 He, F. 何昉 (2020). The 40 Years of Prosperity and Development of Landscape Heritage and Innovation風景園林傳承創新,大鵬展翅騰飛卌年. *Architectural Practice*建築實踐, (11), 168–175.

[157] Liu, J.Q. 刘家麒(2010). The innovative development of landscape construction program in Shenzhen深圳园林绿化建设的发展创新之路. *Landscape Architecture*风景园林, (05), 58–59; Yang, L.L. 杨赉丽 (2010). A Memory of Shenzhen记忆深圳. *Landscape Architecture*风景园林, (05), 60–62; Zhang, X.S. 张信思 (2010). The Landscape Architecture Profession in Shenzhen Will Have a Brighter Future A Celebration of the 30th Anniversary of the Founding of Shenzhen Special Economic Zone深圳园林明天会更好　纪念特区成立30周年的回顾与展望. *Landscape Architecture*风景园林, (05), 63; Zhu, W.H. 朱伟华 (2010). A Dream of the City of Parks The 30 Years' Development of Parks in Shenzhen为了"公园之城"的梦想 深圳公园30年. *Landscape Architecture*风景园林, (05), 64–66.

[158] Li, Z.S. 李錚生 (1994). Planning of green space system in Shenzhen深圳市綠地系統規劃. *Urban Planning*城市規劃, 1994(01), 53–55.

[159] Fu, K.Q., Lai, S.W., & Wang, F. 傅克勤,賴樹翁,王鋒 (2006). Brief notes on the planning of Shenzhen Garden Expo Park深圳園博園規劃小記. *Chinese Landscape Architecture*中國園林, 2006(10), 60–63.

[160] Editorial Board of Guangdong Landscape Journal廣東園林期刊編輯部(2017). Zhang, X.S.張信思. *Guangdong Landscape Architecture*廣東園林, 39(05), 4–9.

[161] He, F., & Ye, F. 何昉,葉楓 (2003). Creating a "living green natural museum" in the urban center of Shenzhen: Overall planning ideas for Lianhua Mountain Park創建深圳城市中心區"活的綠色自然博物館"——蓮花山公園總體規劃思想. *Chinese Landscape Architecture*中國園林, (06), 11–15, 6–7, 10, 23.

[162] David T., Zhang, S.Y., Henry D., et al. 大衛·湯普森,張樹儀,亨利·道爾頓,等 大衛·湯普森,張樹儀,亨利·道爾頓,等 (2013). Shenzhen Oct Bay深圳華僑城歡樂海岸. *Landscape Architecture*風景園林, 2013(03), 68–76; David T., Zhang, S.Y., Henry D., et al.大衛·湯普森,張樹儀,亨利·道爾頓,等 (2012). North Lake Wetland Park in Happy Coast of Shenzhen Oct深圳華僑城歡樂海岸北湖濕地公園. *Landscape Architecture*風景園林, (02), 86–91.

[163] Chen, T., Li, G., & Liang, Z. 陳濤,李剛,梁正陽 (1999). Species selection and demonstration forest construction of ecological landscape forests in Shenzhen深圳生態風景林樹種選擇與示範林營造. *Chinese Landscape Architecture*中國園林, (04), 44–46.

### 2.3.2 Researches by Overseas Scholar

In the late twenty-first century to the early years of the new millennium, foreign scholars conducted historical research on urban parks, with Galen Cranz's work having a particularly wide influence. Cranz studied the relationship between the construction of parks in the United States from 1850 to 1982 and their political context. She viewed urban parks as part of social reform strategies, arguing that understanding them as preserved natural land is naïve.[164] Alexander von Hoffman researched how the space and function of Boston's Franklin Park changed during the Progressive Era at the turn of the nineteenth and twentieth centuries, ultimately deviating from the original intentions of the designers.[165] Cynthia Zaitzevsky conducted detailed research on the design and construction process of the Boston park system based on changes in Boston's urban landscape and the Boston park movement.[166] Roy Rosenzweig and Elizabeth Blackmar reflected New York's urban social history through the history of Central Park, connecting park planning and design to the relationships between merchants, politicians, Irish workers, immigrants, and African American residents.[167] Hillary A. Taylor also studied the spatial design and meaning construction of urban parks in the latter half of the nineteenth century as sites of political, cultural, social, and economic conflict.[168] Terence Young studied San Francisco's Golden Gate Park, arguing that urban parks, like cities, underwent modernization in the nineteenth century.[169] John Crompton's research revealed the role of New York's Central Park, built in the nineteenth century, in raising property prices in the surrounding area.[170] C.W. Thompson criticized nineteenth century American urban parks for often carrying an elitist disdain despite their promotion of philanthropy and democracy.[171] Furthermore, after Japan actively embraced Western urban and urban park construction ideas following the Meiji Restoration, Japanese scholars also conducted research on the history of urban parks. Yosaburo Shirahata traced the origins and development of German parks and

[164] Cranz, G. (1982). *The politics of park design: A history of urban parks in America*. Cambridge, MA: MIT Press. (pp. 61).

[165] Von Hoffman, A. (1988). " Of Greater Lasting Consequence": Frederick Law Olmsted and the Fate of Franklin Park, Boston. *The Journal of the Society of Architectural Historians*, 47(4), 339–350.

[166] Zaitzevsky, C. (1982). *Frederick Law Olmsted and the Boston Park System*. Cambridge, MA: Belknap Press. (pp. 96).

[167] Rosenzweig, R., & Blackmar, E. (1992). *The park and the people: A history of Central Park*. Cornell University Press.

[168] Taylor, H. A. (1995). *Urban public parks, 1840–1900: Design and meaning*. Garden History, 201–221.

[169] Young, T. (1995). Modern urban parks. *Geographical Review*, 535–551.

[170] Crompton, J. L. (2001). The impact of parks on property values: A review of the empirical evidence. *Journal of leisure research*, 33(1), 1–31.

[171] Thompson, C. W. (2002). Urban open space in the 21st century. *Landscape and urban planning*, 60(2), 59–72.

argued that parks not only met the demands of urban dwellers but also served the interests of the ruling class, with the latter being more important.[172] Makiko Ishikawa discussed the formation of green spaces in Japan from the Meiji era to World War II based on the development of European and American urban parks, addressing issues such as the evolution of feudal estates and urban expansion.[173] During this period, research mainly focused on the relationship between urban parks since the nineteenth century and capitalist industrialized cities of the early twentieth century. The important characteristic of the research strategy was using urban parks as a lens to examine urban history, although the focus remained on urban parks themselves.

In the new millennium, using urban parks as a lens to examine urban history remains a primary research strategy. However, the focus has shifted from urban parks themselves to a more diverse and expansive urban historical context, with urban parks serving as entry points into broader issues. For example, Timothy J. Gilfoyle extensively explores the social and historical context surrounding the birth of Millennium Park in Chicago, before delving into a description of the park itself.[174] Sonja Dümpelmann focuses on Italian cities during the Mussolini regime in the twentieth century, arguing that the planning and design of urban parks during this period reflected fascist ideology. Fascism politicized citizens' demands for parks and established a connection between the "Third Rome" and the ancient Roman imperial power through spatial symbolism.[175] Attorre et al. analyze the historical changes of the tree species used in the avenues of Rome, suggesting corresponding relationships between these species and the city's history and politics.[176] Thorbjörn Anderson challenges the notion of parks as isolated oases within cities. He studies the park system in Stockholm, elucidating its political objectives, which include seeking the interests of the people in a utopian manner between the market economy and planned economy. His research also discusses the impact of park construction on individual freedom in certain aspects.[177]

[172] Yozaburo, S. 白幡洋三郎 (2014). *A History of Modern Urban Parks: The Origin of Europeanization*近代都市公園史:歐化的源流. (Trans. by Li Wei and Nan Cheng李偉 南誠). Beijing: New Star Press北京: 新星出版社.

[173] Ishikawa, M., 石川幹子 (2014). *City and Green Space: Striving to Create a New Urban Environment*城市與綠地——為創造新型城市環境而努力. (Trans. by Lei, Y. et al.雷芸 等). Beijing: China Architecture & Building Press北京:中國建築工業出版社.

[174] Timothy, J. G. (2006). *Millennium Park: Creating a Chicago Landmark*. Chicago, IL: University of Chicago Press.

[175] Dümpelmann, S. (2005). La battaglia del fiore. Gardens, Parks and the City in Fascist Italy. *Studies in the history of gardens & designed landscapes*, 25(1), 40–70.

[176] Attorre, F., Francesconi, F., Pepponi, L., Provantini, R., & Bruno, F. (2003). Spatia-temporal analyses of parks and gardens of Rome. *Studies in the History of Gardens & Designed Landscapes*, 23(3), 293–306.

[177] Andersson, T., Conan, M., & Whangheng, C. (2008). Swedish mid-century utopia: Park design as a tool for societal improvements. In Conan, M., and Wangheng. C. (Eds.) *Gardens, City Life and Culture: A World Tour*. Washington, DC: Dumbarton Oaks Research Library and Collection, 157–172.

After 2010, numerous studies have directly addressed the relationship between urban social systems and urban parks, aiming to break free from the binary opposition between cities and parks that dominated previous research strategies. For instance, Denisa Badiu studied the transformation of urban green spaces in Bucharest, Romania, following the end of the socialist regime. Through an examination of legal frameworks, her research suggests that changes in Bucharest's social system directly influenced the quantity and distribution of green spaces, indicating that green spaces are not isolated from the social and historical context of the city.[178] Joan Henderson investigated the development patterns of urban parks in Singapore, emphasizing how the emphasis on order and authority in this city-state is reflected in its urban parks, which serve as tools for physical planning and economic policies.[179] Andrew Smith studied the planning process of London's Olympic Park, revealing how it became a theme park during the period from 2004 to 2014 amidst negotiations between commercial pressures and sustainability concerns.[180] Kevin Loughran explored Washington Park and Jackson Park, both established in the nineteenth century, highlighting the unique aspects of urban parks and greenery: (1) their normative values surrounding nature, (2) power dynamics closely related to dominant landscape aesthetics, and (3) their economic implications through land value associations.[181] Nate Gabriel similarly studied nineteenth-century urban parks, arguing through visual image analysis that discourses about parks were largely constructed through photography, thus artificially shaping the binary opposition between cities and parks.[182] Furthermore, some recent studies have situated cities within the broader context of globalization when discussing their relationship with parks. For example, Hilary Angelo observed that urban greening has become a global policy trend, viewing urbanization as a process of global historical change. Based on this perspective, she conducted research on the greening history of the Ruhr region in Germany.[183] Tal Alon-Mozes and Shirili Gilad-Ilsar examined the planning and design of Yarkon Park in Tel Aviv, Israel, during the 1960s and 1970s within a global knowledge context, highlighting its close association with Jewish nationalism and patriotism, which influenced the park's planning and design.[184]

[178] Badiu, D. L., Onose, D. A., Niță, M. R., & Lafortezza, R. (2019). From "red" to green? A look into the evolution of green spaces in a post-socialist city. *Landscape and urban planning*, 187, 156–164.

[179] Henderson, J. C. (2013). Urban parks and green spaces in Singapore. *Managing Leisure*, 18(3), 213–225.

[180] Smith, A. (2014). From green park to theme park? Evolving legacy visions for London's Olympic Park. *Arq: Architectural Research Quarterly*, 18(4), 315–323.

[181] Loughran, K. (2020). Urban parks and urban problems: An historical perspective on green space development as a cultural fix. *Urban Studies*, 57(11), 2321–2338.

[182] Gabriel, N. (2011). The work that parks do: towards an urban environmentality. *Social & Cultural Geography*, 12(02), 123–141.

[183] Angelo, H. (2021). *How green became good: Urbanized nature and the making of cities and citizens*. University of Chicago Press.

[184] Alon-Mozes, T., & Gilad-Ilsar, S. (2020). Modern park for a modern city: planning Tel Aviv's Yarkon Park during the 1960s–1970s. *Studies in the History of Gardens & Designed Landscapes*, 40(1), 80–94.

Raffaella Sini situated urban parks in Singapore within the historical context of its rapid development in the mid-twentieth century, arguing that the ideology of racial rejuvenation influenced the planning and design of the park system, while also revealing the state's efforts to reconcile local history with global trends.[185]

### 2.3.3 Trend and Gap

As can be seen from the above, the historical research on urban parks in China began to take off in the 1990s. During this period, the research was primarily aimed at inspiring planning and design practices, with many researchers having backgrounds in design. From the 1990s to the early 2000s, researchers mainly came from the humanities and social sciences fields, with almost no motivation for inspiring planning and design practices. The focus of the research was mainly on modern urban parks, with a strong historical orientation. However, in terms of theoretical frameworks, urban parks were simply linked to social issues such as nationalism and citizen life, neglecting the connection between the natural elements of parks and social issues. During this period, research in the built environment domain regarding urban parks mostly consisted of recommendatory studies of foreign parks and discussions on the management system of domestic urban parks and green spaces, rather than strict historical research. After 2010, there was a significant increase in historical research on urban parks in the built environment domain. Except for a few scholars, most of the research still had a focus on inspiring planning and design practices, with discussions mainly revolving around the form and style of urban parks. This discourse structure was shaped by the reality of large-scale infrastructure construction in China since the 1990s, emphasizing the normative. While it has its rationality, at a more rigorous level, historical research should pay more attention to the relationship between context and objects, constructing meaning through specific theoretical perspectives to complement practice rather than directly guiding it. Although there were many studies during this period, few of them meticulously delineated the interaction between context and objects. This standardized discourse structure implies a linear historical view and teleology: that is, inducing past practical experiences, summarizing patterns, and guiding current planning and design practices, while ignoring how planning and design practices are shaped by specific contexts. It is necessary to recognize that historical research has its significance independent of planning and design practice. A detailed understanding of how specific social and historical contexts interact with urban parks is sufficient to constitute historical research. Since 2020, urban parks constructed after the reform and opening up in the built environment field have begun to receive

[185] Sini, R. (2023). Singapore's postcolonial ethnic revival in parks design. *Studies in the History of Gardens & Designed Landscapes*, 43(3), 225–250.

attention from historical research. In addition, some emerging studies have noted the social attributes of natural elements in urban parks but have not directly addressed this issue in theoretical frameworks.

Since the 1980s, historical research on urban parks initiated by foreign scholars has exerted a considerable influence. Before the turn of the millennium, the focus was primarily on urban parks built in the nineteenth century, with discussions centering around historical events with urban parks as the focal point. After the turn of the millennium, research began to shift toward a contextual approach, focusing on the interaction between context and objects. Post-2010, many studies have attempted to break through the binary opposition between city and park in theoretical frameworks, which is essentially a reflection of the dichotomy between society and nature in the built environment domain. Simultaneously, theoretical perspectives from global history have started to be absorbed and applied by researchers.

## Bibliography

### *Periodical*

Friedmann, J., & Wolff, G. (1982). World city formation: An agenda for research and action. *International Journal of Urban and Regional Research, 6*(3), 309–344.

Yeung, Y. M., Lee, J., & Kee, G. (2009). China's special economic zones at 30. *Eurasian Geography and Economics, 50*(2), 222–240.

### *Monographs and Collected Papers*

Cronon, W. (1992). *Nature's metropolis: Chicago and the Great West*. W.W. Norton.

# Chapter 3
# The Landscape History of Shenzhen City and Its Urban Parks

## Contents




**Abstract** This chapter provides an overview of how urban landscapes within the administrative area of Shenzhen have emerged within the context of global political-economic dynamics and the geographical-topographical framework of the Pearl River Delta. Building on this foundation, the study examines the basic distribution patterns of urban parks across Shenzhen and explains the factors underlying their formation.



**Keywords** Urban landscape · Shenzhen administrative area · Topography · Globalization




X. Liu, *Fabricating Nature*, The Urban Book Series,
https://doi.org/10.1007/978-981-95-8208-2_3

Topography is not just a slowly changing inert substance in geology, but seeps into the everyday world of human beings, blending into the practices of human agency.[1,2] Such practices of human agency feature a wide range of types, including emotion, memory, and physical experience from a phenomenological perspective. From the perspective of political economy, farming, logging, urbanization, industrial development, and globalization are included in topographic practices.

This chapter will probe into the landscape history of Shenzhen from this perspective and explain three questions: (1) How does the urban topography of Shenzhen blend with the globalization in the middle to late twentieth century and produce the corresponding urban landscape? (2) What are the morphological characteristics of urban parks (as an important part of the urban landscape) built in different historical periods in Shenzhen? (3) How do these morphological characteristics emerge in the historical context of the blending of topography and globalization? These three questions will be detailed in three sections of this chapter.

## 3.1 Landscape History of Shenzhen

### *3.1.1 Shenzhen in Topography*

Both Ji Cheng's (計成, 1582–1642) conscious emphasis of "siting for garden foundation"[3] in the traditional gardening and Humphry Repton's (1752–1818) unconscious design of almost all his landscape gardens within the soothing sedimentary rocks of eastern England[4] show that the topographic features of the site are important factors affecting the design and construction of gardens. In fact, the construction of gardens is closely related to topography, so is the construction of buildings and cities. After all, they are "always built somewhere."[5]

---

[1] Chen, J. P. 陳潔萍 (2011). *Place-Writing: The Study on Topography in the Expanded Field in Contemporary Architecture, Urbanism, and Landscape Design*書寫場地:當代建築、城市、景觀設計中擴展領域的地形學研究. Nanj ing: Southeast University Press南京:東南大學出版社. (pp. 1–3).

[2] Liu, D. Y. 劉東洋 (2012). Seeing Topography from the Window behind My House從我家後窗看見地形學. In Cousins, M., Chen, W., Li, H., Ge, M. 卡森斯, 陳薇, 李華, 葛明 (Eds.). *Architectural Research 02: Topography & Mental Space*建築研究02:地形學和心理空間. Beijing: China Architecture & Building Press北京:中國建築工業出版社. (pp. 73–106).

[3] Ji, C. (Author) 計成 原著, & Wang, S. Z.王紹增 (Eds.). (2013). *The Craft of Gardens* Reader園冶讀本. Beijing: China Architecture & Building Press北京:建築工業出版社. (pp. 11–21).

[4] Appleton, J. (1986). Some thoughts on the geology of the picturesque. *The Journal of Garden History*, 6(3), 270–291.

[5] Leatherbarrow, D. (2015). *Topographical stories: studies in landscape and architecture*. University of Pennsylvania Press. (pp. 9).

#### 3.1.1.1 Regional and City-Wide Topographies and Their Development History

Located in the southern Guangdong Province, the PRD (珠江三角洲) is known for the topographical features of "surrounded by mountains on three sides, and facing the open sea on the fourth side, confluence of three rivers, and diversion by eight estuaries."[6] Shenzhen is in the southeast of the PRD, as shown in Fig. 3.1, most of

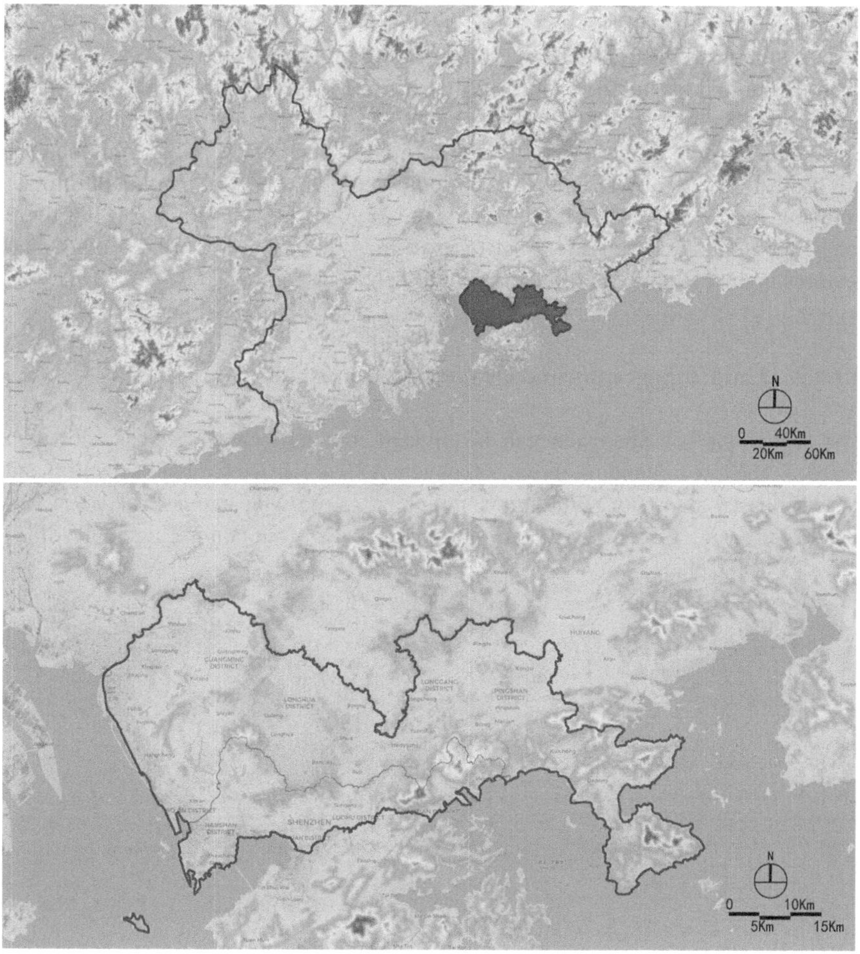


**Fig. 3.1** Topography of the PRD metropolitan area (珠三角区域) and Shenzhen (The base map downloaded from: https://en-us.topographic-map.com/, redrawn by the author)

[6]Guangdong Provincial People's Government廣東省人民政府(2014, December 11). *Notice of the General Office of the People's Government of Guangdong Province on Issuing the Integrated Planning of Ecological Security System in the Pearl River Delta Region* 廣東省人民政府辦公廳關於印發珠江三角洲地區生態安全體系一體化規劃 (2014–2020). Retrieved from http://www.gd.gov.cn/gkmlpt/content/0/143/post_143525.html#7

its landforms are low mountains, gentle platforms, and terrace hills. Dapeng (大鵬) and Kuiyong (葵涌) sub-districts in the southeast are predominantly low mountains; the central and northwest are mainly hilly, and low mountains are scattered to sandwich large alluvial plains; Shajing (沙井), Fuyong (福永), and Xixiang (西鄉) sub-districts in the southwest are dominated by large coastal alluvial plains. According to the survey published by Guangzhou Institute of Geography (廣州地理研究所) in 1984, Shenzhen can be roughly divided into 4 topographic zones, 8 topographic regions, and 26 topographic units. The overall features include[7] (1) high terrain in the southeast and low in the northwest; (2) abundant coastline types; and (3) significant differences in eastern and western landforms.

To better understand the causes of Shenzhen's topographic features, we must go deep into the development of topography in the PRD metropolitan area and Shenzhen. For tens of millions of years without human intervention, such features are basically driven by climate and periodic geological activities, including plate collision, orogenesis, transgression, and regression, as shown in Table 3.1. It is these long-time geological activities that have formed the rich and varied topographic features in the PRD metropolitan area and Shenzhen.

#### 3.1.1.2 Land Topography and Drainage Features

The topography of Shenzhen can be divided into four categories, as shown in Fig. 3.2. The low mountains above 500 meters ASL are distributed in the central and eastern coastal areas (i.e., Wutong Mountain (梧桐山), Meishajian Peak (梅沙尖), and Ta Kwu Ling (打鼓嶺)), the northeast part (Bijia Mountain in Baguang Village

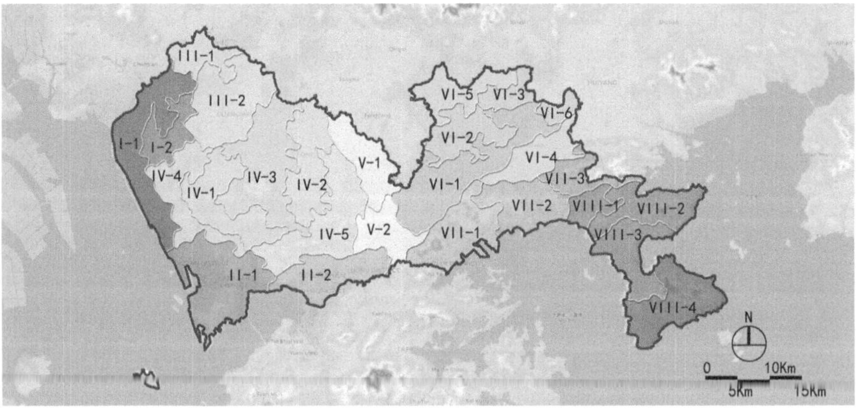


**Fig. 3.2** Topographic zoning of Shenzhen (Guangzhou Institute of Geography廣州地理研究所 (1983). *Shenzhen Landforms*深圳地貌. Guangzhou: Guangdong Science and Technology Press 廣州:廣東科技出版社. (pp. 56))

[7]Guangzhou Institute of Geography廣州地理研究所(1983). *Shenzhen Landforms*深圳地貌. Guangzhou: Guangdong Science and Technology Press廣州:廣東科技出版社. (pp. 3–7).

**Table 3.1** Development of topography in the PRD metropolitan area and Shenzhen ("Shenzhen Geological" Compilation Team《深圳地質》編寫組. (2009). *Geology of Shenzhen*深圳地質. Beijing: Geological Publishing House北京:地質出版社, 51–52)

| Period | Development of topography in the PRD metropolitan area | Development of topography in Shenzhen |
|---|---|---|
| Eocene (53–36.5 million years ago) | The basin near the southern coastal zone of Guangdong descended, and the Pearl River estuary basin received fluvial lacustrine and marine sediments | The crust uplifted, and the intermontane basin in the southeast region descended and received lacustrine sediments |
| Oligocene (34–23 million years ago) | The Chinese mainland, including Guangdong, was mostly ascending, and the South China Sea entered a period of geohistorical development, with mantle uplift and seafloor spreading. The continental slope fracture formed a stepped depression, evolving the Pearl River estuary basin | In the first episode of the Himalayan movement (喜馬拉雅運動), the South China Sea basin further expanded, and the fault depression occurred along the northwest direction of the eastern Mirs Bay (大鵬灣), which gradually separated the Dapeng peninsula (大鵬半島) from Hong Kong. After the middle and late Miocene, the activity basically ceased, and formed the outline of the bay of Shenzhen today |
| Miocene (23–5.33 million years ago) | The Pearl River estuary and other basins formed neritic marine basins | |
| Pliocene (5.3–2.588 million years ago) | The second episode of the Himalayan movement determined the basal landform of the PRD. The differential lifting movement caused parts of the site to uplift, and some fault blocks were subsiding, forming a checkerboard basal landform | The inland area is intermittently uplifted to form a three-order planation surface of mountains and hills (300–350 m, 200–250 m, 100–150 m), representing the height of high and low hills in Shenzhen |
| Pleistocene (2,588,000–11,700 years ago) | The edge of the PRD was an ascending denuded zone, while the inland plain was basically a depressed area. The intermittently frequent crustal movement caused the sea level eustasy, three transgressions and regressions occurred, and the fluvial-delta facies alternated in the PRD | In the third episode of Himalayan movement, high platforms (60–80 m) were formed at the beginning of the early Pleistocene, middle platforms (30–45 m) from the early Pleistocene to middle Pleistocene, and class IV pluvial-alluvial terraces in coastal inland; low platforms (10–25 m) were formed from the middle Pleistocene to the late Pleistocene; the class II river alluvial terraces were formed in the early and middle late Pleistocene, and the second-order pluvial-alluvial terraces were formed in the inland valley of Shenzhen in the late late Pleistocene |
| Holocene (since 11,700 years ago) | | The class II pluvial-alluvial terraces were still developed in the early Holocene, and in the early middle Holocene, the class I pluvial-alluvial terraces were formed in Shenzhen |

(壩光筆架山), Shiren Ridge (石人嶺), Tianxin Mountain (田心山)), and the Dapeng Peninsula (Paiya Mountain (排牙山) and Qiniang Mountain (七娘山)). The high and low hills 100–500 meters ASL are mainly distributed in Jigongtou Peak (雞公頭), Tanglang Mountain (塘朗山), Yangtai Mountain (羊臺山), Diaoshen Mountain (吊神山), Fenghuang Mountain (鳳凰山), Danan Mountain and Xiaonan Mountain (大小南山), and Antuo Mountain (安托山) in the west, around Qinglinjing Reservoir (清林徑水庫) in the north, and in Sha Tau Kok Back Hill (沙頭角後山), Yantian Back Hill (鹽田後山), Dashanjian Mountain (大山尖), Libi Mountain (犁壁山), Qiushui Ridge (求水嶺), and Paiya Mountain in the east. The platforms with an altitude of 10–80 meters are mainly distributed on both sides of the valley plains of Guanlan River (觀瀾河), Shawan River (沙灣河), Longgang River (龍崗河), Maozhou River (茅洲河), and Pingshan River (坪山河) as well as the central and western coastal areas, followed by the middle and upper reaches of Xixiang River (西鄉河) and Dasha River (大沙河), and even Tiegang Reservoir (鐵崗水庫) and Xili Reservoir (西麗水庫) areas.[8]

The diluvial terraces of river channels are mostly distributed in the middle and east coast gully outlets and on both sides of Pingshan River, Longgang River, Guanlan River, Shenzhen River (深圳河), and Maozhou River. Marine deposition terraces are mostly distributed in the middle of Nantou Peninsula (南頭半島), Baishizhou (白石洲), Sha Tau Kok (沙頭角), Mirs Bay, Dameisha, and Xiaomeisha (大小梅沙), while alluvial and coast plains are mainly distributed in Shajing and Fuyong sub-districts southeast of Dongbao River (東寶河) in the western part, and Wangmu River Estuary (王母河口) in Dapeng sub-district north and east of Shenzhen Bay (深圳灣).[9]

Apart from the land topography, Shenzhen's water system is also an important component of the urban topography. Shenzhen's rivers are divided into three river systems, with coast ranges and Yangtai Mountain as the main watershed: (1) There are more than 120 rivers in the bay water system, of which 8 major rivers flow into Shenzhen Bay, Mirs Bay, and Daya Bay (大亞灣); (2) there are more than 40 rivers in the Pearl River Estuary system, of which two major rivers flow into the Lingdingyang Bay (伶仃洋) of Pearl River Estuary; and (3) the Dongjiang River system consists of three major rivers that flow into the Dongjiang River and its tributaries. The distribution of these rivers is heavily influenced by the geological structure and landform. As one of the important watersheds, the coast ranges have the water system on their south slope flowing into the bay in the shape of a comb, characterized by short sources and impetuous torrents. Rivers on the north slope have to flow through the hills and terraces, known for long rivers and many tributaries. They are completely different from the water system on the south slope. The northern hills and valleys radiate from Yangtai Mountain as the center and into the three major drainage systems, respectively.[10]

[8] "Shenzhen Geological" Compilation Team《深圳地質》編寫組 (2009). *Geology of Shenzhen*深圳地質. Beijing: Geological Publishing House北京:地質出版社, 31–32.

[9] "Shenzhen Geological" Compilation Team《深圳地質》編寫組 (2009). *Geology of Shenzhen*深圳地質. Beijing: Geological Publishing House北京:地質出版社, 32–33.

[10] Guangzhou Institute of Geography廣州地理研究所(1983). *Shenzhen Landforms*深圳地貌. Guangzhou: Guangdong Science and Technology Press廣州:廣東科技出版社. (pp. 140–142).

#### 3.1.1.3 Production and Life in the Topography

Seen from the PRD region, the tenth century was the turning point of agriculture in the Lingnan area (嶺南地區), to the extent that the agricultural production used to be dominated by slash-and-burn cultivation and dry farming, followed by the introduction of the rice planting techniques of regions south of the Yangtze River into the Lingnan area and the PRD region. In the Ming Dynasty, the PRD region employed the most land-intensive cultivation pattern and became so productive that it could support a very high population density.[11]

In the case of Bao'an County, the predecessor of Shenzhen, its salt industry and oyster industry enjoy a long history. From the late Qing Dynasty to the Republic of China (民國) era, the population was shrinking, and the production of agriculture, fishing, and aquaculture gradually declined due to the wars. After the founding of the PRC, especially in the early 1950s, farmers actively reclaimed wasteland to bring more land under cultivation (Table 3.2).

In the 1950s, farmers in Bao'an County even remediated the cold-bottom fields, sandy arable land, clay fields, and salty acid fields in the mountainous and hilly areas throughout the county to increase agricultural yield. Besides, forestry and fruit growing, husbandry, fishery, and salt industry were developed significantly in the 1950s after the founding of the PRC.[12] The presence of all these agriculture-related industries must depend on the natural topographic conditions developed over a long geological time: both alluvial plains and mountains and hills, a rich river network, and a vast coastline.

**Table 3.2** Statistics of the arable area in Bao'an County in different years (Bao'an County Local Chronicles Compilation Committee寶安縣地方誌編纂委員會(1997). *Chronicles of Bao'an County*寶安縣誌. Guangzhou: Guangdong People's Publishing House廣州: 廣東人民出版社. (pp. 164))

| | Arable area (*mu*) | | |
|---|---|---|---|
| Year | Total | Paddy field | Dry land |
| Late 1957 | 454,288 | 369,279 | 85,009 |
| Late 1963 | 552,472 | 436,174 | 116,298 |
| Late 1971 | 549,806 | 426,867 | 122,939 |
| Late 1978 | 543,216 | 417,737 | 125,479 |
| Late 1986 | 302,441 | 227,953 | 74,488 |

[11] Marks, R. (1998). *Tigers, rice, silk, and silt: Environment and economy in late imperial South China*. Cambridge University Press. (pp. 100–115).

[12] Bao'an County Local Chronicles Compilation Committee寶安縣地方誌編纂委員會(1997). *Chronicles of Bao'an County*寶安縣誌. Guangzhou: Guangdong People's Publishing House廣州: 廣東人民出版社. (pp. 164–167; 158–237).

## 3.1.2 Shenzhen in Globalization

### 3.1.2.1 Shenzhen in Global and Chinese Transformation

The "Keynesian Era" after World War II brought nearly three decades of prosperity in Europe and the United States, which required mass consumption and also large-scale industrial production to avoid stagflation. By the 1970s, its side effects were concentrated, and various state policies suffered diminishing efficiency in regulating the market, and the industrial production of Fordism, which accompanied the Keynesian welfare state model, was similarly unsustainable.[13] To alleviate this crisis, European and American capitalist countries turned their political and economic practices to neoliberalism and shifted their modes of production organization from Fordism to post-Fordism, which focused on flexible accumulation, with emphasis on order-based contractual relations and production process outsourcing. As a result, blue-collar jobs in industrial production have gradually moved to peripheral countries. This expansion of economic space and the international division of labor allowed East Asian countries and regions (e.g., South Korea, Singapore, China Taiwan, and Hong Kong) to start an economic take-off, and even brought new opportunities and inspirations to the Chinese Mainland at a historical turning point in the late 1970s, since China was likely to use the economic crisis of capitalist countries to integrate into the world market and thus boost its domestic economy.[14]

At the moment, however, the Chinese Mainland's weak economy and shortage of foreign exchange severely limit this possibility. The striking contrast of Gu Mu's visits to Western Europe, Hong Kong, and Macao in April to May 1978 and Xi Zhongxun's visit to Bao'an County, Guangdong in July 1978 reinforced the commitment to reform at the national level. In the meantime, the China Merchants Group of the Ministry of Transport in Hong Kong (交通部駐香港招商局) wanted to build an industrial zone in Bao'an, which has aroused great attention at the national level. In January 1979, the Guangdong Provincial Committee of the Revolutionary Committee (廣東省革命委員會) and the Ministry of Transport jointly submitted a report to the State Council (國務院), proposing to establish an industrial zone of China Merchants Group in Shekou, Bao'an. It was eventually identified as the Shekou Industrial Zone (蛇口工業區) and became an outpost of the SSEZ.[15]

External opportunities and internal challenges drive change. In August 1979, the State Council issued the *Provisions on Several Issues Concerning Vigorously*

[13] Palley, T. I. (2005). From Keynesianism to Neoliberalism: Shifting Paradigms in Economics. In Saad Filho, A., & Johnston, D. (Eds). *Neoliberalism: a critical reader*. London; Ann Arbor, Mich.: Pluto Press. (pp. 20–29).

[14] Harvey D. (2005). *A brief history of neoliberalism*. New York: Oxford University Press. (pp. 120).

[15] Shenzhen Municipal Chronicles Office深圳市史志辦公室(Eds.). (1997). *Establishment and Development of China's Special Economic Zones: Shenzhen Volume*中國經濟特區的建立和發展:深圳卷. Beijing: Communist Party History Publishing House北京:中共黨史出版社. (pp. 50).

*Developing Foreign Trade and Increasing Foreign Exchange Earnings* (關於大力發展對外貿易增加外匯收入若幹問題的規定). It is expressly stated that "Special export zones may be piloted…in Shenzhen and Zhuhai first…."[16] Obviously, the original intention of establishing special economic zones is closely related to the development of foreign trade and the increase of foreign exchange earnings. In fact, this coincides with the initial purpose of the reform and opening-up at the national level at that time.

Shenzhen, a city established in 1979 and a special economic zone identified in 1980, is located in the Pacific Rim Economic Circle, in the middle of Southeast Asia and Northeast Asia, and adjacent to Hong Kong, with unique geographical conditions. In the international landscape at that time, it became a test field for China's Guangdong Province to make "One Step Ahead" in the reform and opening-up nationwide.[17] In 1980, Shenzhen was a small border city in South China with a built-up area of only about three square kilometers and a total population of only 330,000. By 2020, its built-up area reached 973.5 square kilometers with a population of 17.56 million.[18] The establishment of the SSEZ has opened the prelude of urbanization and become the epitome of cities in contemporary China, because it's strategically exemplified to be a mock-up for the modernization of domestic cities, in addition to showing China's ability and determination to reintegrate into the world.[19]

Between 1965 and 1990, the per capita GNP growth rate of 23 East Asian countries and regions was significantly higher than that of other regions in the world,[20] according to the World Bank. This economic take-off is also reflected in the urban landscape, in particular the Chinese Mainland. By 2010, the Chinese Mainland had significantly more urban land and population than other East Asian countries and regions, as shown in Fig. 3.3. In addition, the Chinese Mainland occupies 600 of the 869 cities in East Asia, and its total area is more than two-thirds of the total urban land area in East Asia. The PRD metropolitan area even surpassed the Tokyo metropolitan area in terms of area and population to become the largest metropolitan area

[16] State Council of the PRC issued "Regulations on Foreign Trade and the Increase of Foreign Exchange Earnings" 國務院頒發《關於大力發展對外貿易增加外匯收入若干問題的規定》, In Wang Z.C. 王振川 (Eds.), *Yearbook of China's Reform and Opening-Up New Era*中國改革開放新時期年鑒. Press of Chinese Democratic Legal System北京: 中國民主法制出版社. (pp. 662–663).

[17] Vogel, E. F. (1989). *One step ahead in China: Guangdong under reform (Vol. 2)*. Harvard University Press. (pp. 60–66).

[18] Chen, Y.X. 陳一新 (2022). *The History of Urban Planning in Shenzhen*深圳城市規劃簡史. Beijing: China Social Sciences Press北京: 中國社會科學出版社. (pp. 6–7).

[19] Bach, J. (2017). Shenzhen: From Exception to Rule In O'Donnell et al. (Eds.) *Learning from Shenzhen : China's post-Mao experiment from special zone to model city*. Chicago: The University of Chicago Press. (pp. 23–38).

[20] World Bank (2003). *The East Asian miracle: Economic growth and public policy*. New York, N.Y: Oxford University Press.

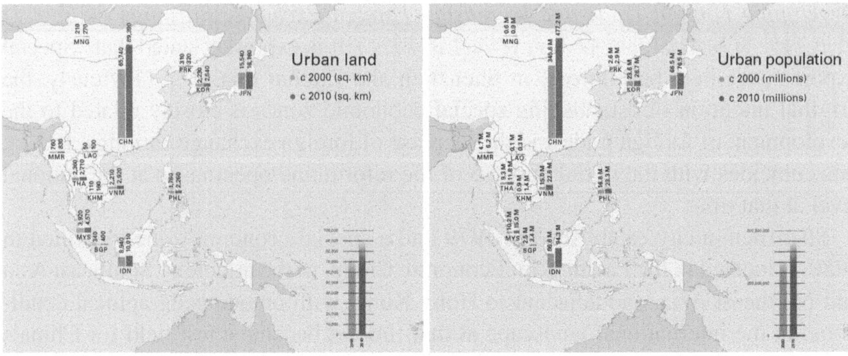


**Fig. 3.3** Urban land and population in East Asia, 2000 and 2010

in the world.[21] As a representative city in the PRD metropolitan area, Shenzhen has contributed a lot to the delta in its urban development and construction over the 40 years after 1980. This is obviously tied to the spatial resources and their allocation in Shenzhen and even the PRD metropolitan area.

#### 3.1.2.2 The Production of Urban Landscape in Shenzhen

For Henri Lefebvre, the production of urban landscape is, in fact, the representation of specific power relations, which are imbued with ideological characteristics.[22] This "ideological" character of Shenzhen can be understood on two levels that are both related and not exactly overlapping: First, the expansion of economic space in developed countries since the 1970s and the division of labor of neoliberal globalization promoted the urban space production of Shenzhen; second, China's state forces responded to neoliberalism and guided the local urban construction in Shenzhen.

At the first level, the emergence of flexible accumulation and post-Fordism has allowed industrial mass production to be broken down into several processes that are outsourced and assembled globally, while absorbing cheap labor. Since the 1980s, the influence of external forces on Asian cities has become more apparent, and cheap costs have attracted the supply chains of multinational companies.[23] The neoliberal turn in the developed countries thus produced a spatial restructure on a

---

[21] World Bank Group (2015). *East Asia's Changing Urban Landscape: Measuring a Decade of Spatial Growth*. Urban Development;. Washington, DC: World Bank. https://openknowledge.worldbank.org/handle/10986/21159 License: CC BY 3.0 IGO.

[22] Lefebvre, H.亨利·列斐伏爾 (2015). *Space and Politics*空間與政治. (Trans. By Li C. 李春) Shanghai: Shanghai People's Publishing House上海: 上海人民出版社. (pp. 1–16;50–57).

[23] Yeung, Y. M. (2011). Rethinking Asian cities and urbanization: four transformations in four decades. *Asian Geographer*, 28(1), 65–83.

global scale,[24] and the early economic take-off of East Asian countries and regions started with the construction of a large number of industrial parks, evidenced by Ulsan Industrial Park (蔚山工業區) in South Korea, Jurong Industrial Estate (裕廊工業區) in Singapore, Changhua Industrial Park (彰化工業區) in Taiwan, and industrial buildings throughout Hong Kong. The same is true of the SSEZ. The earliest document dealing with urban spatial planning is the *Program for Urban Development of Special Economic Zone of Shenzhen Municipality (1980)*, under which 655 ha. of planned industrial land (282 ha. in Huanggang District (皇崗區), 305 ha. in Shangbu District (上步區), and 68 ha. in Luohu District (羅湖區)) would account for 13.32% of the total planned land. In addition to residential areas, industrial land would occupy the highest proportion of land types.[25] According to the *Master Plan of Shenzhen Special Economic Zone (1986)* (深圳經濟特區總體規劃 (1986)), 15 industrial zones were planned, where a total of 1848 ha. of industrial land accounted for 15.1% of urban construction land, second only to 18.2% of residential land.[26] In subsequent editions of the city's master plan, the strategies of developing the financial, high-tech, and modern service industries have promoted Shenzhen to enter the international global city network, which makes the new urban landscape such as Futian Central Business District (福田中心區), Nanshan Science and Technology Park (南山科技園), Shenzhen Bay Super Headquarters Base (深圳灣超級總部基地), and Qianhai Shenzhen-Hong Kong Modern Service Industry Cooperation Zone (前海深港現代服務業合作區) emerge in the climate mentioned above.

David Harvey interprets neoliberalism as a political project aimed at the realignment of international capitalism.[27] Although China has never fully embraced the theory and practice of neoliberalism, economic theories with neoliberal overtones have been widely discussed in academia, especially in economics, since the late 1970s. The discussion has also clearly influenced the urban planning academic community in China.[28] Especially in the context of Chinese cities after the 1990s, "urban renewal (舊城改造)," "free trade zone (保稅區)," "urban marketing," and

---

[24] Harvey, D. (2005). *Spaces of neoliberalization: towards a theory of uneven geographical development (Vol. 8)*. Franz Steiner Verlag. (pp. 7–54).

[25] *Program for Urban Development of Special Economic Zone of Shenzhen Municipality*深圳市經濟特區城市發展綱要 (1980) (Discussion Draft討論稿), preserved at the archive office of Shenzhen Planning and Natural Resources Bureau.

[26] *Master Plan of Shenzhen Special Economic Zone (1986)*, preserved at the archive office of Shenzhen Planning and Natural Resources Bureau.

[27] Harvey D. (2005). *A brief history of neoliberalism*. New York: Oxford University Press. (pp. 5–38).

[28] Zhang, T. 張庭偉, & LeGates, R. (2009). Changing Planning Paradigm in China in the Post-Neoliberal Era後新自由主義時代中國規劃理論的範式轉變. *Urban Planning Forum*城市規劃學刊, 05, 1–13; Zhang, T. 張庭偉 (2004). Neoliberalism, City Management, City Governance, and City Competitiveness in China自由主義·城市經營·城市管治·城市競爭力. *City Planning Revie*城市規劃, 05, 43–50.

"government-business joint venture (官商聯營)" fully demonstrated the historical and spatial results of the localization practice of neoliberalism.[29]

At the second level, the Chinese government is involved in both the distribution and production of the "cake." On the one hand, the government owns urban land, and on the other hand, it has full control over the financial, taxation, and planning approval related to land development. Therefore, the government must also be deeply involved in the allocation of urban space resources and cannot stay out of land development and infrastructure construction.[30] Thus, China's government behavior cannot be explained by a framework based on Western classical economics; in the ideal scenario of Western classical economics, the market is self-regulating, and the government only works to protect citizens' property and never interferes with market behavior.[31] On this basis, it is not difficult to understand that the urban construction of the SSEZ is clearly driven by the government efforts, because the establishment of the SSEZ has benefited from the clear political and policy orientation from the national level—"the big contract fiscal system under the unified leadership of the central government (中央統壹領導下實行大包幹)."[32]

In June 1992, the Shenzhen Municipal Government (深圳市政府) issued the *Interim Provisions on Rural Urbanization in the Shenzhen Special Economic Zone* (關於深圳經濟特區農村城市化的暫行規定), commencing the Unified Acquisition(統征), in an effort to nationalize the rural collective land in the SSEZ for urban construction. A total of 46,000 permanent farmers, fishermen, and oyster breeders in the SSEZ from 68 village committees and Shahe Overseas Chinese Farm (沙河華僑農場) became urban residents all at once. The government quickly completed the urbanization transformation of rural areas in the SSEZ by means of administration. In June 2004, a total of 270,000 villagers from 218 village committees of 18 towns, and 956 square kilometers of collective land in Bao'an and Longgang Districts were nationalized, due to the issuance of the *Measures on Urbanization Land Management in Bao'an and Longgang Districts of Shenzhen Municipality* (深圳市寶安龍崗兩區城市化土地管理辦法) and the nationalization of collective land outside the SSEZ. This process is also abbreviated as "Unified Transformation(統轉)."[33] The resulting urban villages in the city became the

[29] Wu, Q. Y. 吳啟焰 (2011). The Critic of the Neoliberalism Urban Studies in China新自由主義城市空間重構的批判視角研究. *Scientia Geographica Sinica*地理科學, 31(07), 769–774.

[30] Lan, X. H. 蘭小歡 (2021). *Immersed in Affairs: The Chinese Government and Economic Development*置身事內 中國政府與經濟發展. Shanghai: Shanghai People's Publishing House上海:上海人民出版社. (pp. 85–86).

[31] Huang, C.P. 黃宗智 (2022). *The Dual Unity of State and Society*國家與社會的二元合一. Guilin: Guangxi Normal University Press桂林:廣西師範大學出版社. (pp. 238–275).

[32] Zhong, J., & Guo, M. J. 鐘堅 郭茂佳(2010). *Documentation and Materials on Chinese Special Economic Zones (Vol. 1)* 中國經濟特區文獻資料(第1輯). Beijing: Social Sciences Academic Press北京:社會科學文獻出版社. (pp. 18).

[33] Wang, J. B. 王江波 (2022). *Research on the Transformation of Shenzhen's Land System*深圳土地制度變遷研究 (1979–2020). Beijing: China Financial and Economic Publishing House北京:中國財政經濟出版社. (pp. 37–40).

"special zones in the SSEZ (特區中的特區)." They provided living space for migrant workers and low-income groups, but in the meantime created new problems for the government's urban governance.[34] When the agriculture-related land, such as cultivated land, oyster cropland, and mountainous forest land, was used to promote urban construction, farmers in dozens of agricultural production brigades, oyster breeders, and fishermen in coastal areas in SSEZ were gradually leaving the places where they had lived and worked for generations. The resultant space is gradually used to build factories, houses, urban parks, expressways, high-tech industrial parks, and headquarters bases.

### *3.1.3 The Production of Urban Landscape: The Integration of Topography and Globalization*

Undoubtedly, both topography and society are the foundation of construction.[35] If we just focus on the former as matter, the meaning context of the topography will be cut off; if we focus only on the latter, it will be difficult to generate inspiring thinking in the field of built environment. The two must be integrated to study the vein of Shenzhen urban landscape shaping. How does the integration of topography and globalization shape the urban landscape of Shenzhen? This integration manifests itself in at least three important ways.

#### 3.1.3.1 Topography and Boundaries: Second Line

The demarcation of 327.5 square kilometers of the SSEZ is closely related to its topographic characteristics. As we all know, the special economic zone is not originally created by China. They are a concept of economic geography and have long played a role in international trade. In the late Cold War period, special economic zones proliferated around the world, especially in Asia and other developing countries, primarily to attract investment from multinational corporations in labor-intensive manufacturing and to integrate themselves into globalization through export-led growth policies.[36] SSEZ was built for reasons. On the one hand, despite economic weakness in the late 1970s, China had a large population and a vast territory, which made it difficult to enter the global industrial chain quickly and

[34] O'Donnell, M. A. (2017). Laying Siege to the Villages: The Vernacular Geography of Shenzhen In O'Donnell et al. (Eds.) *Learning from Shenzhen : China's post-Mao experiment from special zone to model city*. Chicago: The University of Chicago Press. (pp. 107–123).

[35] Leatherbarrow, D. (2015). Topographical stories: studies in landscape and architecture. University of Pennsylvania Press. (pp. 19–20).

[36] Farole, T., Akinci, G. (2011). *Special economic zones: progress, emerging challenges, and future directions (English).* Washington, D.C.: World Bank Group.

comprehensively; on the other hand, the development of special economic zones is full of mixed results and challenges on a global scale.[37] Both affirmed the need to demarcate boundaries and contain risks locally.

In order to define the scope of the "test field" of market economy in this socialist environment and highlight the "special" of the SSEZ, the Guangdong Provincial People's Congress delimited the scope of the SSEZ in *Interim Provisions on the Administration of Shenzhen Special Economic Zone* (深圳經濟特區行政管理暫行規定) in December 1981, as shown in Fig. 3.4.

*It starts from Beizaijiao Beach (背仔角) in Dapeng Bay in the east, extends southwest to Shekou and Yijia Village of Nantou Commune (南頭公社), and northwards until Wutong Mountain, Dalinggu (大嶺古) of Yangtai Mountains, Dagu Ridge (打鼓嶂), Zhangding Peak (嶂頂), Jiuweiding Peak (九尾頂), Yaji Mountain (雅髻山), Dayangtou (大洋頭), as well as Shawan Village (沙灣村), Dushu Village (獨樹村), and Baimang Village (白芒村), with a total area of 327.5 square kilometers.*[38]

If important political and economic differences cannot be maintained by both cultural inertia and spatial distance, establishing boundaries is the best way to do so. This is true of the Berlin Wall to separate East and West Germany, the 38th Parallel between North and South Korea, and the US–Mexico Border Wall. For the SSEZ, the mountainous and hilly features evolved by the crustal movements over a long period of time provided a basis for the demarcation of boundaries in the 1980s. The boundaries came to be known as the "Second Line (二線關)," consisting of 84.6

**Fig. 3.4** Second line and its topographic features

[37] Akinci, G., Crittle, J. (2008). *Special economic zone: performance, lessons learned, and implication for zone development (English)*. Washington, D.C.: World Bank Group.

[38] Zhong, J., & Guo, M. J. 鐘堅, 郭茂佳 (Eds.). (2010). *Documentation and Materials on Chinese Special Economic Zones (Vol. 2)* 中國經濟特區文獻資料(第2輯). Beijing: Social Sciences Academic Press北京:社會科學文獻出版社. (pp. 154–157).

kilometers of barbed wire running east to west, patrol roads, and 16 border checkpoints along the route, as shown in Fig. 3.5.

In terms of physical space, the SSEZ is separated from the mainland only by a "Second Line" with mountains as natural geographical barriers. However, it made the SSEZ a special political and economic space that was neither "domestic" nor "overseas" for China at that time. First, the "Second Line" blocked those people and goods attempting to enter/leave the SSEZ illegally. The deeper impact lies in the different laws and regulations implemented by the Shenzhen Municipal Government on the north and south sides of the Second Line. In the nearly 30 years (1982–2010), the land system, administration, infrastructure, industrial distribution, real estate prices, and even traffic fines policies varied on the north and south sides.[39] The boundaries defined by natural mountains have constructed the "Second Line" in a physical, political, and economic way, which has had an important effect and a profound impact on the industrial and spatial development of Shenzhen.

**Fig. 3.5** Second line in progress (Shenzhen Urban Daily Comprehensive Report深圳都市報綜合報導 (2018, January 16). *Historical Overview of Second-Line*二線關歷史. Shenzhen Urban Daily 深圳都市報.(A02))

[39] Emma X. M. and Blackwell A. (2017). The Political Architecture of the First and Second Lines In O'Donnell et al. (Eds.) *Learning from Shenzhen : China's post-Mao experiment from special zone to model city*. Chicago: The University of Chicago Press. (pp. 124–137).

#### 3.1.3.2 Topography and Uncertainty: Cluster Structure

The strip-cluster urban pattern in SSEZ is also determined based on the significant premise of "adapting to the geographical and topographic conditions of Shenzhen." The flexible accumulation of global capital restructuring made Shenzhen in the 1980s, and even China, face uncertainties from social, political, economic, and other aspects. Post-Fordism under the ideology of neoliberalism implements the make-to-order strategy based on specific customer requirements, making its production more personalized than Fordism. The production outsourcing relationship between capital and the contractual relationship between capital and labor have greatly impacted the stable employment relationship and production organization.[40] This more flexible way of production organization brings great uncertainty to urban space, too.

According to the urban structure defined in the *Master Plan of Shenzhen Special Economic Zone (1986)*, the strip-shaped special economic zone is divided into six clusters. Shangbu-Luohu cluster focuses on business and trade, industry and administration, Overseas Chinese Town cluster on tourism, Nantou cluster on industry, culture, and education, and Shekou cluster on industry and ports, presenting minimal mutual interference.[41] In the meantime, five isolation belts are sandwiched between six clusters: Danan Mountain and Xiaonan Mountain isolation belt, Dasha River isolation belt, Huangniulong isolation belt, Futian River 800-meter isolation belt, and Wutong Mountain isolation belt from west to east. These isolation belts formed by rivers or mountains and other topographic conditions separate the originally banded urban space, forming a spatial pattern of "strip-cluster" with independent functions but mutual connections, as shown in Fig. 3.6. After the 1990s, the urban space expanded beyond the Second Line, and this strip structure of "clusters" formed a "network-cluster" in the topographic conditions of hills, valleys, and basins outside the Second Line. This is reflected in the *Urban Master Plan of Shenzhen Municipality (1996–2010)* and laid the city-wide spatial structure. For example, the central cluster is located in the north of the Tanglang Mountain-Danaoke (大腦殼)-Jigong Mountain (雞公山) range, and is divided into four parts (Guanlan, Longhua, Pinghu, and Buji) by the Laohuyan (老虎岩)-Niuweiling (牛尾嶺)-Baishaling (白砂嶺) mountain series, as shown in Fig. 3.7.

During the rapid expansion of Shenzhen's urban space, few people can clearly predict Shenzhen's future size and prospects, and even large deviations in the predictions of population and land area in previous urban master plans, as shown in Table 3.3. However, such deviations have not caused serious consequences in the process of urban development and construction, because the flexibility of the cluster structure makes the development and construction of Shenzhen not rely on

[40] Harvey, D. (1989). *The condition of postmodernity : an enquiry into the origins of cultural change.* Blackwell. (pp. 141–172).

[41] Shenzhen Institute of Building Design Co., Ltd. 深圳市建築設計研究總院有限公司 (2019). *Shenzhen Forty Years: Industry and City*深圳四十年:產業與城市. Beijing: China Architecture & Building Press北京:中國建築工業出版社. (pp. 63).

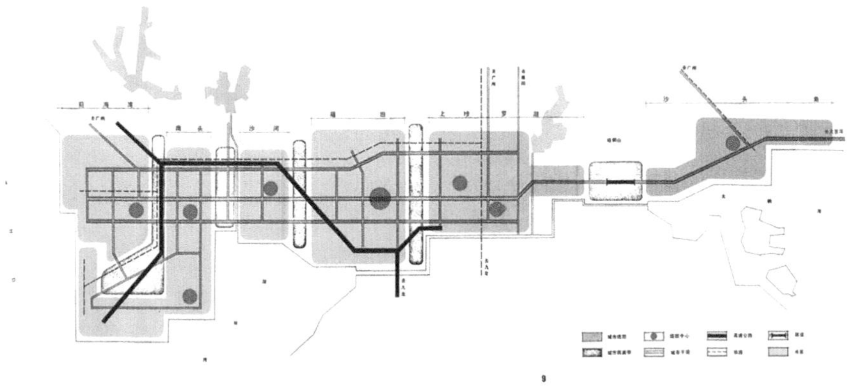

**Fig. 3.6** Urban structure in the *Master Plan of SSEZ (1986)* (Source from: *Master Plan of Shenzhen Special Economic Zone*深圳經濟特區總體規劃(1986), preserved at the archive office of Shenzhen Planning and Natural Resources Bureau)

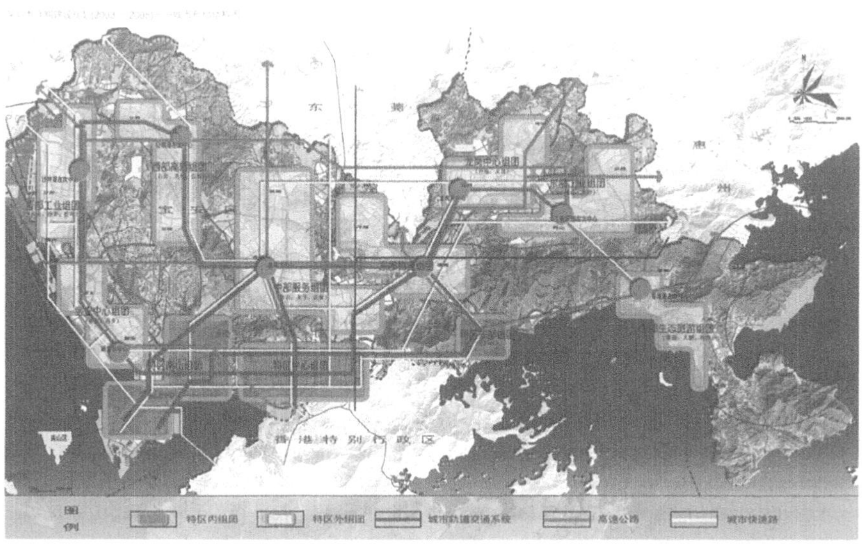

**Fig. 3.7** Zonal cluster and network cluster structure (Source from: Urban Master Plan of Shenzhen Municipality (1996–2010))

**Table 3.3** Envisaged vs. actual sizes in previous planning efforts (Source from: Multiple urban master plans formulated in Shenzhen since its establishment as a city)

| Document name | Scope | Planned population (in 10,000) | Actual resident population (in 10,000) | Planned land for construction ($km^2$) | Actual land for construction ($km^2$) |
|---|---|---|---|---|---|
| *Program for Urban Development of Special Economic Zone of Shenzhen Municipality (1980)* | SSEZ | 30(1985)<br>60(1990) | 46.98(1985)<br>100.98(1990) | 49(1990) | 69.34(1990) |
| *Master plan of Shenzhen special economic zone*《深圳經濟特區總體規劃》*(1986)* | SSEZ | 80(2000) | 205.30(2000) | 123(2000) | 189.62(2000) |
| *Urban Master Plan of Shenzhen Municipality (1996–2010)* | SSEZ | 173(2000)<br>180(2010) | 205.30(2000)<br>682.94(2010) | 130.25(2000)<br>160.90(2010) | 189.62(2000)<br>247.21(2010) |
| | The City Scope | 400(2000)<br>430(2010) | 701.24(2000)<br>1037.20(2010) | 377.77(2000)<br>478.73(2010) | 620.09(2000)<br>1040.35(2010) |
| *Urban Master Plan of Shenzhen Municipality (2010–2020)* | The City Scope | 1100(2020) | 1763.38(2020) | 890(2020) | 1022.26(2019) |

particularly accurate predictions of the city's size.[42] Besides, it can well adapt to the uncertainty of urban development under the conditions of market economy, which is conducive to the development and construction in phases and zones according to the changes in the situation.[43]

There is no evidence that such urban structure is consciously designed by policy-makers or planners in view of the uncertainty of Shenzhen's development, and topographical conditions are the main reason for the emergence of this spatial structure. It can be imagined that without topographical restrictions, Shenzhen is likely to develop into an urban structure for large-scale contiguous development along main trunk roads such as Guangzhou-Shenzhen Expressway or Shenzhen-Huizhou Expressway due to agglomeration effect. Subject to topography, Shenzhen forms a cluster structure to deal with the uncertainty of urban spatial development. It can be seen that the topographic characteristics of Shenzhen have a good adaptability to the uncertain development of industrial space in the global economy. From this point of

[42] Zhao, Y. J. 趙燕菁(2004). Rapid Development and Spatial Extension: A Case Study by Shenzhen 高速發展與空間演進——深圳城市結構的選擇及其評價. *City Planning Review*城市規劃, 06, 32–42.

[43] Zou, B. 鄒兵 (2017). The Evolution of Urban Spatial Structure in Shenzhen and the Utility Evaluation of Its Planning深圳城市空間結構的演進歷程及其中的規劃效用評價. *Urban & Rural Planning*城鄉規劃, 06, 69–79.

view, the topography "resist" and "allow" the urban spatial development of Shenzhen.[44]

#### 3.1.3.3 Topography and "Space–Time Compression": Manufacturing and Shipping

Labor-intensive manufacturing is the priority of the early industrial construction of the SSEZ, and its development is closely related to the topography. In the 1980s, the SSEZ became one of the most suitable receptors to the transfer of labor-intensive industries from developed countries and regions. In the initial phase, the SSEZ had a weak industrial foundation, and a large amount of capital and means of production were raised by foreign investors and businessmen. The SSEZ mainly provided land, factories, living facilities, transport, post, and telecommunications infrastructure. As we all know, industrial development requires the large-scale construction of plants, living facilities, and transport and freight lines, which must be built on flat topography as far as possible to reduce costs and construction difficulties. This is an important factor affecting industrial location selection. As a result, the platforms (79.61 $km^2$), terraces (9.8 $km^2$), and plains (85.61 $km^2$) within the SSEZ have been used in a way that can obtain more economic benefits than the agricultural production before. In the flat coastal platforms and plains south of the Second Line, Shangbu Industrial zone (上步工業區), Shahe Industrial Zone (沙河工業區), Shekou Industrial Zone, Bagualing Industrial Zone (八卦嶺工業區), and Shuibei Industrial Zone (水貝工業區) rose one after another to focus on the development of electronics, light industry, building materials, machinery, and textile industries. All of them are located on these open, flat, and gently sloping platforms and foreslopes with a good geological foundation. Except for some areas, most areas can be constructed after slight leveling. In addition, after the land for construction is saturated in the SSEZ, the vast hinterland in the north of the Second Line including large areas of platforms (343.05 $km^2$), terraces (91.3 $km^2$), and plains (332.56 $km^2$) can still support the development of manufacturing.

In addition to developing these manufacturing industries themselves, Shenzhen needs a strong logistics system to truly develop export-oriented industries and become a global manufacturing base. Thanks to Shenzhen Port, Shenzhen has become a window for China to connect with the world. It also has a lot to do with the topography.

In the 1980s, the urban construction of the SSEZ exhibited a "point-like" layout.[45] In addition to the Shangbu-Luohu cluster relying on the old town, the urban

---

[44] Leatherbarrow, D. (2016). Not Anywhere, Not Only Here. In Girot, C., & Imhof, D. (Eds.). *Thinking the contemporary landscape (First edition.)*. Princeton Architectural Press. (pp. 199–207).

[45] Shenzhen Institute of Building Design Co., Ltd. 深圳市建築設計研究總院有限公司 (Eds.). (2019). *Shenzhen Forty Years: Industry and City*深圳四十年:產業與城市. Beijing: China Architecture & Building Press北京:中國建築工業出版社. (pp. 16).

construction began at Shekou-Nantou and Sha Tau Kok, which are qualified for building ports, in favor of the industrial matchmaking between the SSEZ and Hong Kong or even abroad. The geological movement for a long period of time separated Dapeng Peninsula from Hong Kong and formed Mirs Bay. As the coast closest to the Mirs Bay within the SSEZ, Sha Tau Kok is located in a flat valley between Wutong Mountain and Maluan Mountain and is thus conducive to the construction of ports. In the west, the Shekou Peninsula (蛇口半島) has also become a site for port construction due to its topographic characteristics. Located in the terrace plain at the lower reaches of Xixiang River/Dasha River, Shekou Peninsula does not have sufficient water depth at the southern tip, but the earth's rotation makes the Pearl River Estuary always scoured in the east and silted in the west, and the dredged dock basins are free from serious backsilting.[46] In the later development, Shekou Peninsula built Shekou, Chiwan (赤灣), Mawan (媽灣), and other port areas; and Sha Tau Kok built the Yantian Port (鹽田港).[47] Inside the city, the port logistics service has greatly affected the layout of urban space, forming a new relationship between port and city.[48]

The development of international shipping is a necessary means to shorten the time and space distance between Shenzhen and international markets.[49] For Shenzhen, a city committed to an export-oriented economy, the possibility of building its port area has been nurtured in geological movements over tens of millions of years. Benefiting from a total of 229.96 km of coastline (154.69 km in the east and 75.27 km in the west),[50] Shenzhen is endowed with a number of port areas, making it an important deep-water port city in the Chinese Mainland and even in the Asia-Pacific region. David Harvey reconsiders the concept of time and space planned by the Enlightenment and integrates the flexible accumulation of post-Fordist modes of production with the new concept of time and space, proposing the concept of "time-

---

[46]Yan, F. Y. 顏鳳迎 (2018). Special Zone Window: Three Lives and Three Worlds of Shekou Port特區視窗:蛇口港的三生三世. In Y. H. Xu & W. Liu許永軍 劉偉 (Eds.), *Shekou: The Place Where Dreams Begin - Paying Tribute to 40 Years of Reform and opening-up*蛇口 夢開始的地方——致敬改革開放40年. Beijing: People's Daily Press北京:人民日報出版社. (pp. 225–236); Sun, S. X. 孫紹先 (1984). Port Construction in Shekou Industrial Zone蛇口工業區的港口建設. *Port & Waterway Engineering*水運工程, Z1: 78–81.

[47]Shenzhen Municipal Chronicles Compilation Committee深圳市地方誌編纂委員會. (2014). *Chronicles of Shenzhen: Volume on the Tertiary Industry*深圳市志·第三產業卷. Beijing: Fangzhi Publishing House北京: 方志出版社. (pp. 96).

[48]Zhang, W., & Le, Y. C. 張偉,樂宜春 (2019). Shenzhen Port City: Spatial Relationship Transformation Process and Coordinated Development Strategy深圳港城空間關係變遷歷程與協調發展策略. In Urban Transportation Planning Academic Committee of China Urban Planning Society中國城市規劃學會城市交通規劃學術委員會(Eds.). *Proceedings of the 2019 China Urban Transportation Planning Annual Conference* 2019年中國城市交通規劃年會論文集. (pp. 10).

[49]Lin, Z. J. 林祖基(Eds.). *Essays on Shenzhen - An International City*深圳-國際性城市論文集. Beijing: China Economic Publishing House北京:中國經濟出版社. (pp. 8).

[50]Huang, Z. G. 黃鎮國 (1984). *Shenzhen's Natural Resources and Economic Development*深圳自然資源與經濟開發. Guangzhou: Guangzhou Institute of Geography廣州:廣州地理研究所. (pp. 18).

space compression."[51] The physical topography does change slowly, but the geographic space does not. Before and after the construction of ports, changes in shipping time and costs reshaped the space and time distance between Shenzhen's urban industry and the world. The construction of Shenzhen's port areas is right a node in the compressed global space-time network.

In short, the global and national efforts are found behind the rise of Shenzhen after 1980, both of which are intricate and produce a new urban landscape on Shenzhen's natural terrain. This urban landscape is undoubtedly very different from the agricultural landscape that once existed on the east bank of the Pearl River Estuary. In this process, the original surface morphology was reshaped, so were the production and living patterns interwoven therein. Urban parks emerge in the basic context of such landscape history.

## 3.2 Changing City and Parks

Henri Lefebvre criticized the situation in which specialized activities become detached from everyday life, and even specialized activities are developed, refined, and advanced outside of everyday life.[52] In terms of components, parks are nothing more than plants, rocks, terrain, water, and buildings, which can be summarized in a few words and can be described by other increasingly advanced techniques. However, if put into human practice, urban parks will vividly reflect the development and changes in urban social life. Where the difference lies in its expression compared to urban public spaces, such as streets and squares, is that it is based on the "signifiers of nature"—plants, mountains, rocks, topography, and water bodies—to unfold the construction of meaning. Simultaneously, it allows these "natural" objects to permeate with layers of historical and societal significance.

With the change of history and society, the shift in human needs will reshape its relationship with physical topography again and again, forming a temporal landscape lamination.[53] Since 1980, the gradual globalization of Shenzhen has changed the relationship between man and the natural environment and has given birth to a large number of urban parks. In this section, we will understand how this happened in the historical and social context, and meanwhile present an explanatory view on the morphological description of the previous section.

---

[51] Harvey, D. (1989). *The condition of postmodernity : an enquiry into the origins of cultural change*. Blackwell. (pp. 284–307).

[52] Lefebvre H. *Critique of everyday life,volume I*,Introduction; (Trans. by John Moore). Verso, London, New York 1991. (pp. 87); Liu, H.Y. 劉懷玉(2003). Henri Lefebvre and Several Trends of Critique of Everyday Life in the 20th Century West. 列斐伏爾與20世紀西方的幾種日常生活批判傾向. *Seeking Truth*求實, (05), 44–50.

[53] Corner J. (2014). The landscape imagination : collected essays of James Corner, 1990–2010 (First edition). New York: Princeton Architectural Press: 101–103.

### 3.2.1 Forests and Arable Land into Parks

Although parks or arable and forest land can be considered "natural," their meanings are quite different. Even so, it still defies common sense to think the two are unrelated.

On a regional scale, in the tenth century, rice planting techniques from regions south of the Yangtze River were introduced into the Lingnan area and the PRD region, which changed the previous slash-and-burn farming method and became the starting point for the extensive cultivation of arable land. From the fourteenth century to the eighteenth century, the population in Lingnan area increased, and the reclamation of arable land became the primary reason for the decrease of forest land.[54] By the early twentieth century, most of the forests in the Lingnan area had disappeared, and the mountainous area of northern Guangdong were largely "barren hills, overgrown with weeds."[55] In the first half of the twentieth century, the war, corruption, and economic depression made the development of forestry in Guangdong continue to decline. After 1949, the local afforestation in Guangdong Province was still unable to meet the needs of socialist industrialization for wood.[56] The habitat destruction caused by long-term "trees felled without reafforestation" made the survival rate of afforestation in Bao'an County between the founding of the PRC and the early period of reform and opening-up not even 60%.[57] Around 1960, Bao'an County "could described as 'deserted'… with sparse vegetation and bare surface,"[58] as shown in Fig. 3.8.

From the perspective of Shenzhen, it has gradually become a new manufacturing base for Hong Kong and the global economy since 1979. Changes at the global, national, regional and urban scales have exposed Bao'an County, a local agricultural economy under socialism, to global capitalism and the urban process it drives. This is reflected in at least three aspects: (1) The area of arable land and forest land decreased; (2) the proportion of the output value of the primary industry was decreasing year by year; and (3) the population employed in the primary industry was rapidly decreasing.

---

[54] Marks, R. (1998). *Tigers, rice, silk, and silt: Environment and economy in late imperial South China*. Cambridge University Press. (pp. 35–46;319–327).

[55] Marks, R. (1998). *Tigers, rice, silk, and silt: Environment and economy in late imperial South China*. Cambridge University Press. (pp. 35–46;319–327).

[56] Zaizhi Z. and Chokkalingam U. Success and sustainability: Lessons from Guangdong Province In Chokkalingam, U. et al. (Eds.) *Learning lessons from China's forest rehabilitation efforts: national level review and special focus on Guangdong Province*. CIFOR. (pp. 69–134).

[57] Guangzhou Institute of Geography廣州地理研究所 (Eds.) (1985). *Atlas of Natural Resources and Economic Development in Shenzhen*深圳市自然資源與經濟開發圖集. Beijing: Science Press北京:科學出版社. (pp. 109).

[58] He, F. 何發. Counting the Trees on Which Hill Even with Closed Eyes閉著眼睛都能數出哪個山頭種了什麼樹. In Shenzhen Urban Planning and Natural Resources Bureau, Shenzhen Forestry Bureau深圳市規劃和自然資源局 深圳市林業局 (Eds.). *Memory Forest: Oral History of Shenzhen Forestry*記憶森林 深圳林業口述史 (1980–2020). Unpublished internal document.

**Fig. 3.8** Shenzhen reservoir (深圳水庫) under construction and Wutong Mountain with sparse vegetation (1959) (Zhang, Y. F. 張燕方. (2014). *Spatiotemporal Images: Works from Zheng Zhongjian Photography Exhibition*影像時空:鄭中健攝影展作品集. Shenzhen: Haitian Publishing House深圳:海天出版社. (pp. 107))

First, according to the survey of Guangzhou Institute of Geography, in 1983, at the beginning of reform and opening-up, arable land and forest land in Shenzhen accounted for 27.55% and 53.47% of the city's land area,[59] respectively. Rapid urbanization made large areas of contiguous arable land being occupied for urban construction in the early 1980s. By the end of 2019, Shenzhen's arable area was more than 2844.74 ha., accounting for only 1.44% of the total land citywide, mainly distributed in Guangming, Bao'an, and Pingshan districts.[60] In the decades since the SSEZ was established, the agricultural land has been shrinking from 1010.68 $km^2$ in 1995 to 890.12 $km^2$ in 2015, with an average annual decrease of 6.03 $km^2$, in which the arable area has been sharply declined from 65.32 $km^2$ to 39.77 $km^2$, with an average annual decrease of 1.28 $km^2$.[61]

In 1983, the forest land of Shenzhen accounted for 53.47% of the city's land area.[62] By 2019, the forest area of Shenzhen still accounted for 32.5% of the city's

[59] Guangzhou Institute of Geography廣州地理研究所 (1985). *Atlas of Natural Resources and Economic Development in Shenzhen*深圳市自然資源與經濟開發圖集. Beijing: Science Press北京:科學出版社. (pp. 128).

[60] Shenzhen Urban Planning and Natural Resources Bureau深圳市規劃和自然資源局(2022, April 26). *Major Data Bulletin (Charts) of the Third National Land Survey in Shenzhen*深圳市第三次全國國土調查主要資料公報(圖表). Retrieved from https://pnr.sz.gov.cn/gkmlpt/content/9/9744/post_9744274.html#4290

[61] Shenzhen Urban Planning and Land Development Research Center深圳市規劃國土發展研究中心 (2019). *Shenzhen Land Resources*深圳市土地資源. Beijing: Science Press北京:科學出版社. (pp. 124–125).

[62] Huang, Z. G. 黃鎮國(1984). *Shenzhen's Natural Resources and Economic Development*深圳自然資源與經濟開發. Guangzhou: Guangzhou Institute of Geography廣州:廣州地理研究所. (pp. 109).

land area.[63] The spatial distribution of forest land in Shenzhen is mainly compatible with mountains and hills, which are not conducive to rapid urban development and construction.[64] This is one of the main reasons for the rapid decline of arable land and the slow decline of forest area caused by urban construction in the past 40 years, as shown in Table 3.4.

Second, the primary, secondary, and tertiary industries in 1979 contributed to 37.0%, 20.5%, and 42.7% of Shenzhen's GDP, respectively. It can be seen that agriculture was still of great importance at this time. With the industrial upgrading and transformation, by 2020, the primary, secondary, and tertiary industries accounted for 0.1%, 37.8%, and 62.1% respectively,[65] and the output value of the primary industry, such as agriculture, forestry, animal husbandry, and fishery had already been negligible.

**Table 3.4** Statistics of arable land and forest land in Shenzhen over the years (Date of 1995–2015 come from: Shenzhen Urban Planning and Land Development Research Center深圳市規劃國土發展研究中心(2019). *Shenzhen Land Resources*深圳市土地資源. Beijing: Science Press北京:科學出版社. (pp. 124–125); Date of 2016–2019 come from: *Land change survey data* (土地變更調查) of various years released on the official website of the Shenzhen Urban Planning and Natural Resources Bureau)

| Year | Arable land area (ha.) | Forest land area (ha.) | Year | Arable land area (ha.) | Forest land area (ha.) | Year | Arable land area (ha.) | Forest land area (ha.) |
|---|---|---|---|---|---|---|---|---|
| 1980 | 32,673 | – | 1993 | 6033 | – | 2006 | 4089 | 58,203 |
| 1981 | 30,653 | – | 1994 | 4573 | – | 2007 | 3847 | 57,344 |
| 1982 | 29,273 | – | 1995 | 6532 | 72,009 | 2008 | 3816 | 56,987 |
| 1983 | 28,660 | – | 1996 | 6465 | 71,529 | 2009 | 3160 | 58,578 |
| 1984 | 26,760 | – | 1997 | 6443 | 71,322 | 2010 | 3057 | 57,974 |
| 1985 | 22,627 | – | 1998 | 6416 | 70,692 | 2011 | 3022 | 57,987 |
| 1986 | 21,940 | – | 1999 | – | – | 2012 | 3019 | 58,228 |
| 1987 | 20,653 | – | 2000 | 6372 | 64,478 | 2013 | 4096 | 58,215 |
| 1988 | 19,640 | – | 2001 | 6147 | 61,576 | 2014 | 4045 | 57,890 |
| 1989 | 18,833 | – | 2002 | 6014 | 60,510 | 2015 | 3977 | 57,787 |
| 1990 | 18,673 | – | 2003 | 4694 | 59,503 | 2016 | 3837 | 57,873 |
| 1991 | 15,113 | – | 2004 | 4522 | 59,004 | 2017 | 3715 | 57,617 |
| 1992 | 10,333 | – | 2005 | – | – | 2018 | 3618 | 57,393 |
| | | | | | | 2019 | 2845 | 64,323 |

[63] Shenzhen Urban Planning and Natural Resources Bureau深圳市規劃和自然資源局 (2022, April 26). *Major Data Bulletin (Charts) of the Third National Land Survey in Shenzhen*深圳市第三次全國國土調查主要資料公報(圖表). Retrieved from https://pnr.sz.gov.cn/gkmlpt/content/9/9744/post_9744274.html#4290

[64] Shenzhen Urban Planning and Land Development Research Center深圳市規劃國土發展研究中心(2019). *Shenzhen Land Resources*深圳市土地資源. Beijing: Science Press北京:科學出版社. (pp. 118).

[65] Shenzhen Statistical Bureau深圳市統計局 (2021). *Shenzhen Statistical Yearbook 2021*深圳統計年鑒2021. Beijing: China Statistics Press北京:中國統計出版社. (pp. 25–27).

Third, we must see changes in people who manage these lands. In 1982, Shenzhen had a total population of 354,500,[66] and the number of those engaged in the primary, secondary, and tertiary industries accounted for 57.15%, 17.67%, and 25.18%, respectively. With a large number of farmers engaged in other industries, only 13,035 of the 12,922,900 industrial workers were engaged in the primary industry in 2020, accounting for 0.01% of the total industrial workers.[67]

From three aspects, globalization since the 1970s has brought great changes to the agricultural economy of Bao'an County. Export-oriented industrial and spatial development policies mean less agricultural and ecological land, and even a dramatic change in the social relations attached to these lands, that is, in the way human interacts with nature. An important manifestation of this change lies in the fact that the nature (arable land and forest land) originally used for agricultural products is transformed into another cultural form of nature (urban parks). In the initial stage of reform and opening-up, on the one hand, capital looked for room for regeneration on a global scale, and on the other hand, hundreds of millions of surplus labor force in rural China were being left idle. The two initially found a joint point in the space and market of Shenzhen,[68] and the global space production, along with the local urban process, re-collaged the urban landscape of Shenzhen. The self-sufficient agricultural economy is rapidly being drawn into the process of globalization, leading to the conversion of arable land and forests into factories, residences, and urban parks that serve industrialization and urbanization—these uses are more economically beneficial.

Compared with the former forest land and arable land, plant planting has emerged as a new identity of urban greening in the heart of this emerging city. As the heart, Luohu was almost entirely arable land in 1980, with the exception of three square kilometers of built-up area. With the gradual start of Shenzhen's local urbanization, as shown in Table 3.5, the early parks in Shenzhen, as a symbolic nature to improve the environment for foreign investment, were mainly built on the former forest land and arable land. The context will be elaborated in Chap. 4.

In fact, most of the parks built in the 1980s within the Second Line and the parks built outside the Second Line following the "territory-wide development (全境開拓)" in 1990 are based upon the original arable land and forest land. Although the transformation of urban parks shares the same political and economic context as factories and houses, the difference is that before and after this transformation, the land is presented as "nature"—whether it is arable land, forest land, or urban parks.

[66] Shenzhen Special Economic Zone Yearbook Editorial Committee 深圳經濟特區年鑒編輯委員會 (1986). *Shenzhen Special Economic Zone Yearbook 1985*深圳經濟特區年鑒1985. Shenzhen: Shenzhen Special Zone Yearbook Press深圳:深圳特區年鑒社. (pp. 581).

[67] Wang, S.W. 王世巍(2014). *Population Changes in Shenzhen*深圳市人口變遷研究. Shenzhen: Haitian Publishing House深圳:海天出版社. (pp. 237).

[68] Xiao, J., & Cao, K. 肖競,曹珂 (2014). Nowhere to Place Our Field: Study on the Invasion About Cultivated Land in China During the Globalization of Space Production Era我們無處安放的農田——空間生產全球化格局下我國耕地資源侵佔問題研究. *Urban Planning International*國際城市規劃, 29(02), 59–67.

**Table 3.5** Land sources and uses of parks built in SSEZ in 1980s (Information comes from oral accounts of multiple eyewitnesses and internal data from the Shenzhen Park Management Center. A en dash ("–") indicates that information is yet to be obtained)

| Name | Built-up area (ha.) | Year of completion | Main source of land requisition | Land use | Location |
|---|---|---|---|---|---|
| Lychee Park | 18.60 | 1985 | Caiwuwei brigade (蔡屋圍大隊) | Arable land, fruit-bearing forest | Shangbu-Luohu |
| Honghu Park | 59.15 | 1987 | Shuibei and Tianbei brigades (水貝、田貝大隊) | Arable land | |
| Children's Park | 5.95 | 1987 | Dongyuan Flower & Wood Market (东园花木场) | Flower & Wood Market | |
| People's Park | 12.95 | 1986 | Cainong brigade (菜農大隊) | Arable land, vegetable plot | |
| Donghu Park | 233.30 | 1984 | Huangbeiling Village (黃貝嶺村)<br>Bao'an County fish farm (寶安縣魚苗場);<br>Wutong Mountain nursery farm (梧桐山苗圃場) | Arable land, fruit-bearing forest<br>Fish pond<br>Nursery | |
| Cemetery of Revolutionary Martyrs | 5.20 | 1987 | – | – | |
| Fairy Lake Botanical Garden | 592.00 | 1988 | Wutongshan Village (梧桐山村)<br>Shenzhen Forest Farm (深圳林場) | Arable land<br>Forest land | |
| Zhongshan Park | 53.80 | 1987 | – | – | Shekou-Nantou |
| Sihai Park | 16.79 | 1987 | – | – | |
| Bihai Park | 18.60 | 1985 | Sea reclamation | – | Sha Tau Kok |

As Hunter's "ideal landscape," the urban parks are picturesque, while the arable land and forest land, as a "functional landscape," seem primitive and rough in contrast.[69] However, such aesthetic cognition has blurred the logic of power behind global space production, as well as the competing and even possibly conflicting views of nature—both space and natural production are carried out in interaction with social forces.

[69] Hunter, J.M. (1985). *Land into landscape*. London: George Goodwin. (pp. 3–5).

### 3.2.2 *Plains to Mountain Land*

From the description and calculation in Sects. 3.2.2 and 3.2.3, it can be seen that the urban parks built in Shenzhen in the past 40 years tend to be gradually distributed to areas with large slopes and high altitudes, which makes the spatial forms of parks built in different periods greatly different. For urban parks in Shenzhen, changes in slope and elevation over time are not natural but intrinsic to the interrelationship between topography and urbanization: On the one hand, the topography of parks is shaped by the broader urban topography, and on the other hand, its location is also structured in the urban process under a specific historical and social context.

The oil shocks of the 1970s caused economic slowdown and the rise of protectionism in most Western industrial countries, and Hong Kong's manufacturing industry, relying on foreign raw material supply and markets, was thus frustrated by the international market, and the role of manufacturing in Hong Kong's economy had gradually declined. By the end of the 1990s, Hong Kong's industrial structure had successfully shifted from "industry" to "service economy," such as finance, insurance, and real estate.[70] The manufacturing industry of Hong Kong shifted to the Chinese Mainland on a large scale, mainly to the PRD metropolitan area, which reconstructed the urban landscape of the PRD cities represented by Shenzhen. In Shenzhen, this reconstruction takes place in the SSEZ within the Second Line, and also in the rapid change of urban landscape outside the Second Line.[71]

When the SSEZ was initially established, the State Council agreed to demarcate the scope of 327.5 square kilometers,[72] in addition to the steep mountains not suitable for construction, the planned available area was 160 square kilometers, of which 123 square kilometers were to be developed.[73] By 1989, 109 square kilometers of land had been allocated in the SSEZ, leaving little of the area originally intended for development and construction.[74]

In response to the shortage of land in SSEZ, the *Urban Development Strategy in Shenzhen* (深圳市城市發展策略) formulated in December 1989 pointed out the overall strategy to "seek the territory-wide development with key points and steps by centering on the special zone." The territory-wide development is to put the

[70] Feng, B. Y. 馮邦彥 (2002). *A Study on the Industrial Structure of Hong Kong*香港產業結構研究. Beijing: Management Publishing House北京:經營管理出版社. (pp. 22–24).

[71] Shenzhen Construction Bureau深圳市建設局(1989). *Urban Development Strategy of Shenzhen*深圳市城市發展策略. Unpublished. (pp. 16–17), preserved at the library of the School of Architecture, Southeast University.

[72] Zhong, J., & Guo, M. J. 鐘堅 郭茂佳 (2010). *Documentation and Materials on Chinese Special Economic Zones (Vol. 2)* 中國經濟特區文獻資料(第2輯). Beijing: Social Sciences Academic Press北京:社會科學文獻出版社. (pp. 154–157).

[73] Urban Planning and Natural Resources Bureau of Shenzhen深圳市規劃和自然資源局*Memories of Land System Reform in Shenzhen*深圳土地制度改革記憶. Unpublished. (pp. 78).

[74] Shenzhen Construction Bureau深圳市建設局(1989). *Urban Development Strategy of Shenzhen*深圳市城市發展策略. Unpublished. (pp. 16–17).

Bao'an County[75] outside the Second Line and the SSEZ within the Second Line under unified planning and construction.[76] In February 1994, the Shenzhen Municipal Government issued the *Notice on Ceasing the Approval of the "Three-processing and One Compensation " and Other Projects in the Special Economic Zone* (關於經濟特區停止審批"三來壹補"等項目的通知), making the (greater) Bao'an and (greater) Longgang[77] outside the Second Line the main base to receive the industrial transfer of "Three-processing and One Compensation." This directly turned the idea of "territory-wide development" into a reality. In fact, this idea was only given legal approval in the *Urban Master Plan of Shenzhen Municipality (1996–2010).* Sure, the transfer of "Three-processing and One Compensation "industries is only the initial driving force for the urban development outside the Second Line, and since then, multiple industrial upgrades outside the Second Line have gradually promoted the construction and development of urban space. The agricultural land was first converted to industrial land, then supplemented by supporting public facilities and requirements, and finally into the city. This is the general path of urbanization outside the Second Line.[78] In any case, there is no doubt that the spatial strategy of "territory-wide development" after the 1990s opened the rapid expansion of urban space outside the Second Line.

The topography outside the Second Line must be considered in the "territory-wide development" of urban construction. In fact, the *Urban Development Strategy in Shenzhen (1989)* did take note of this problem, to the extent that it divided the space outside the Second Line into eastern, central, and western areas, and preliminarily assessed the advantages and disadvantages that the topographic conditions of each area might bring to urban construction, as shown in Table 3.6. The U*rban Master Plan of Shenzhen Municipality (1996–2010)* also divided the city into eastern, central, and western zones. Although the topographic effects were not explicitly mentioned, the planned urban structure was superimposed on top of the terrain. It can be found that there is an obvious, interlocking relationship between the urban structure and topography outside the Second Line in the master plan, as shown in Fig. 3.9. All six clusters in the eastern, central, and western areas outside the Second Line were sandwiched by a series of mountains and hills, interweaving the city with hills.

---

[75] In January 1993, Bao'an County was dissolved and divided into (Greater) Bao'an and (Greater) Longgang districts.

[76] Shenzhen Construction Bureau深圳市建設局(1989). *Urban Development Strategy of Shenzhen*深圳市城市發展策略. Unpublished. (pp. 16–17).

[77] (Greater) Bao'an District includes the present Bao'an District, Guangming District, and Longhua District; (Greater) Longgang District includes the present Longgang District, Pingshan District, and Dapeng New District.

[78] Wen, W.文維(2021). Building Shenzhen into a National and Global Maritime Center City and Hub City將深圳打造成國家乃至全球海洋中心城市和樞紐城市 In. Shenzhen Urban Planning and Natural Resources Bureau深圳市規劃和自然資源局(Eds.). *Forty Years of Shenzhen Special Economic Zone: Oral History Interviews on Planning and Land Use*深圳經濟特區四十年 規劃國土口述歷史訪談錄. Unpublished. (pp. 330).

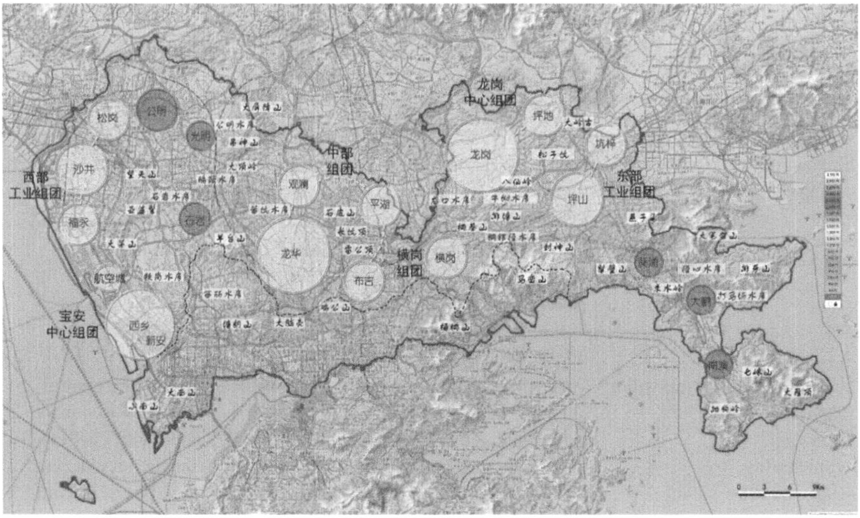


**Fig. 3.9** The urban structure outside the Second Line intersecting with the topography (Redrawn from the Urban Structure Diagram (城市結構圖) in the Urban Master Plan of Shenzhen Municipality (1996–2010))

According to the *Urban Master Plan of Shenzhen Municipality (1996–2010)*, Shenzhen's area of land for urban construction in 2010 would be 47,875 ha. In fact, according to the land use survey published by Shenzhen, Towns and Villages and Land for Mining and Industry(城鎮村及工礦用地) alone reached 78,342 ha. in the city-wide land for construction in 2010,[79] as in Table 3.7.

This shows that the scale of construction land envisaged in the *Urban Master Plan of Shenzhen Municipality (1996–2010)* has been seriously broken. These out-of-plan metrics flowed largely to the cluster isolation belts that originally limited the size of these clusters. These mountains and hills, making up the isolation belts, were not strongly protected but were invaded by extensive development and construction.[80] As a result, the boundaries of the planned clusters were gradually blurred, and the clusters became morphologically continuous. The land for urban construction broke through the originally flat basins and valleys and intersected with

[79] Shenzhen Urban Planning and Natural Resources Bureau深圳市規劃和自然資源局 (2014, September 19). Bulletin on the *Main Data Results of the Land Change Survey in Shenzhen for the Years 2010–2012* 深圳市2010–2012年度土地變更調查主要資料成果的公報. Retrieved from http://pnr.sz.gov.cn/xxgk/sjfb/tjsj/content/post_5837064.html

[80] Wen, W.文維(2021). Building Shenzhen into a National and Global Maritime Center City and Hub City將深圳打造成國家乃至全球海洋中心城市和樞紐城市 In. Shenzhen Urban Planning and Natural Resources Bureau深圳市規劃和自然資源局(Eds.). *Forty Years of Shenzhen Special Economic Zone: Oral History Interviews on Planning and Land Use*深圳經濟特區四十年 規劃國土口述歷史訪談錄. Unpublished. (pp. 330).

**Table 3.6** Assessment of city-wide topography in *Urban Development Strategy of Shenzhen* (1989)

| | | Western | Central | Eastern |
|---|---|---|---|---|
| Urban development strategy of Shenzhen (1989) | Favorable factors | Good conditions of land use, flat terrain, free from natural topographic thresholds | Relatively concentrated development, prone to the aggregated economy effect | Many types of shoreline are conducive to comprehensive development |
| | Adverse factors | Relatively dispersed structure of the city, not conducive to the organization of urban life | Many mountainous areas, difficult traffic organization, and high construction costs of new transport facilities | Many mountainous areas and high development difficulty |
| Urban Master Plan of Shenzhen Municipality (1996–2010) | Composition | Western industrial cluster + bao'an central cluster +3 independent towns: SSEZ by way of xin'an, Xixiang, Fuyong, Shajing to Songgang, connecting the west of Dongguan | Central cluster + Henggang cluster: Composed of Longhua, Buji, Pinghu and Guanlan, connecting the eastern part of Dongguan | Longgang central cluster + eastern industrial cluster +3 independent towns: Composed of Buji, Henggang, Longgang, Pingdi, Pingshan, and Kengzi, connecting Huizhou |
| | Positioning | One of the transportation hubs and an important industrial belt on the future Shenzhen-Hong Kong-Huizhou international city belt; Bao'an central cluster and the western industrial cluster have become important industrial intensive areas of the city | The peripheral supporting clusters of the SSEZ central cluster, the logistics corridor connecting the Beijing-Kowloon Railway and Beijing-Guangzhou-Kowloon Railway to Hong Kong and Shenzhen's East and West Ports, and an important auxiliary region for Shenzhen to play its role in regional and international exchanges in the future | It is the gateway of the PRD linking to the eastern Guangdong, and developed into an important industrial base in the PRD region under the drive of Yantian Port. Longgang central cluster and eastern industrial cluster will become the most potential growth points of the city in the twenty-first century |

**Table 3.7** Planned land for construction in 2010 vs. the actual land for construction in 2010 (Area of land for construction in 2010 (planned) is derived from the *Urban Master Plan of Shenzhen Municipality* (1996–2010); Area of land for construction in 2010 (actual) from: Shenzhen Urban Planning and Natural Resources Bureau深圳市規劃和自然資源局 (2014, September 19). *Bulletin on the Main Data Results of the Land Change Survey in Shenzhen for the Years 2010–2012* 深圳市2010–2012年度土地變更調查主要資料成果的公報. Retrieved from http://pnr.sz.gov.cn/xxgk/sjfb/tjsj/content/post_5837064.html)

| Area of land for construction in 2010 (planned) | | | | Area of land for construction in 2010 (actual) | | |
|---|---|---|---|---|---|---|
| Spatial location and cluster name | | | Land for construction (ha.) | Towns, villages and land for mining and industry (ha.) | Transportation (ha.) | Water and water conservancy facilities (ha.) |
| OSL | Western | Western industrial cluster | 4144 | 59,860 | 6615 | 12,839 |
| | | Bao'an center cluster | 4942 | | | |
| | Central | Central cluster | 9013 | | | |
| | | Henggang cluster | 2044 | | | |
| | Eastern | Longgang central cluster | 4155 | | | |
| | | Eastern industrial cluster | 3375 | | | |
| | Independent towns | Gongming | 1134 | | | |
| | | Guangming | 379 | | | |
| | | Shiyan | 662 | | | |
| | | Kuiyong | 613 | | | |
| | | Dapeng | 820 | | | |
| | | Nan'ao | 504 | | | |
| | Subtotal | | 31,785 | | | |
| ISL | Nanshan cluster | | 7170 | 18,482 | 2706 | 3533 |
| | Central cluster | | 7442 | | | |
| | Eastern cluster | | 1478 | | | |
| | Subtotal | | 16,090 | | | |
| Total | | | **47,875** | **78,342** | 9321 | 16,372 |

mountains and hills in space,[81] as shown in Fig. 3.10. Therefore, "More than 50% of the urban space outside the SSEZ is barbarously grown, and about 50% is controlled, a 50/50 split between the controlled and uncontrolled space."[82]

The above situation of urban construction encroaching on mountains and hills was curdled after the implementation of the *Regulations on the Management of the Basic Ecological Control Line in Shenzhen City* (深圳市基本生態控制線管理規定) in November 2005,[83] as shown in Fig. 3.11. The Regulations clearly define the scope of the basic Ecological Control Line that cannot be used for general urban construction, including:

*(1) First-grade water protection zones, scenic areas, nature reserves, concentrated basic farmland protection areas, forest and country parks; (2) mountains, forest land with slopes greater than 25%, and highlands more than 50 meters ASL within the Second Line and more than 80 meters ASL outside the Second Line; (3) main rivers, reservoirs and wetlands; (4) ecological corridors and green spaces that maintain ecosystem integrity; (5) islands and littoral land with ecological protection value; and (6) other areas requiring basic ecological control.*[84]

It is worth noting that some types of construction are permitted within the basic Ecological Control Line:

**Fig. 3.10** The city outside the Second Line intertwining with hills (Longhua-Dalang area) (Photo taken by the author on Yangtai Mountain, December 10, 2022)

[81] Zou, B. 鄒兵(2017). The Evolution of Urban Spatial Structure in Shenzhen and the Utility Evaluation of Its Planning深圳城市空間結構的演進歷程及其中的規劃效用評價. *Urban & Rural Planning*城鄉規劃, 06, 69–79.

[82] Wen, W.文維(2021). Building Shenzhen into a National and Global Maritime Center City and Hub City將深圳打造成國家乃至全球海洋中心城市和樞紐城市 In. Shenzhen Urban Planning and Natural Resources Bureau深圳市規劃和自然資源局(Eds.). *Forty Years of Shenzhen Special Economic Zone: Oral History Interviews on Planning and Land Use*深圳經濟特區四十年 規劃國土口述歷史訪談錄. Unpublished. (pp. 330).

[83] Zou, B. 鄒兵 (2017). The Evolution of Urban Spatial Structure in Shenzhen and the Utility Evaluation of Its Planning深圳城市空間結構的演進歷程及其中的規劃效用評價. *Urban & Rural Planning*城鄉規劃, 06, 69–79.

[84] People's Government of Guangdong Province廣東省人民政府. (March 10, 2022). *Regulations on the Management of Shenzhen's Basic Ecological Control Line*深圳市基本生態控制線管理規定. Retrieved from: http://www.gd.gov.cn/zwgk/wjk/zcfgk/content/post_2532255.html?eqid=a032e0d200011b5a00000006643f56a0

**Fig. 3.11** Scope of the basic Ecological Control Line in Shenzhen

*(1) Major road traffic facilities; (2) municipal public utilities; (3) tourism facilities; (4) urban parks; and (5) agricultural, educational, and scientific research facilities suitable for ecological and environmental protection.*[85]

After the slope and elevation of the site selection of construction land are clearly defined, the general urban construction, such as houses and factories, can no longer encroach on the mountainous forest land, but urban parks are an exception, and their siting is not legally bound by slope or elevation.

Second, there is another important reason to promote the shifting of park sites to large slopes and high elevations. Before the promulgation of the *Regulations on the Management of the Basic Ecological Control Line in Shenzhen City*, Shenzhen officially put forward the "four difficulties"[86] faced by the city from the aspects of land space, population, resources, and environmental carrying capacity. After the delineation of the basic Ecological Control Line, the 974 square kilometers of land area within its scope cannot be used for general urban construction, which accounts for nearly half of Shenzhen's total urban area. Within the remaining 1000 square kilometers available for development, no less than 270 square kilometers were

[85] People's Government of Guangdong Province廣東省人民政府. (March 10, 2022). *Regulations on the Management of Shenzhen's Basic Ecological Control Line*深圳市基本生態控制線管理規定. Retrieved from: http://www.gd.gov.cn/zwgk/wjk/zcfgk/content/post_2532255.html?eqid=a032e0d200011b5a00000006643f56a0

[86] Sun, R. F., & Zhao, J. 孫榮飛,趙傑 (2005, August 12). *"Calling for the Effectiveness of the 'Four Difficulties' in Shenzhen."* "四個難以為繼"呼喚效益深圳. First Financial Daily第一財經日報. (A06).

demarcated for industrial land, and the remaining land must harmonize a series of urban functions such as residence, commercial services, transport facilities, and public service facilities. On the one hand, the government should provide relatively flat and easily accessible construction land for a series of urban functions, and on the other hand, it must build urban parks to meet the needs of citizens in a city with an overspeed population growth. Under such circumstances, the site selection and construction of parks in the large slopes and high elevations becomes the common solution to the contradiction between the two aspects.

Third, even so, the government was challenged by land property rights to build parks outside the Second Line before 2004. At the establishment, the land available for development and construction in the SSEZ was basically the local farmers' collective owned land. Consequently, the government embarked on urban development by acquiring land from farmers, parcel by parcel. With the increasing scale and speed of construction, this approach is difficult to adapt to the demand. In June 1992, the Shenzhen Municipal Government issued the *Interim Provisions on Rural Urbanization in the Shenzhen Special Economic Zone* to nationalize the rural collective land in the SSEZ at one time. The same was true for land outside the Second Line before the *Measures on Urbanization Land Management in Bao'an and Longgang Districts of Shenzhen Municipality*[87] was promulgated in June 2004. According to the *Land Administration Law of the People's Republic of China* (中華人民共和國土地管理法) at that time, rural collective land belonged to the collective of village farmers rather than the state. Only after the collective land for construction was expropriated as state-owned land could it be transferred to entities or individuals for other purposes. For such land, the government can only acquire land use rights by "parcel by parcel requisition." As far as the government is concerned, this hindered the rapid progress of urban construction and even made it difficult to control the use of land outside the Second Line, resulting in a number of industrial zones based on rural villages without governmental permission. This is obviously not conducive to the implementation of the "territory-wide development" strategy at the municipal level. As a result, the government hoped to quickly nationalize the rural land outside the Second Line. The *Measures on Urbanization Land Management in Bao'an and Longgang Districts of Shenzhen Municipality* specified to nationalize 956 square kilometers of collective land in the (greater) Bao'an, (greater) Longgang Districts outside the Second Line, actually involving 241 square kilometers in the land compensation.[88] Article 5 of this Measures reads:

[87] Guangdong Provincial People's Government廣東省人民政府 (2020, March 24). *Notice on Issuing the "Shenzhen Bao'an and Longgang Urbanization Land Management Measures" by the Shenzhen Municipal People's Government*深圳市人民政府關於印發《深圳市寶安龍崗兩區城市化土地管理辦法》的通知. Retrieved from http://www.gd.gov.cn/zwgk/wjk/zcfgk/content/post_2939390.html

[88] Urban Planning and Natural Resources Bureau of Shenzhen. (Eds.) 深圳市規劃和自然資源局編. *Memories of Land System Reform in Shenzhen*深圳土地制度改革記憶. Unpublished. (pp. 91–93).

*For the collectively owned land converted to state-owned land in the two districts, the governments of the two districts shall give appropriate compensation in stages and batches to the successor entities or economic organizations and individuals of the original collective organizations, except for the following land:*

*(1) Built-up area; (2) land for construction approved but not yet been used; (3) mountainous forest land; and (4) parks with slopes greater than 25 degrees and not used as the land for construction.*

In this way, the government acquired the right to use all mountainous forest land and fields with slopes greater than 25 degrees outside the Second Line. They can be included in the government's unified planning for "territory-wide development," including the construction of urban parks.

It can be seen that there are at least three policy and institutional reasons for the situation observed in Sect. 3.2: (1) The slope and elevation of urban parks built after the 1990s are significantly higher than those in the 1980s, because the urban construction built in the 1980s is mainly in Nanshan District, Futian District, eastern and central Luohu District and central Yantian District, which are flat, low-lying areas. (2) Although residential buildings, factories, commercial areas, and other urban construction purposes can no longer be sited at large slopes and high elevations after 2005, urban parks were exempt. Even outside the Second Line dominated by mountainous and hilly landscapes, the urban park construction was climaxed only in the period after 2005, with a large average area of a single park.

### 3.2.3 Sea Reclamation for Oasis (Fig. 3.12)

Based on the previous section, we know that Shenzhen has been short of land almost throughout its short history of development. It is not difficult to understand that sea reclamation has become an important strategy for urban space development in Shenzhen, a coastal city with 260 kilometers of natural coastline. In addition to developing manufacturing itself, Shenzhen in the 1980s required a strong logistics industry to support its transportation, warehousing, and distribution, and other links in order to truly become a global manufacturing base. Shekou, Chiwan, and Yantian were selected as ports, thanks to the topographic features of a wide sea surface, deep water with few waves, less siltation, and no deep-sea undercurrent.[89,90] The site selection of airports was also subject to geographical and topographical conditions.

---

[89] Li, H. J., Huang, H. H., & Ye, Y. R. 李華傑,黃涵蓀,葉煜榮 (1982). The Rise of Shekou Industrial Zone—An Investigation into the Establishment of Shekou Industrial Zone by China Merchants Group 蛇口工業區的崛起——對招商局創辦蛇口工業區的調查. *Academic Research*學術研究, (01), 13–18.

[90] Gu, X.J. 辜曉進 (Eds.). (2018). *In the Name of News III: Selected Works of News Graduation Design from Shenzhen University - Frontier Observations* 新聞的名義III:深圳大學新聞畢業設計作品選·前沿觀察. Beijing: Renmin University of China Press 北京:中國人民大學出版社. (pp. 46).

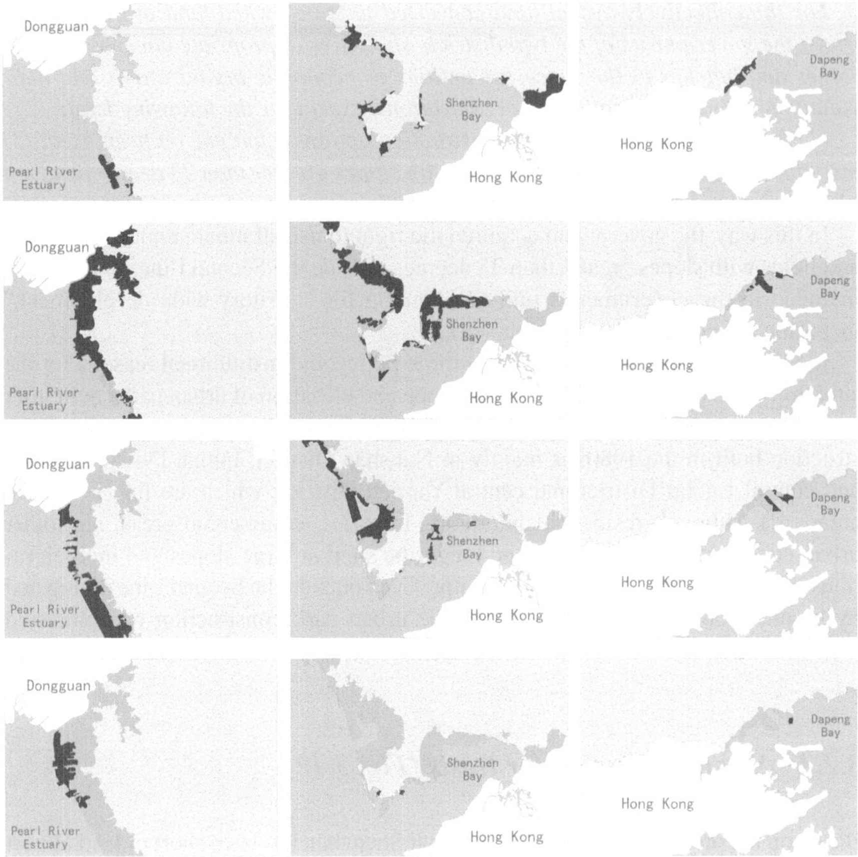


**Fig. 3.12** Shenzhen's extent of reclamation in four different periods (Redrawn based on high-resolution satellite images from USGS and historical satellite images from Google Earth)

The airport was whittled down from an initial list of 18 sites to three, among which Deep Bay was rejected for being too close to Wutong Mountain and the bird sanctuary and too close to the urban area, and Minzhi (民治) was also rejected because it was too close to Yangtai Mountain and the clouds were hardly dispersed. In this way, Huangtian Beach (黄田海灘) along the Pearl River Estuary on the west side of Bao'an District, with a mean depth of 7–8 m silt, became the final site for the airport (Table 3.8).[91]

[91] Shenzhen Municipal Political Consultative Conference Cultural深圳市政協文化文史和學習委員會. Historical, and Learning Committee (Eds.) (2010). *The Rise of Shenzhen's Four Main Industries: High-Tech*深圳四大支柱產業的崛起·物流. Beijing: China WenShi Publishing House 北京:中國文史出版社. (pp. 129).

**Table 3.8** Large-scale reclamation projects in Shenzhen and the total estimated area of reclamation in different periods (Collected from various sources, for reference only)

| Period | Along the Pearl River Estuary | | Shekou Peninsula and surrounding areas | | Yantian Port | | Total reclamation area (estimated) |
|---|---|---|---|---|---|---|---|
| | Project | Area ($km^2$) | Project | Area ($km^2$) | Project | Area ($km^2$) | |
| 1980–1992 | Huangtian Airport | 2.96 | Shekou Port; Chiwan Port; Futian Free Trade Zone | 2.32 3.55 1.35 | Yantian Port | 2.32 | 12.50 |
| 1993–2004 | Shajing/ Fuyong | 22.57 | Bao'an Center; Houhai; West Access, North of Binhai Avenue | 15.0 4.04 5.11 1.76 | | 2.39 | 50.87 |
| 2005–2012 | Airport Runway 2 | 11.0 | Qianhai; Dachan Bay | 7.70 10.2 | | 1.69 | 30.59 |
| 2013–2020 | New Marine City | 7.44 | – | – | | 0.26 | 7.70 |

Reclamation from the early 1990s to the mid-2000s was interwoven in two layers of rationale, with the early stage of reclamation still for the development of manufacturing industry, while the later stage has begun to prepare for industrial upgrading. The reclamation area is mainly distributed in three places: first, Shajing-Fuyong. With the mass migration of manufacturing and the suspension of "Three-processing and One Compensation" projects inside the Second Line after 1994, a large number of manufacturing industries moved to outside the Second Line. Here became the preferred destination for manufacturing migration. Second, Bao'an Central District (寶安中心區). At the end of 1992, Bao'an County was converted into a district, in need of a new urban center to lead the development of the district. By 2005, 15 square kilometers of Bao'an Central District was formed here.[92] At this time, Shenzhen's reclamation idea was not to increase land supply for low-end manufacturing, but to provide space for industrial upgrading thereafter.[93,94] Third, Shenzhen Bay is divided into two parts: the area north of Binhai Avenue and Houhai.

After the mid-2000s, reclamation is no longer open to low-end manufacturing. Since then, there have been two large-scale reclamation projects: the expansion of Airport Runway 2 and the Qianhai Center. As Shenzhen Bao'an International

[92] He, J. 何俊 (2022, February 25). *Thirty Years: The Changes and Rise of Bao'an Central Area*三十年，寶安中心區的變遷與躍升. Southern Metropolis Daily南方日報. (BD02).

[93] Shenzhen Urban Planning Bureau深圳市規劃局 (2007). *Shenzhen 2030: Toward a pioneer metropolis based on sustainable development*建設可持續發展的全球先鋒城市 深圳2030城市發展策略. Beijing: China Architecture & Building Press北京:中國建築工業出版社. (pp. 26).

[94] Editorial Committee of the " Plan for the Coordinated Development of the PRD metropolitan area." 《珠江三角洲城鎮群協調發展規劃》編委會 (2007). *Pearl River Delta Urban Agglomeration Coordinated Development Plan*珠江三角洲城鎮群協調發展規劃: 2004–2012. Beijing: China Architecture & Building Press北京:建築工業出版社. (pp. 30–31).

Airport (深圳寶安國際機場)[95] became increasingly saturated with flights, the reclamation project commenced in December 2005 for Runway 2 of this airport. To implement the national planning for the PRD metropolitan area and to build a new urban center for Shenzhen near the Pearl River Estuary, the Qianhai reclamation project kicked off in September 2006.

From the spatial distribution of the reclamation area, Shenzhen's reclamation is clearly focused on the west, mainly in the northwest and southwest coastal plains; formed by reclamation in the east, Yantian Port cannot be compared with the reclamation in the west in terms of overall scale. There are at least two reasons for this situation: (1) The eastern coastal area is mountainous, not conducive to traffic and urban development. Even the reclamation of Mirs Bay in the east has struggled to create a vast land depth, while the original topography of the western coastal area is plain, conducive to the construction of cities, and reclamation can further expand the urban space. (2) More importantly, the west borders the Pearl River Estuary, conducive to the linkage development with other cities in the Pearl River Estuary and the formation of the PRD metropolitan area (珠三角城市群). In fact, cities in the PRD metropolitan area have a trend of developing toward the sea and thus strengthening their linkage with other surrounding cities, such as Nansha District in Guangzhou, Daya Bay in Huizhou, Yinzhou Lake in Jiangmen, Doumen in Zhuhai, Torch Industrial Zone in Zhongshan, and Humen Port in Dongguan.[96]

The competition and cooperation among the cities in the PRD metropolitan area are the key drivers for reclamation in western Shenzhen. Since the 1980s, the PRD has developed based on an export-oriented economy, which has resulted in other cities, except Guangzhou and Shenzhen, not being very dependent on the service functions of their own city centers, because Hong Kong was the real center of service and tertiary industry at that time. With the further liberalization of reform and opening-up, especially after China's accession to the World Trade Organization, Shenzhen is no longer the only city to enjoy the open policy. In the context of cooperation and integrated development, cities in the PRD metropolitan area compete with each other, gradually producing higher requirements for their service functions of city centers. The service function of the city center will largely determine the quality of urban development as a single entity. The weak service function may cause the transfer of high-end industries that could be settled locally to other surrounding cities. Under such circumstances, all the cities in the PRD metropolitan area have adopted the development idea of enhancing the service function of the city center.[97] Therefore, Shenzhen must build a coastal city center in its west and at

---

[95] On September 18, 2001, Shenzhen Huangtian International Airport was officially renamed Shenzhen Bao'an International Airport.

[96] Shenzhen Urban Planning and Land Resources Commission深圳市規劃和國土資源委員會. (2011). *Guiding Urban Transformation through Transformation Planning - Shenzhen Urban Master Plan*轉型規劃引領城市轉型——深圳市城市總體規劃 (2010–2020). Beijing: China Architecture & Building Press北京:中國建築工業出版社. (pp. 64).

[97] Shenzhen Urban Planning and Land Resources Commission深圳市規劃和國土資源委員會 (2011). *Guiding Urban Transformation through Transformation Planning - Shenzhen Urban*

the east bank of the Pearl River Estuary to be integrated into the PRD and enhance its service function as a city center.

The industrial development of the PRD metropolitan area has experienced the stage of accepting the transfer of Hong Kong and international manufacturing industry, the stage of domestic demand-oriented localized industry growth, and is gradually moving toward the stage of independent innovation with more complete categories and functions.[98] In Shenzhen, a series of urban centers on the east bank of the Pearl River Estuary, such as Shenzhen Bay Super Headquarters Base (深圳灣超級總部基地), Qianhai Shenzhen-Hong Kong Modern Service Industry Cooperation Zone (前海深港服務業合作區), Bao'an Center, International Exhibition Center (國際展覽城), New Marine City, and even slightly further Guangming Science City (光明科學城), have gradually risen to meet the needs of high-end independent innovation industries. However, the land along the Pearl River Estuary in places such as Shajing and Fuyong has long been occupied by low-end manufacturing and the ensuing residential and commercial services were transferred in the 1990s. As a result, reclamation has become the only option to accommodate new urban centers, which is the motivation for Shenzhen to spend huge amounts of money and environmental costs in the west for decades.

For Shenzhen, especially within the Second Line, the construction of urban parks is closely related to the construction of city centers. Before the 1990s, the SSEZ received investment from both Hong Kong and international investors. During this period, the service functions of the city were developed near the old town of Luohu, evidenced by Lychee Park, People's Park, Honghu Park, and Donghu Park, all built in or near the old town of Luohu. With the saturation of Luohu's urban functions, it was proposed in the *Urban Master Plan of Shenzhen Municipality (1996–2010)* to move the new city center westward to Futian District as the future administrative, cultural, and business center. Therefore, during this period, Shenzhen Central Park (深圳中心公園), Bijiashan Park, Huanggang Park (皇岗公园), Meilin Park, and Civic Center Square (市民中心廣場) were built successively in the central area of Futian District. The *Urban Master Plan of Shenzhen Municipality (2010–2020)* defines the spatial layout of Qianhai-Houhai center and Futian-Luohu center, during which the western Shenzhen Bay Park, Talent Park, Binhai Cultural Park (濱海文化公園), and other parks have been built and opened. When the *Master Plan for Shenzhen's Territorial Space (2020–2035)* (深圳市國土空間總體規劃(2020–2035)) (publicity version) clearly considers Shekou Peninsula and its surrounding areas as the only "comprehensive core area for innovation" in the city, Nanshan District has become a "hub of internet-famous parks" and a "district of 100

*Master Plan*轉型規劃引領城市轉型——深圳市城市總體規劃 (2010–2020). Beijing: China Architecture & Building Press北京:中國建築工業出版社. (pp. 64).

[98] China Development Research Foundation中國發展研究基金會. (2022). *Research on the Integration of City Clusters in China*中國城市群一體化研究. Beijing: China Development Press 北京: 中國發展出版社. (pp. 155–156).

parks."[99] It can be seen that the city center of Shenzhen has been moving westward for decades, and each move has driven the construction of several large parks in and around the city center. This westward shift of the central area is benefiting from the integrated development of the PRD metropolitan area.[100] After 2010, the city center moved to the Qianhai-Houhai area, and urban parks built on reclaimed land emerged in such context.

In addition to parks, large public service facilities such as large theaters, concert halls, libraries, auditoriums, and exhibition centers were moving west along with the Shenzhen Central Business District. This also shows from the side that although urban parks are always presented as green and natural, they are not independent of the urban process, but shaped by the same social and economic context as other public service facilities.

The metropolitan area is a new spatial form emerging in the global information society, which develops in a variety of social and geographical contexts. This spatial form is constructed around the flow of capital, information, technology, people, images, sounds, and other elements. The physical space expresses the needs of the technology-finance-management elite that occupies the leading position in society, which, in turn, is closely related to the construction of the milieux of innovation.[101] For the local government, to strengthen the regional connection with the cities in the PRD, the city center should, even at the cost of land reclamation, be transformed into a "space of flow" that can express the needs of elites in the global information society. It must contain residential and leisure-oriented spaces, gather the dominant urban functions, and have easy access to the art, culture, and entertainment.[102]

The city is shaped by greening and urban parks in the double context of space and society. As a cultural form of nature, urban parks are the nature located in the city, and even the nature shaped by the rationale of urbanization—the urban process dominates the construction and expression of this nature. The construction of urban parks since the nineteenth century is still regarded today as an identifiable and acceptable aesthetic and social practice.[103] The same is true in Shenzhen. It has been positioned as a core innovative city in the PRD metropolitan area by national and regional policies. Government departments, planners, architects, and landscape architects believe that the urban environment can be made more comfortable and

[99] Nanshan District People's Government of Shenzhen深圳市南山區人民政府(2021, February 18). *Nanshan Becomes a Gathering Place for 'Internet Celebrity Parks,' Truly Living Up to the Name of 'The District of a Hundred Gardens'* 南山成"網紅公園"聚集地 "百園之區"名副其實. Retrieved from http://www.szns.gov.cn/xxgk/qzfxxgkml/bmdt/content/post_8555918.html

[100] Zhu H.B. 朱惠斌 (2013). Urban Administrative Center Location Selection and Migration Type: A Case Study of Shenzhen-Dongguan-Huizhou Metropolitan Area城市行政中心區位選擇與遷移類型——以深莞惠都市區為例. *Tropical Geography*熱帶地理, 33(05), 527–532.

[101] Scott, A. J. (2012). A *world in emergence: cities and regions in the 21st century*. Edward Elgar Publishing. (pp. 64–78).

[102] Castells, M. (2011). *The Rise of the Network Society: The Information Age: Economy, Society, and Culture Volume I (2nd ed., Vol. 12)*. West Sussex: Wiley-Blackwell. (pp. 440–452).

[103] Angelo, H. (2021). *How green became good: Urbanized nature and the making of cities and citizens*. University of Chicago Press. (pp. 1–7).

sustainable through the construction of urban parks. Therefore, a "green face" is particularly required to shape its milieux of innovation.[104]

As important infrastructure to enhance urban service functions, parks built on the basis of reclamation are placed in the context of regional co-opetition in such a way as to understand and analyze. Their construction is related to the construction of urban milieux of innovation, and represents the imagination of technology, finance, and management elites for the urban space environment.

## 3.3 Summary

Through research on the physical topography of the Shenzhen city scope, it was found that the SSEZ located south of the second line is predominantly characterized by flat plains, while outside the second line, mountainous and hilly topography prevails. Before the era of reform and opening-up, the residents of various villages and towns in Bao'an County engaged in agricultural-centered production and life practices intertwined within such topographic conditions. Since the 1970s, neoliberal globalization has profoundly transformed these indigenous practices. Driven by national and urban policies, a large amount of rural collective land within and beyond the second line has gradually been converted into state-owned land controlled by local governments. Simultaneously, this has generated new urban landscapes. Whether it is the demarcation of the second line, the cluster structure, or the deep-water port areas of Shekou and Yantian, the topography, as a non-human element, exhibits a similar agency in shaping urban spatial structures as the globalization propelled by human endeavors. It significantly shapes the relationship between Shenzhen and the world; it is not merely a passive, inert element subject to transformation but rather an active participant in the process.

In the new urban landscape, urban parks also emerge abundantly with a verdant natural appearance. The spatial and temporal distribution of these urban parks within the city's scope may seem to have formed inadvertently through countless site selections, but in reality, they are traceable. Their size, slope, elevation, and the evolution of land use within their scope are all intrinsic to Shenzhen's unique urbanization process, intertwined with both local topography and the broader socio-historical context of globalization. In terms of park area, the number of small parks is increasing while the number of large parks is decreasing. Additionally, the quantity and area of park construction are gradually shifting from within the second line to outside. Regarding the average slope of parks, urban park construction is increasingly inclined to be distributed in areas with relatively steep slopes. Similarly, concerning the average elevation of parks, urban park construction gradually tends to be distributed in areas with relatively higher elevations. In terms of

---

[104] Sze, J. (2015). *Fantasy islands: Chinese dreams and ecological fears in an age of climate crisis*. University of California Press. (pp. 14).

land use, approximately one-fourth of the parks built and opened in the past forty years were established on the arable land of 1980, while about half were established on forest land in the same year. From the period between 1993 and 2004 onward, the proportion of urban park construction occupying forest land began to rise significantly. By the fourth period, the proportion of urban park construction occupying water bodies in 1980 noticeably increased. In terms of terrestrial spatial evolution, parks constructed on reclaimed land are mostly located in the Qianhai-Houhai area and were predominantly built during the fourth period. For this study, these phenomena necessitate explanations within the social and historical context of Shenzhen.

First, the transformation of urban landscapes reflects changes in industries. Between 1980 and 2020, the area of arable land, the output value of the primary industry, and the number of people employed in the primary industry in Shenzhen all experienced sharp declines. The agricultural economic system under the planned economic system was almost completely replaced by the globalization-driven economic system. Regardless of whether within or beyond the second line, almost all types of urban construction, including urban parks, were built on arable land, forest land, fish ponds, orchards, nurseries, and other agricultural land. Therefore, from the perspective of land use, it is not difficult to understand that the sites chosen for parks constructed over the past forty years were primarily located on arable land and forest land from 1980. Beginning in the period between 1993 and 2004, as urban space expanded into mountainous and hilly areas beyond the second line, the proportion of park construction on forest land noticeably increased. Although the process of urban park construction represents a transformation of appearance from natural to natural, the practices of production and life, as well as the socio-historical context before and after the transformation, are entirely different.

Furthermore, with the development of urban space within the SSEZ, the land resources of the flat plain in the Luohu area were nearly exhausted by the early 1990s. Subsequently, the city center developed to Futian and finally expanded to land reclamation in the Qianhai-Houhai area. This process, combined with the industrial upgrading within the SSEZ, explains the necessity for high-quality spatial resources in the Qianhai-Houhai area. This area must strengthen its service function as a new city center, consolidating Shenzhen's position as the core city in the PRD metropolitan area. Constructing urban parks on reclaimed land enhances the overall quality of urban space, aligns with the environmental imagination of urban elites, and reflects its important political and economic significance. Moreover, the timing of urban spatial development also explains why park construction on water bodies only began in the fourth period. Lastly, we seek to understand several spatio-temporal distribution characteristics of urban parks built over the past 40 years in terms of area, average slope, and average elevation: (1) Park construction has gradually shifted toward areas with steeper slopes and higher elevations. (2) Park construction has tended toward dispersed small-scale parks rather than constructing large-scale individual parks. In fact, this reflects the reality of Shenzhen facing tight land resources, where it is necessary to meet the recreational needs of citizens while also freeing up land for industrial development. This brings us back to the initial definition and understanding of urban parks: they are a shared solution of bottom-up demand and top-down governance practices.

Urban parks, unlike other "gray infrastructure" in urban areas, undergo a transformation from natural to natural. Agriculture dominates the former, while urban social life shapes the latter. Although both are not purely natural and are cultural form of nature, they carry significantly distinct connotations. In Shenzhen's urban social life, urban parks do indeed provide recreational spaces for citizens; however, at the level of urban governance, urban parks have a more structural significance. This structural significance, which differs from that of "gray infrastructure," lies in its construction around the production of nature: whether they are the Shangbu-Luohu Park Group on the plains, the Meilin-Gangxia Group on mountainous hills, or the Shenzhen Bay Park Group on reclaimed land. The varying topographic conditions result in urban parks having different spatial forms and natural landscape resources, which not only showcases the richness of the relationship between cities and nature but, more importantly, makes the interaction between these urban parks and urban social life distinct. This is also the reason for choosing topography as the entry point for discussion in this study; it will lay the groundwork for the case studies in Chaps. 4, 5, and 6. In historical research, we should not interpret natural topography as a passively modified inert object or as purely natural. Instead, it is critical to carefully observe how urban social life accepts the facilitation or restriction imposed by topography.

# Chapter 4
# Shangbu-Luohu Park Group: Nationalism, Market Economy, and Public Security Governance

## Contents




**Abstract** This chapter focuses on the park group constructed in the Shangbu-Luohu plain area during the 1980s, examining how the development and management of urban parks relate to city marketing, public security management, environmental conservation, and the concept of the nation-state within the context of neoliberal globalization.



**Keywords** Shangbu-Luohu · Plain · Urban marketing · Nation-state





X. Liu, *Fabricating Nature*, The Urban Book Series,
https://doi.org/10.1007/978-981-95-8208-2_4

## 4.1 The Landscape History in Shangbu-Luohu Area

### *4.1.1 Before Reform and Opening Up*

The Shangbu-Luohu Area, which developed out of the "Shenzhen Fair(深圳墟)," served as the social, political, and economic center of Shenzhen City throughout the 1980s. It was also the primary location where China's reform and opening-up policies were implemented. However, its historical significance extends far beyond this time frame. The earliest documented reference to the "Shenzhen Fair" could possibly be discovered in the *Xin'an County Chronicle* (新安縣誌) authored in 1688 by Jin Wenmo (靳文謨, Year of birth and death unknown). This chronicle officially recognized a total of 28 marketplaces. The "Shenzhen Fair" was one of these periodically held markets.[1] The primary cartographic representation of the "Shenzhen Fair" in its entirety is the *Guangdong Coastal Map* (廣東沿海圖)[2] during the Qianlong period of the Qing Dynasty (1736–1795). Figure 4.1 depicts the geographical location and historical evidence that trace the origins of the "Shenzhen Fair" back to the reign of Emperor Kangxi of the Qing Dynasty or even earlier.

In 1911, the Kowloon-Canton Railway (KCR) was fully completed and made accessible to the public, and the final station on the Chinese side, Shenzhen Fair

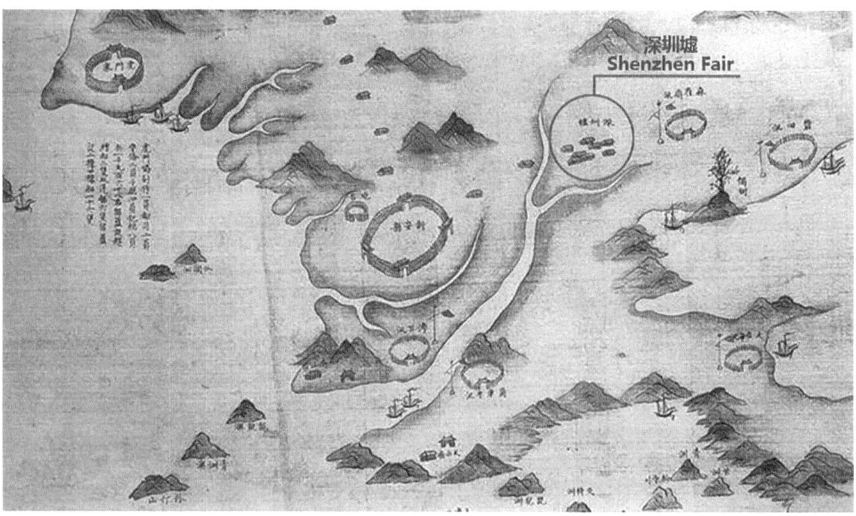


**Fig. 4.1** "Shenzhen Fair" in *Guangdong Coastal Map* (Liang, E. P. 梁二平(2022, July 16). *Rarely Seen "Shenzhen Fair" on Ancient Maps*古地圖上難得一見的"深圳墟." Shenzhen Evening News 深圳晚報, A07)

[1] Zhang, Y. B. (Eds.) 張一兵 校(2006). *Three Editions of Old Chronicles of Shenzhen*深圳舊志三種. Shenzhen: Haitian Publishing House深圳:海天出版社. (pp. 268).

[2] Liang, E. P. 梁二平(2022, July 16). *Rarely Seen "Shenzhen Fair" on Ancient Maps*古地圖上難得一見的"深圳墟." Shenzhen Evening News深圳晚報, A07.

Railway Station (深圳墟站), was set up in close proximity to Shenzhen Fair. In subsequent decades, Shenzhen Fair emerged as the region with the most efficient transit infrastructure in the entire county, leading to tremendous growth in its commerce. In the Republican era, the Shenzhen Fair extended its reach to the south, encompassing the Luohu Bridge (羅湖橋) and the railway station area.

This region became a hub for a wide variety of stores that catered to people from all social backgrounds. These stores included photo studios, parks, churches, Guandi Temple (關帝廟), Tin Hau Temple (天后廟), Ancestral Hall, and others. Several main streets concentrated these stores. The Shenzhen Fair was thriving at this time, frequently engaging in trade with Sheung Shui (上水), Sha Tau Kok (沙頭角), and Yuen Long (元朗) in Hong Kong, as well as Dongguan (東莞), and Huiyang (惠陽) in Guangdong province.[3] Amidst the Anti-Japanese War, Shenzhen Fair endured significant bombardment but steadily recuperated following the triumph of the anti-Japanese forces.

In 1950, the Shenzhen Fair was promoted to the status of Shenzhen Town. The process of municipal development commenced gradually when the Bao'an County Government relocated from Nantou to Shenzhen Town in 1953. The authorities expanded and merged four streets to create Jiefang Road (解放路), and restored and enhanced Renmin Road (人民路). Furthermore, the authorities improved other streets like Huancheng Road (環城路), along with various cultural and living facilities such as schools, hospitals, theaters, cultural palaces, and bookstores. The urban area of Bao'an County in 1978 covered approximately three square kilometers,[4] as depicted in Fig. 4.2.

In 1979, the provincial authorities of Guangdong province declared their intention to rename Bao'an County as Shenzhen and later promoted Shenzhen to the rank of a city. In the same year, the provincial authorities of Guangdong province founded Luohu District as the inaugural district and expanded it beyond its initial confines, encompassing Luohu Village, Luohu Bridge, and Luohu Hill in the southern region of the original Shenzhen Town. Luohu District was established as a county-level administrative division, stretching from Dapeng Bay (大鵬灣) in the east to Chegong Temple (車公廟) in the west.

Luohu District is characterized by a predominantly mountainous and hilly topography, except for the coastal alluvial plain in the southwest region known as Shangbu-Luohu Area (上步-羅湖片區). The upper part of the soil has a mixture of soil and sand particles, while the lower section consists of glacial-fluvial deposits, which is a common geological composition found in alluvial plains. The subtropical marine climate's geological characteristics contribute to the soil's fertility, enabling the cultivation of crops with minimal terrain adjustments. In ancient times, the capacity to sustain agricultural activities resulted in a rise in the human population and the growth of farming communities.

---

[3] Liao, H. L. 廖虹雷 (2016). Historical Evolution of Shenzhen Markets深圳墟歷史沿革. *Shenzhen Historical Records*深圳史志, (52), 63–65.

[4] Bao'an County Local Chronicles Compilation Committee寶安縣地方誌編纂委員會(1997). *Chronicles of Bao'an County*寶安縣誌. Guangzhou: Guangdong People's Publishing House 廣州: 廣東人民出版社. (pp. 318–319).

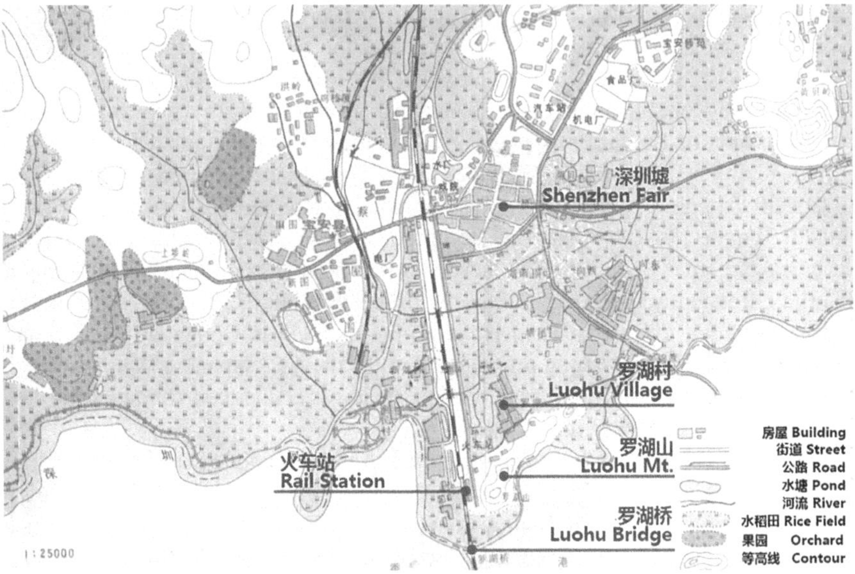


**Fig. 4.2** Map of Bao'an County town from 1953 to 1978 (Bao'an County Local Chronicles Compilation Committee 寶安縣地方誌編纂委員會(1997). *Chronicles of Bao'an County*寶安縣誌. Guangzhou: Guangdong People's Publishing House 廣州: 廣東人民出版社. (pp. Illustration Page插圖頁))

Based on Fig. 4.3, the Shangbu-Luohu Area has undulating topography in the northern, eastern, and southern sections. Nevertheless, the majority of the inland area is primarily level, with sporadic fluctuations in height caused by small hills such as Huangbeiling Hill (黃貝嶺) and Luohu Hill (羅湖山). The Kowloon-Canton Railway passes through this area, linking the northern and southern halves as it heads toward Hong Kong. The region's transport infrastructure primarily comprises an integrated network of rural roads. The Buji River (布吉河) and Shawan River (沙灣河), along with their minor tributaries, originate from the mountains in the north. Furthermore, the Shenzhen River functions as the demarcation line between Shenzhen and Hong Kong, coursing in an east-to-west direction. Moreover, there are other villages scattered across extensive agricultural areas, predominantly dependent on farming as their economic foundation. Notable villages in the vicinity include Shenzhen Fair, Luohu Village (羅湖村), Xiangxi Village (向西村), Hubei Village (湖貝村), Shuibei Village (水貝村), Sungang Village (筍崗村), Caiwuwei Village (蔡屋圍村), Huangbeiling Village (黃貝嶺村), and several others. By glancing into their aboriginal genealogy, one can infer the geography and environmental features of these villages, which date back hundreds of years.

*The Luohu lineage originates from Wentang, the inhabitants of which region migrated from Jiangxi province. The ancestors of this family are renowned for their prominence in the city, which is attributed to the exceptional individuals. While the location benefits from its natural surroundings, its prosperity is primarily a result of*

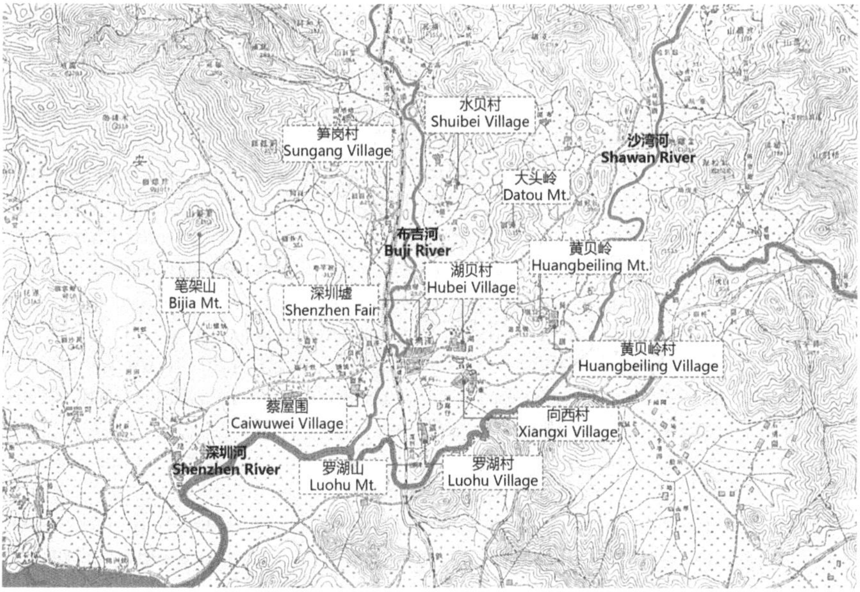


**Fig. 4.3** Map of Shangbu-Luohu Area in 1930 (Source from the *Central Research Institute (Taiwan) Digital Map Collection Integration and Query System*. Data code: map_loc_033061)

*human activities. For what reason? JiuLongchuan Ridge and Jinpai Mountain are encompassed by interconnected bamboos and trees while being flanked on the left side. Beidi and Liantang Fields, known as Yuanwu, are guarded by towering bamboos and trees on the right side.*[5]

*The Wutongshan mountain spits out the moon to reflect the Luo River, and the silk ribbons flutter in the west. A few fishing rods passed by, and the high singing made the white clouds lower.*[6]

*The ancestors chose Chikang, located on Xin'an East Road, as their home in Dajing, Dongguan. It has been based here since the Yuan Dynasty, which is the former site of Yuegang Tun. Gen Xiangkun and Yinshen were both born in his hometown. He was under the ancient temple on Beishan Mountain at the back. There was a big pond on the right. There was a pond in front of the township, and a sacred mountain with towering green trees. The river was on the left, Bijiashan Hill was on the right, and there was Shenzhen in the northeast.*[7]

*The luxuriant forest in Beiling is a hundred-foot-tall parasol, the sun is beautiful, the wind is clear and the colorful phoenixes are singing, the beautiful water in the*

[5] Peng, Q. 彭全民 (2015). *Shenzhen anecdotes: A casual discussion*深圳掌故漫談. Shenzhen: Shenzhen Press Group Publishing House深圳:深圳報業集團出版社. (pp. 49).

[6] Peng, Q. 彭全民 (2015). *Shenzhen anecdotes: A casual discussion*深圳掌故漫談. Shenzhen: Shenzhen Press Group Publishing House深圳:深圳報業集團出版社. (pp. 50).

[7] Caiwuwei Genealogy Compilation Committee蔡屋圍族譜編纂委員會. (2005). *Caiwuwei Genealogy of Shenzhen*深圳蔡屋圍族譜. Internal document. (pp. 36).

*stone pool is full of golden waves, and the state of the world is dancing like a dragon.*[8]

According to the illustration in Fig. 4.4, the Shangbu-Luohu Area was previously characterized by extensive agricultural land, encompassed by mountains, bodies of water, and ponds. Over the course of several centuries of urban development, particularly after the implementation of the Reform and Opening-up policy, landmarks such as Wutong Mountain (梧桐山), Bijia Mountain (筆架山), Huangbeiling Hill (黃貝嶺), Shenzhen River, Buji River, Shawan River (沙灣河), and others have managed to survive till present. However, the ground ponds that were once scattered throughout this region have completely vanished. The previous pastoral panorama of this area remains alive solely in the writings of our ancestors.

Rice cultivation has been the primary mode of production due to the geographical and natural conditions. Consider the rice production of four villages in the years 1955, 1965, and 1975 as illustrated in Table 4.1.

The contemporary urban environment of the Shangbu-Luohu Area, characterized by its abundance of tall structures, was not created from a "blank slate." Previously, there was wide agricultural land encompassed by mountains and bodies of water. For centuries, the terrain and natural environment have played a significant role in the agricultural society and way of life of the Yuan (袁), Zhang (張), Cai (蔡), and other clans. In 1980, the government officially designated it as a SSEZ to attract both domestic and foreign investments for the development of the "Three-Processing and One Compensation" industry.

**Fig. 4.4** Shenzhen River and Shenzhen Fair in 1927 (Image from Luohu Archive, Shenzhen)

[8]Peng, Q. M.彭全民(2019). *Shenzhen Huangbeiling Village Zhang Family Genealogy*深圳黃貝嶺村張氏族譜. Shenzhen: Haitian Publishing House深圳:海天出版社. (pp. 241).

**Table 4.1** Rice production in four villages in 1955, 1965, and 1975 (*Annual Rice Production in Fucheng Commune from 1953 to 1975*(1953–1975年附城公社歷年水稻產量). Catalog Number A13;Archive Number C1;Volume Number 8- Item Number 1(檔案全宗號C1; 目錄號A13; 卷宗號8; 件號1). Preserved at the Shenzhen Luohu Archives)

| Year | Village | Area (mu[a]) | Yield per mu (dan[b]) | Total (dan) |
|---|---|---|---|---|
| 1955 | Huangbeiling | 4020 | 385 | 15,494 |
| | Hubei | 4980 | 220 | 10,236 |
| | Sungang | 3636 | 388 | 14,139 |
| | Caiwuwei | 4500 | 275 | 12,375 |
| 1965 | Huangbeiling | 2956 | 402 | 11,902 |
| | Hubei | 3664 | 464 | 16,929 |
| | Sungang | 4447 | 362 | 16,113 |
| | Caiwuwei | 2783 | 373 | 10,398 |
| 1975 | Huangbeiling | 3123 | 416 | 12,927 |
| | Hubei | 1220 | 423 | 5160 |
| | Sungang | 4100 | 437 | 17,915 |
| | Caiwuwei | 2678 | 446 | 11,959 |

[a]Mu is a traditional unit of area in China, 1 mu equals 666.67 square meter
[b]Dan is a traditional unit of weight in China, 1 dan equals 50 kilograms

## 4.1.2 Three-Processing and One Compensation Period

> During the initial phase of construction, the SSEZ was comprised of a deserted agricultural plain encircled by desolate mountains. In order to attract internal and external investments, it is imperative to develop infrastructure and effectively execute the "Three Supplies and One Leveling (三通一平)" strategy, which involves providing water, power, road access, and leveling the ground. By doing so, we can establish a favorable investment climate for both foreign and domestic firms.[9]
>
> —Zhou Ding

Zhou Ding (周鼎, 1927–2019) held the position of deputy mayor, with responsibility for supervising the urban construction of the SSEZ. His statement above indicated that the urban construction during that period involved extensive modification of the topography. During the early 1980s, urban construction mostly took place in three specific locations inside the SSEZ, namely the Shatoujiao area (沙頭角), Shekou-Nantou area (蛇口-南頭), and Shangbu-Luohu Area. The Shangbu-Luohu Area, which encompassed the original Shenzhen Town and the new district built in the 1980s, was the economic and political heart of the city. The Shangbu-Luohu Area, serving as the political and economic center, is a focal point of urban construction and represents the initial development zone. It encompasses Shenzhen Town, formed based on the original Shenzhen Fair, and the newly constructed district in the 1980s.

[9]Wang, J.王靜(2018, December 13). *Zhou Ding: It's Unimaginable to Achieve Today's Success Without Good Policies*周鼎:如果沒有好的政策，是無法想像能有今天. Retrieved from http://www.china.com.cn/cppcc/2018-12/13/content_74258098.htm

By early 1983, the process of adding soil to the lower end and smoothing out the eroded area, known as the "Three filling and One Leveling" project, was successfully implemented in the Shangbu-Luohu Area. As a result, the road network in the area was essentially established. The rehabilitation and expansion of roads in the original Shenzhen Town were primarily conducted by building upon the existing foundation. In contrast, the roads in the new region were predominantly designed in a square grid pattern, as depicted in Fig. 4.5. The original Shenzhen Town is located in the southeastern part of the Shangbu-Luohu Area, with the Shenzhen River to the south and Luofang Mountain (罗芳山)-Tangpai Mountain (唐排山) to the east. However, due to these geographical constraints, there is limited space for expansion in these two directions. Consequently, undertaking extensive urban development has become an unavoidable decision to facilitate expansion into the western and northern regions of the original Shenzhen Town. Prior to 1980, the Futian area, situated to the west of the original Shenzhen town, was predominantly comprised of agricultural land and uncultivated terrain. Similarly, the region extending as far north as Buji, with the exception of a few established villages like Shuibei, Tianbei, and Sungang, consisted mainly of barren hills.[10]

However, the construction of the Shangbu-Luohu Area initially began in the Luohu Village (羅湖村)area due to its proximity to the port. Without this

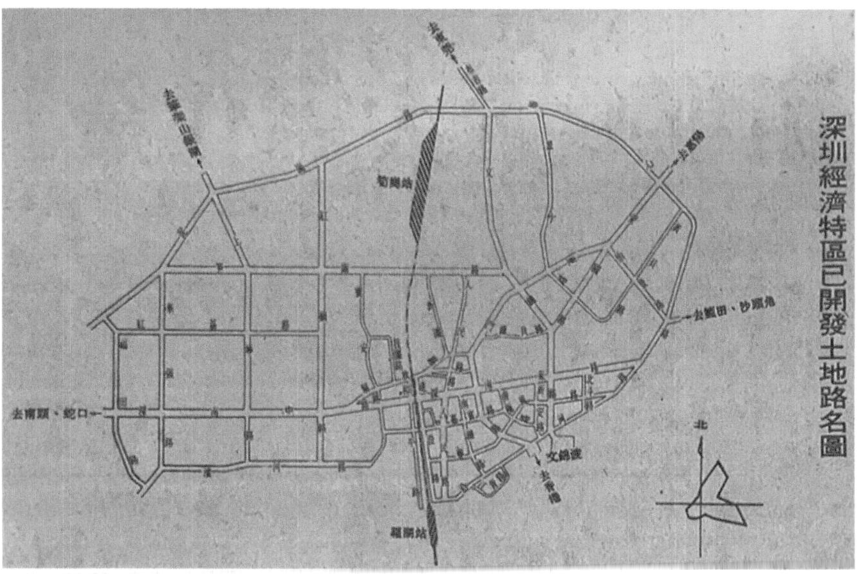


**Fig. 4.5** Road network map of the Special Economic Zone in 1983 (*Shenzhen Economic Special Zone Developed Land Road Map*深圳經濟特區已開發土地路名圖. Shenzhen Special Zone Daily深圳特區報, 1983-02-28(05))

[10] Shenzhen Municipal Chronicles Compilation Committee深圳市地方誌編纂委員會(2009). *Chronicles of Shenzhen: Volume on Infrastructure Construction*深圳市志·基礎建設卷. Beijing: Fangzhi Publishing House北京: 方志出版社. (pp. 531).

advantageous location, the Luohu Village area would not have been suitable for extensive urban development compared to the Shangbu area, as the presence of Luohu Hill (羅湖山) restricts available space for future expansion. Conversely, this region is replete with ponds and experiences frequent occurrences of flood.[11] There has been a contentious discussion on the development of Shangbu, a flat area located far from the port, or the Luohu Village area, which is disadvantageous but adjacent to the port. This matter has even caught the attention of high-ranking officials in the Central Government. The Luohu Village area was ultimately chosen for urban construction due to its proximity to the national gate and its significance in shaping the image of Shenzhen and even China as a whole. The resolution of the topographical challenges in the Luohu Village area was the initial and significant undertaking in the urban development of the SSEZ during that period. Within a year after the creation of the SSEZ, Luohu Hill, located on the eastern side of the Canton-Kowloon Railway and on the northern side of the Shenzhen River, was demolished. According to Figs. 4.6 and 4.7, the excavation of 1.3 million cubic meters of earth raised the elevation of the vicinity of Luohu Village by 1.07 meters over 0.8 square kilometers, achieving "Three Supplies and One Leveling" and improving water hazards to some extent.[12]

On a larger scale, the significant changes in the urban structure of the Shangbu-Luohu Area primarily took place throughout the 1980s, as depicted in Fig. 4.8: (1) Over the course of a decade, the urban area of Shenzhen Town rapidly expanded, extending from its original location in a flat area surrounded by hills to encompass a larger region in the south, north, and east directions. (2) A significant amount of previously cultivated land and some forested areas were converted into urban construction sites, resulting in the disappearance of natural features like rivers, lakes, and hills due to human activity. (3) Although the four large parks' spatial range has been maintained to preserve their overall natural appearance, human activity has led to artificial alterations in the topography.

The primary catalyst for the extensive alteration of the landscape is the process of urbanization in SSEZ, which may be attributed to the following factors:

One cause is the advancement of the industrial sector through the implementation of the "Three-Processing and One Compensation" policy. The establishment of a worldwide industrial division of labor has provided SSEZ with the potential to cultivate and expand their manufacturing industries. In the 1980s, SSEZ constructed several industrial zones in the Shangbu-Luohu Area, such as the Bagualing Industrial Zone (八卦嶺工業區), Shangbu Industrial Zone (上步工業區), and Shuibei Industrial Zone (水貝工業區), along with the Sungang Storage Area (筍崗倉儲區) and Qingshuihe (清水河倉儲區) Storage Area. As of the end of 1989, the built-up

[11] Li, Y. T. 李詠濤 (2010). *Da Dao 30: National Memories on Shennan Avenue*大道30 深南大道上的國家記憶. Shenzhen: Shenzhen Press Group Publishing House深圳:深圳報業集團出版社. (pp. 29).

[12] Tao, Y. T., & Lu, Z. G. 陶一桃 魯志國 (2008). *Historical Discourse on China's Economic Special Zones*中國經濟特區史論. Beijing: Social Sciences Literature Press北京:社會科學文獻出版社. (pp. 40–42).

**Fig. 4.6** Excavation of Luohu Hill in progress (1980) (來自羅湖檔案館. From Luohu Archive, Shenzhen)

**Fig. 4.7** Filling of the large pond in Luohu Village (1980) (He, H. Y. 何煌友(2019, March 23). *LuoHu Village Area (from south to north)* 羅湖村區域(由南往北), 1980. Retrieved from http://www.szlh.gov.cn/ssznzt/jy/lszpdb/content/post_5027271.html)

area of the SSEZ totaled 60.1 square kilometers, with the industrial land area covering 7.3 square kilometers, accounting for 11.97%.[13]

Another factor is the development of residential neighborhoods. According to a study on population changes in Shenzhen City, the total permanent population grew from 332,900 to 881,500 between 1980 and 1985 and then sharply climbed to 1,677,800 by 1990.[14] The influx of individuals migrating to Shenzhen City has

[13] Shenzhen Real Estate Yearbook Editorial Committee《深圳房地產年鑒》編輯委員會(1991). *Shenzhen Real Estate Yearbook 1991*深圳房地產年鑒1991. Shenzhen: Haitian Publishing House 深圳:海天出版社. (pp. 16).

[14] Wang, S.W.王世巍(2014). *Population Change in Shenzhen,1979–2010*深圳市人口變遷研究1979–2010.深圳:海天出版社Shenzhen: Haitian Press. (pp. 33).

**Fig. 4.8** Contrastive plan of Shangbu-Luohu Area between 1980 and 1990 (Source: The left diagram is created based on US Geological Survey (USGS) historical imagery, Document No.: D3C1216-300662F047(Sep. 1980) and *Shenzhen Place Names Gazetteer*(深圳地名志). The right diagram is created based on the *Shenzhen Transportation and Sightseeing Map*(深圳交通遊覽圖) issued by Guangdong Provincial Map Publishing House in April 1990)

resulted in a significant increase in housing demand. The construction of housing areas at Mutoulong (木頭龍), Hongwei (紅圍), Tongxinling (通心嶺), and Nanyuan (南園) commenced in 1983. Subsequently, the construction of residential areas such as Yuanling Residential Area (圓嶺居住區), Bibo Residential Area (碧波居住區), Baishaling Residential Area (白砂嶺居住區), and Binhe Community (濱河社區) has started in the SSEZ of China.[15] By the end of 1989, the total land area designated for habitation measured 24.8 square kilometers, or 40.66% of the developed area within the SSEZ.[16]

The third reason is to enhance public service infrastructure and develop an urban center. The exponential increase in population necessitated the need for easily accessible public service amenities. Furthermore, the urban development plan of "attracting external and connecting internal investments" requires that the SSEZ possess a contemporary urban aesthetic to enhance its appeal to potential investors. In late 1982, the Shenzhen Municipal Government made the decision to construct "eight significant cultural facilities," which comprised a library, museum, newspaper office, TV station, gymnasium, grand theater, science and technology museum, and news and cultural center. Their positions were strategically placed near the government office district in order to establish a contemporary urban ambiance for

[15] Huang, L. M. (Eds.) 黃麗滿 編(1997). *Establishment and Development of the Special Economic Zone in Shenzhen(Shenzhen Volume)* 深圳市中國經濟特區的建立與發展(深圳卷). Beijing: Party History Press of the Communist Party of China北京:中共黨史出版社. (pp. 86).

[16] Shenzhen Real Estate Yearbook Editorial Committee《深圳房地產年鑒》編輯委員會(1991). *Shenzhen Real Estate Yearbook 1991*深圳房地產年鑒1991. Shenzhen: Haitian Publishing House 深圳:海天出版社. (pp. 16).

SSEZ.[17] The building of the four parks, including Lychee Park, People's Park, Children's Park, and Honghu Park, has been initiated in accordance with the same objective, as Table 4.2 shows.[18]

The Shangbu-Luohu Area has a long-standing landscape history that encompasses both the physical characteristics of the natural environment and the historical relationship shaped by the dynamic interplay between human activities and the natural surroundings. Prior to the Reform and Opening, the inhabitants of this region engaged in agricultural pursuits, specifically cultivating rice in this flatland, for several centuries. As a result, a social network emerged, consisting of small-scale farmers engaged in commodity production, represented by Shenzhen Fair, and a system where multiple clans coexisted. This transformation was facilitated by the national policy of Reform and Opening. The agricultural plain landscape swiftly transitioned into a modern urban landscape to accommodate the industrial relocation from Hong Kong and other foreign countries and regions in line with the global division of labor in the industry. Urban parks have emerged as the primary spaces that maintain their natural characteristics among the landscape modifications. This embraces not just apparent alterations in structure but also imperceptible alterations in possession. The changes in meaning surrounding nature through urban parks can only be observed by integrating the two.

## 4.2 Alternations in Ownership and Morphology

### *4.2.1 Land System of China*

Upon the establishment of the People's Republic of China in 1949, the feudal land ownership system was eliminated and replaced with a system that granted equalized private property rights in farmland. Nevertheless, the objective at the state level is to promote socialist industrialization, which means that "land reform is a prerequisite for the advancement of productive forces and national industrialization." Land

**Table 4.2** Basic information for the four parks (Shenzhen Municipal Chronicles Compilation Committee深圳市地方誌編纂委員會(2009). *Chronicles of Shenzhen: Volume on Infrastructure Construction*深圳市志·基礎建設卷. Beijing: Fangzhi Publishing House北京: 方志出版社. (pp. 631–639))

| Park | Area (ha.) | Preparation time | Open time |
|---|---|---|---|
| Lychee Park | 18.60 | 1982–09 | 1985–10 |
| People's Park | 12.95 | 1983–06 | 1986–10 |
| Children's Park | 5.95 | 1983–06 | 1987–06 |
| Honghu Park | 59.15 | 1984–09 | 1987–02 |

[17] According to the interview with Mr. Guo Binghao, 2023-02-16.

[18] According to the interview with Mr. Ye Guidong, 2023-12-30.

reform in mainland China was essentially finished by the end of 1952. However, the implementation of an equalized private land property rights system may potentially result in the gradual consolidation of land rights into enormous real estate holdings and a widening wealth gap between the rich and the poor. Furthermore, the majority of the income generated from the land has been acquired by farmers, making it challenging for the Central Government to collect the equivalent revenue to support socialist industrialization.[19] Hence, the Chinese government during that period held the belief that it was imperative to mobilize farmers in order to secure a greater supply of food and industrial raw materials for the state, as opposed to relying on the system of private land ownership. From 1952 to 1956, mainland China implemented Mutual Assistance and Cooperation in Agricultural Production, which was based on private ownership of land. This was followed by advanced corporatization based on collective land ownership from 1956 to 1958. Finally, from 1958 to 1978, the rural land in mainland China adopted the Collective Ownership of Land, which advocated for "three-level ownership, accounting on a team basis (三級所有，隊為基礎)" (1958–1978),[20] which means that rural land is owned collectively by communes (公社), production brigades (生產大隊), and production teams (生產隊) at three levels, with production teams serving as the fundamental accounting unit.

The establishment of rural collective land ownership in 1956 facilitated the implementation of state ownership of urban land. Urban governments nationalized private housing and private land through mechanisms such as state leasing and public–private partnerships. The urban government grants nationalized urban land to land users without any cost for an indefinite duration. Additionally, it specifies that land use rights cannot be exchanged among land users, effectively establishing state ownership of urban land. The revision of the *Constitution of the People's Republic of China* (中華人民共和國憲法) occurred during the Fifth Session of the National People's Congress (全國人民代表大會第五次會議) in 1982. This revision established the legal ownership of urban land by the state and clarified that rural land is collectively owned by farmers. The Land Administration Law of the People's Republic of China, enacted in 1986, reiterated the principle that urban land is collectively held by the state, representing the entire population. "With the exception of state-owned land as mandated by legislation, land in rural areas and urban suburbs is owned collectively." "With the exception of land owned by the state as mandated by law, land in rural areas and urban suburbs is owned collectively." "The ownership of homestead land, private land, and private hills is shared among a group of individuals".[21] The primary impetus behind the establishment of the Dual System of Urban–Rural Land (城鄉土地二元制), which allowed for the coexistence of

---

[19] Wu, L., & Zhang, Q. 武力,張強(2015). *Four Transformations of the Rural Social Governance Structure*鄉村社會治理結構的四次變革. National Governance國家治理, No. 38(14), 14–25.

[20] Gao, L. L.郜亮亮(2023). Reform Practice, Logic of Change, and Future Evolution Direction of China's Rural Land Property Rights System中國農地產權制度的改革實踐、變遷邏輯及未來演進方向. *China Review of Political Economy*政治經濟學評論, 14(01), 48–76.

[21] Liu, S. Y. 劉守英(2018). *Land System and China's Development*土地制度與中國發展. Beijing: Renmin University of China Press北京:中國人民大學出版社. (pp. 155–157).

state ownership and collective ownership, was the country's desire to advance socialist industrialization in the face of a significantly low overall productivity level. This was done to navigate the international climate during the Cold War.[22]

Consequently, while the Chinese government's land systems in rural and urban areas are both characterized by communist public ownership, there are significant distinctions in the valuation, utilization, and administration of urban and rural land. The construction of the SSEZ in the early 1980s necessitated a substantial allocation of urban land that could be effectively managed and utilized by the government. However, during that period, the majority of the land within the SSEZ was predominantly under collective ownership by rural villages. The acquisition of construction land for urban development in the SSEZ was primarily hindered by the requirement to convert property rights from rural collective land. This obstacle existed for various types of construction, including urban roads, residential areas, public facilities, and urban parks.

### 4.2.2 The Land Source of the Four Parks in Shangbu-Luohu Area

Projects such as Shennan Avenue (深南大道), Renmin South Road (人民南路), Aiguo Road (愛國路), Beihuan Road (北環路), and Shenzhen University, Shenzhen Gymnasium, Shenzhen Library, and Grand Theatre were all accomplished by acquiring rural collective land, and the construction of the four parks is no exception. In February 1982, the Shenzhen Municipal Government constituted the Municipal Basic Engineering Working Group (基礎工程工作組) (hereinafter referred to as the "Basic Working Group (基礎工作組)" to oversee the first development of infrastructure and acquisition of land in the SSEZ. In May of the same year, the Basic Working Group formed a preparatory group for the four main parks: Lychee Park, People's Park, Children's Park, and Honghu Park (洪湖公園). Furthermore, the Municipal Landscape Corporation (市園林總公司) and the Municipal Planning and Design Administration (市規劃設計管理局) were also involved, and the three agencies collaborated to successfully accomplish the necessary tasks. Their respective duties and obligations are as outlined below in Table 4.3.

The issue of land is the problem all three departments are facing and needs to be tackled, so its importance is evident. The government's urban development requires the possession of relevant land use rights. Therefore, the first stage in preparing for the construction of urban parks is to acquire the necessary land. Based on the pertinent historical records, the origins of the land obtained throughout the development of the park group are in Table 4.4.

---

[22] Xia, Z. Z. 夏柱智(2022). *Between the City and the Village Rural Revitalization in the Process of Urbanization*亦城亦鄉 城市化進程中的鄉村突圍. Guilin: Guangxi Normal University Press桂林:廣西師範大學出版社. (pp. 5–7).

The table indicates that the majority of land comes from collectively owned rural land, with the exception of a tiny portion controlled by government-operated institutions. The term "rural collectives" mostly pertains to the four production brigades inside the Fucheng Commune. The table indicates that the majority of land comes from collectively owned rural land, with the exception of a tiny portion controlled by government-operated institutions. The term "rural collectives" mostly pertains to the four production brigades inside the Fucheng Commune, namely Caiwuwei Brigade, Hubei Brigade, Cainong Brigade, Shuibei/Tianbei Brigade (水貝/田貝大隊). The production and revenue conditions of the four brigades, which were the initial sources of land for urban park building, were not promising. Let's consider the Hubei Brigade and the Cainong Brigade as Table 4.5.

Although these lands, initially developed as urban parks, have not yet yielded clear economic advantages, they have emerged as precious natural spaces within this burgeoning city, assuming an unusual function in the urban ecosystem. Prior to comprehending these novel responsibilities, it is crucial to ascertain the process by which the land transitioned into state-owned property and was designated as an urban park.

### 4.2.3 *Nationalization of Collective Land*

To begin, it is important to comprehend the two primary approaches employed by the Shenzhen Municipal Government in the 1980s for compensating land acquisition: monetary compensation and the "land retention" system. In 1979, Guangdong Province elevated Bao'an County to the status of Shenzhen city while concurrently establishing the Municipal Revolutionary Committee (市革命委員會) and the Municipal Capital Construction Committee (市基本建設委員會). In order to acquire property for urban development, the Municipal Revolutionary Committee published two notices: the *Notice on Interim Provisions on Urban Construction Management* (關於城鎮建設管理暫行規定的通告) and the *Notice on Further*

**Table 4.3** Park construction relevant departments division of responsibilities (Compiled based on oral accounts from multiple eyewitnesses involved in park construction and relevant historical materials)

| Department | Responsibilities |
|---|---|
| Municipal Planning and Design Administration | Land allocation, park planning and design plan approval, etc. |
| Municipal Landscape Corporation | Provide technical support, garden appearance management, implement red line scope, handle land acquisition procedures, and organize the preparation of planning and design plans, etc. |
| Municipal Basic Engineering Working Group | Park land acquisition and compensation, fund settlement, project construction management, etc. |

**Table 4.4** Land sources composition for the four Parks (Compiled based on oral accounts from multiple eyewitnesses involved in park construction and relevant historical materials)

| Park | Preparation time | Total area (hectare) | Source of the land | Acre (hectare) | Percentage |
|---|---|---|---|---|---|
| Lychee Park | 1982–10 | 30.18 | Caiwuwei Brigade (蔡屋圍大隊) | 20.68 | 68.54% |
| | | | Acquisited land by BWG | 8.93 | 29.60% |
| | | | Subgrade of Jiefang Rd. | 0.56 | 1.86% |
| People's Park | 1983–06 | 13.86 | Hubei Brigade (湖貝大隊) | 4.34 | 31.28% |
| | | | Cainong Brigade (菜農大隊) | 6.67 | 48.08% |
| | | | Shenzhen waste company (深圳市利廢公司) | 0.64 | 4.58% |
| | | | City three defense office (深圳市三防辦) | 1.67 | 12.02% |
| Children's Park | 1983–06 | 5.95 | East Garden Nursery (東園花木場) | 5.95 | 100% |
| Honghu Park | 1984–09 | 70.44 | Shuibei Brigade | 65.51 | 93% |
| | | | Shenzhen department store, textile company (深圳百貨公司、紡織品公司) | 3.52 | 5% |
| | | | Shenzhen leather products factory (深圳市皮革製品廠) | 1.40 | 2% |

**Table 4.5** Agricultural production information of Hubei and Cainong Brigade (1976 &1977) (*Compilation of Basic Statistical Data on Shenzhen Town, 1977* (1977年深圳鎮基本情況統計資料彙編). [Unpublished manuscript]. Catalog Number A1.3; Archive Number C1; Volume Number 8; Item Number 3 (檔案全宗號C1;目錄號A1.3;案卷號8;件號3). Preserved at the Shenzhen Luohu Archives. *Compilation of Basic Statistical Data on Shenzhen Town,1976* (1976年深圳鎮基本情況統計資料彙編). [Unpublished manuscript]. Catalog Number A1.3; Archive Number C1; Volume Number 10; Item Number 3(檔案全宗號C1; 目錄號A1.3; 案卷號10; 件號3). Preserved at the Shenzhen Luohu Archives)

| Year | Brigade | Total population | Total agricultural output (dan) | | | | | | Per capita annual income per person (Yuan) |
|---|---|---|---|---|---|---|---|---|---|
| | | | Rice | Vegetable | Aqua-culture | Peanut | Sugar Can | Euryale Ferox | |
| 1976 | Hubei brigade | 791 | 5870 | 4060 | 43 | 110 | 140 | 71 | 131 |
| | Cainong brigade | 476 | 378 | 17,343 | – | 9 | 222 | – | 197 |
| 1977 | Hubei brigade | 787 | 6056 | 4000 | 220 | 57 | 200 | – | 166 |
| | Cainong brigade | 493 | 346 | 23,987 | – | 9.6 | 430 | – | 203 |

*Improving Property Acquisition for National Construction* (關於進一步做好國家建設徵用土地的通知). The former forbids the unauthorized occupation of land, while the latter establishes the criteria for compensating land acquisition, as

Table 4.6 shows. In addition, the *Shenzhen Special Economic Zone Land Management Implementation Measures* (深圳經濟特區土地管理實施辦法) enforced in 1983 aimed to procure farmers' communal land and augment urban building land.

Land serves as a fundamental resource for agricultural producers. As per the *Regulations on Land Acquisition for National Construction*(國家建設徵用土地辦法) enacted in January 1958 and the *Regulations on Land Acquisition for National Construction*(國家建設徵用土地條例) issued in May 1982, the government bears the responsibility of relocating farmers affected by land acquisition. In the early 1980s, there were few institutions in the SSEZ hat could accommodate rural labor. Consequently, the Shenzhen Municipal Government used the strategy of "land retention" following property acquisition. In September 1982, the *Shenzhen Municipal Government issued the Interim Regulations on the Land Used for Rural Commune Members to Build Houses in the SSEZ* (深圳市经济特区农村社员建房用地的暂行规定), which established specific guidelines for "land retention." These guidelines stipulated that each household is entitled to a homestead area of 150 square meters with a building base area of 80 square meters. The construction should not exceed three floors, and an additional 15 square meters of industrial land is allocated per person. In 1986, the government implemented regulations to

**Table 4.6** Compensation standards for land requisition from farmers (Urban Planning and Natural Resources Bureau of Shenzhen深圳市規劃和自然資源局(Eds.). *Memories of Land System Reform in Shenzhen*深圳土地制度改革記憶. Unpublished. (pp. 73))

| Compensation type | | Compensation price | Compensation type | | Compensation price |
|---|---|---|---|---|---|
| Corps | Trees | 4–9<br>Yuan per plant | Building | Reinforced concrete structure | 100 yuan per m$^2$ |
| | Bamboo | 10<br>Yuan per plant | | Brick and wood structure | 80 yuan per m$^2$ |
| | Fruit Trees | 10–100<br>Yuan per plant | | Grave relocation | 20 yuan per plant |
| | | | | Pile tower | 5 yuan per plant |
| | Tea Park | Compensation calculation based on total output value of 3 years and current national price | Land | Fishpond (1983)<br>Fishpond (1987) | 4500 yuan per mu<br>5000 yuan per mu |
| | Pineapple and Banana parks | Compensation calculation based on total output value of 1–3 years and current national price | | Water field (1983)<br>Water field (1987) | 3500 yuan per mu<br>4000 yuan per mu |
| | | | | Dry field (1983)<br>Water field (1987) | 1500 yuan per mu<br>2000 yuan per mu |

oversee the allocation of land previously used by farmers for residential construction. This was done through the issuance of the *Notice on Further Strengthening Rural Planning in the SSEZ* (關於進一步加強深圳特區內農村規劃工作的通知). In addition to addressing the housing problem, the government also allocates land specifically for farmers to address concerns related to employment, business, and production. The land acquisition compensation system, which has undergone testing since the early 1980s, was officially stipulated in the Several Provisions on Land Acquisition in SSEZ (關於深圳經濟特區征地工作的若干規定) established in January 1989:

1. *The Municipal Land and Resources Bureau* (市國土局) *will prioritize allocating a piece of land, free of charge, to the land-expropriated village for the purpose of constructing commercial and service buildings to facilitate production. Factors such as the size of the village, the amount of land to be requisitioned, the planned investment by the land-acquired unit, and the urban planning requirements will determine the allocation.*
2. *If the current positioning of individual villages is unsuitable for the development of commercial and service buildings, a limited portion of land can be designated for the establishment of factories adjacent to existing industrial facilities or within the remaining vacant areas of the village, without compromising urban planning.*

Table 4.7 demonstrates that the Central Government, Guangdong Province, and the SSEZ have established institutional frameworks to address the issue of landless farmers, with varying degrees of regulatory depth. The legal foundation for this arrangement was established as early as 1953, with the creation of the People's Republic of China.

In the 1980s, the Caiwuwei Brigade had the largest amount of land acquisition during the development of the four parks. Hence, we will use the Caiwuwei Brigade as a case study to analyze the current state of rural collectives following the loss of their land.

Caiwuwei is an ancient community with a rich history spanning over six centuries.[23] Throughout its various stages of development, agriculture has consistently remained their primary mode of production, as depicted in Fig. 4.9. Since the establishment of the SSEZ in 1980, Caiwuwei has continuously provided over 5000 mu (equivalent to 3335 ha.) of agricultural land to the government and received a total of 15 million yuan in compensation for the land acquisition. The locals did not have complete control over the allocation of land acquisition monies. During that period, the Shenzhen Municipal Party Committee mandated that the monies allocated for the acquisition of land be utilized for the operation of collective enterprises.[24]

[23] Caiwuwei Genealogy Compilation Committee蔡屋圍族譜編纂委員會(2005). *Caiwuwei Genealogy of Shenzhen*深圳蔡屋圍族譜. Internal document. (pp. 41).

[24] Shenzhen Local Chronicles Compilation Committee 深圳市地方誌編纂委員會 (2014). Shenzhen Chronicles: Volume on Reform and opening-up深圳市志·改革開放卷. Beijing: Fangzhi Publishing House北京:方志出版社. (pp. 58).

**Table 4.7** Regulations on land requisition from farmers at different administrative levels

| Administrative level | Laws, regulations and provisions | Implementation time |
|---|---|---|
| Central Government | *Measures for land Acquisition for National Construction* 國家建設徵用土地辦法 | 1953–12–05 |
| | *Measures for land Acquisition for National Construction (revised)*國家建設徵用土地辦法(修訂) | 1958–01–06 |
| | *Regulations on land acquisition for national Construction*國家建設徵用土地條例 | 1982–05–14 |
| | *Land management law of the people's republic of China* 中華人民共和國土地管理法 | 1986–06–25 |
| | *Land management law of the people's republic of China (revised)*中華人民共和國土地管理法(修訂) | 1998–08–29 |
| Guangdong Province | *Guangdong Province land acquisition management regulations* 廣東省征地管理規定 | 1993–07–01 |
| | *Guangdong Province measures for the Administration of Compensation Fees for Acquisition of Rural Collectively owned land* 廣東省徵用農村集體所有土地各項補償費管理辦法 | 1995–01–01 |
| SSEZ | *Interim regulations on land use for housing construction by rural commune members in SSEZ* 深圳市經濟特區農村社員建房用地的暫行規定 | 1982–09–17 |
| | *Notice on further strengthening rural planning work in SSEZ*關於進一步加強深圳特區內農村規劃工作的通知 | 1986 |
| | *Several provisions on land acquisition in SSEZ* 關於深圳經濟特區征地工作的若干規定 | 1989 |

Consequently, the land acquisition amount was not disbursed to the 398 families comprising over 1000 individuals. Instead, it was utilized as the first capital for the advancement of the common economy and the collective joint-stock economy.

Capitalizing on the policies in the SSEZ during that period, the residents of Caiwuwei facilitated the establishment of over 30 enterprises by introducing Hong Kong's "Three-Processing and One Compensation" businesses. Following the acquisition of experience, villagers initiated a series of sole proprietorships, joint ventures, and cooperative enterprises, which effectively facilitated the swift expansion of the collective economy. In 1992, the combined total assets and net assets of Caiwuwei reached 220 million yuan and 202 million yuan, respectively, representing a growth of 95 times and 87 times respectively compared to the figures in 1978. The overall revenue and net profit had significant growth of 86-fold and 68-fold, respectively. Additionally, per capita income surged to 8370 yuan, marking a remarkable 77-fold gain compared to the year 1978.[25]

[25] Feng, Q馮慶(2022, September 1). *Cai Hongliang: From washing feet in the fields to brainwashing in the city, Caiwuwei is a microcosm of rural urbanization in Shenzhen*蔡洪亮:從洗腳上田到洗腦進城，蔡屋圍是深圳農村城市化的縮影.[EB/OL]. Retrieved from https://mp.weixin.qq.com/s?__biz=MzAxMjA1NDk1MA==&mid=2650016078&idx=1&sn=89e72fb24a77d4571cc

**Fig. 4.9** Caiwuwei Village and its surrounding farmland (1929) (41 Caiwuwei Genealogy Compilation Committee蔡屋圍族譜編纂委員會(2005). *Caiwuwei Genealogy of Shenzhen*深圳蔡屋圍族譜. Internal document. (pp. 39))

Caiwuwei exemplifies the process of rural community change within the SSEZ. During that period, numerous rural collectives were experiencing a comparable procedure. During the early 1980s, amidst the enormous urban development of the SSEZ, numerous rural production teams received substantial funding for land acquisition. By utilizing these combined resources, farmers are incentivized to secure finances and make investments to establish an economic organization that promotes the growth of both agriculture and industry. Farmers who have lost their property are now embracing the concept of sustainable development as a means of seeking a new path forward. Collective joint-stock cooperative firms started to grow in rural parts of the SSEZ under this model. The city had formed a total of 1352 joint-stock cooperative firms by 1991. Furthermore, the municipal government actively endorses and fosters these endeavors, having convened a dedicated assembly to further the development of rural communal joint-stock cooperative firms with the aim of accumulating knowledge and coordinating efforts.

Rural collective land ownership plays a crucial part in this process. The primary impetus for the establishment of this system in the 1950s was to thwart the encroachment of property under individual ownership, which would have exacerbated the socioeconomic divide between the affluent and the impoverished. In the early 1980s, the presence of rural collective organizations in the SSEZ ensured that the initial capital of the collective economy remained intact and was not fragmented or lost. The amalgamation of these funds and the "land retention" method empower

ba81fde7026e6&chksm=83b71187b4c09891bacab946cb286519b15e959dbfdd3a55c4df65dcc3ef269e9cb21a5e71f5&scene=27

farmers who have experienced land loss to acquire finances and essential land resources for subsequent advancement. The establishment of collective joint-stock cooperative firms, which operate on the basis of collective land ownership, allows rural collectives to maintain a relatively stable development trajectory even in the absence of land ownership.

The primary challenge encountered in the development of the first set of urban parks is the Dual System of Urban-Rural Land, which has gradually evolved since the formation of the People's Republic of China. Under the regime established by the "Alliance of Workers and Peasants (工農聯盟)," the SSEZ has effectively addressed this issue by implementing institutional measures such as providing financial compensation and retaining land, thereby acquiring land resources for urban development. Simultaneously, the rural communal economy has also acquired resources and advantages that have been made available by urban growth. Inevitably, changes in land ownership led to changes in spatial form as a means of representation.

## 4.3 Nationalism-Oriented Nature: The Reshaping of Cultivated Land and the Representation of the Nation-State

### *4.3.1 Political and Economic Context of the Four Parks*

During the 1970s, there was a significant growth in the global economy, and the international allocation of work enabled the "Four Asian Tigers" to commence their economic advancement. Additionally, this situation provided fresh prospects for mainland China, which was undergoing a pivotal transition in the late 1970s. China possessed the capacity to assimilate into the global market and invigorate its domestic economy by leveraging the economic crises in capitalist nations.[26] Nevertheless, China's endeavor was impeded by a significant deficiency in foreign currency reserves. In August 1979, the State Council issued the *Provisions on Several Issues Concerning Vigorously Developing Foreign Trade to Increase Foreign Exchange Income* (關於大力發展對外貿易增加外匯收入若干問題的規定), which explicitly indicated that "Shenzhen and Zhuhai were the designated pilot cities" for the establishment of special export zones in China.[27] The phrase "special zone (特區)" made its inaugural appearance in China's official policy texts during the 1970s. The

[26] Harvey D. (2005). *A brief history of neoliberalism*. New York: Oxford University Press. (pp. 120).

[27] Wang, Z. C.王振川 (Eds.) (1979). *Yearbook of the New Period of Reform and opening-up in China*中國改革開放新時期年鑒. Beijing: China Democratic Legal System Press北京:中國民主法制出版社. (pp. 662–663).

primary objective behind the establishment of SSEZ was predominantly to "enhance the inflow of foreign exchange earnings (增加外匯收入)."

The development of SSEZ commenced is earnest, despite the absence of sufficient national resources that could be relied upon and allocated for the construction of new cities, due to the consistently unfavorable economic conditions in China during that period. On the contrary, the constructors of SSEZ need to "work on their own and fight their way out" (自己去搞，殺出一條血路).[28] The initial phase of urban development following the foundation of SSEZ focused primarily on achieving investment outcomes. Following the economic triumph of China's inaugural commercial housing endeavor, Donghu Liyuan (東湖麗苑) in SSEZ, policymakers were cognizant of the substantial value of land.[29] Within the "Sixth Five-Year Plan (六五計畫) (1981–1985)" of the SSEZ, there is an enhanced emphasis on "converting vacant land into valuable assets" through land development and urban construction.[30] Thus, according to decision-makers, if an urban park primarily benefits citizens but does not align with the initial aim of establishing the SSEZ, the administrative workflow will not acknowledge it.

During that period, urban development and environmental improvement were significant topics of research closely connected to SSEZ's educational approach—Singapore, a nation situated in East Asia. Since the 1970s, numerous Asian cities have actively engaged in self-promotion to enhance their worldwide competitiveness and appeal, with the aim of attracting multinational corporations to participate in local labor-intensive manufacturing. Singapore aspired to transform itself into a globally renowned oasis in Southeast Asia, distinguishing itself from other developing nations. Since 1965, Singapore, a city-state with limited resources, has thrived due to substantial investments from industrialized nations. Under the leadership of Lee Kuan Yew, the Singapore government adopted the "garden city" idea as a strategic approach to attract international investments. During this period, the Singapore government revived colonial legacies by restoring the Singapore Botanic Gardens. Additionally, new parks were established, including the 4.8-hectare Toa Payoh Town Park, the 5-hectare Ang Mo Kio Town Garden East, and the 248-hectare Jurong Park.[31]

In the summer of 1983, following the completion of the majority of urban parks in Singapore, the government of Shenzhen dispatched a significant delegation of

---

[28] Central Compilation and Translation Bureau of the Communist Party of China中共中央文獻研究室 (2004). *Deng Xiaoping Chronology (1975–1997)* 鄧小平年譜(1975–1997). Central Literature Publishing House北京:中央文獻出版社. (pp. 510).

[29] Shenzhen Innovation and Development Institute (2018). Luo Jinxing: Crab eater who won the "first auction" of land use rights in China. Special Zone Economy, 355(8), 10–11.

[30] Shenzhen Development and Reform Commission 深圳市發展和改革委員會(2017). *Compilation of Shenzhen's Five-Year Plans*深圳市歷次五年規劃彙編. Shenzhen: Haitian Publishing House深圳:海天出版社. (pp. 4).

[31] Zhang T.J., Chen B.Q張天潔,程秉鈴(2021). City Cultural Capital and Local Rebuilding: Analysis of Singapore Park Construction and Renewal Process城市文化資本與地方重塑:新加坡公園建設與更新歷程解析(1968–2020). *Landscape Architecture*風景園林. (06), 56–63.

officials to Singapore. The delegation had two primary objectives: firstly, to gain knowledge and skills in urban planning and the establishment of green spaces; and secondly, to comprehend the investment promotion approach utilized by the Jurong Town Corporation. Examining the historical context of that period reveals the inherent interconnectedness of these two aims. The educational tour in 1983 had a significant influence on the perspective of Shenzhen officials about the implementation of green spaces in the city.[32] The city authorities highly acknowledged Singapore's expertise and proposed their goal of achieving Hong Kong's operational efficiency inside Singapore's urban environment".[33] Singapore's expertise in greening has elevated the importance of promoting greening in Shenzhen to the same level as industrial production efficiency, making Singapore a benchmark for learning and emulation.

The quality of the urban spatial environment in the 1980s had an impact on the confidence and influence of external investors,[34] similar to how other infrastructure affected growing cities in Asia like Shenzhen and Singapore. Urban greening and the creation of the four parks, along with the eight major cultural facilities, serve a commensurate political function in this context. In this regard, the strategies of Luo Changren(羅昌仁,1923–), a deputy mayor responsible for urban construction, and Lee Kuan Yew are quite alike. Both governments prioritize the attraction of foreign entrepreneurs and matching their investment requirements as the main focus of urban construction. The "Sixth Five-Year Plan" (1981–1985) has specified that the primary means of financing the creation of the entire SSEZ is "the influx of foreign investment." According to Luo Changren, "urban construction projects… that primarily cater to foreign businessmen…should adhere to internationally recognized construction standards in order to meet their requirements".[35] Similarly, one of Lee Kuan Yew's strategies is to establish a foundation for business and tourism in the region by attracting businessmen and tourists from developed countries, of all his schemes, he held the belief that the greening plan is the most economically efficient.[36]

Upon the return of the delegation from Singapore, Shenzhen promptly adopted Singapore's relevant practices. This included designating 30 meters of Shenzhen's main roads as green belts and establishing an 800-meter green belt from the

---

[32] National Committee of the Chinese People's Political Consultative Conference, Committee on Culture, History, and Learning全國政協文史和學習委員會 (2009). *Selected Compilation of Historical and Literary Materials, Issue 153*文史資料選輯. 第153輯. Beijing: China Social Sciences Press北京:中國文史出版社. (pp. 27–46).

[33] Urban Master Plan of Shenzhen Municipality (1996–2010), the preface of the text (review draft), is kept in the Futian District Library in Shenzhen.

[34] According to the interview with Mr. Guo Binghao, 2023-02-16.

[35] Luo, C. R. 羅昌仁 (1987). Improving Urban Planning in Shenzhen把深圳市的規劃工作做得更好. In Hai Tian Publishing House海天出版社. *Comprehensive Review of the General Plan of Shenzhen Special Economic Zone* 深圳經濟特區總體規劃評論集. Shenzhen: Hai Tian Publishing House深圳:海天出版社. (pp. 31–35).

[36] National Archives of Singapore新加坡國家檔案館 編. (2015). *Lee Kuan Yew's Governance Strategies*李光耀執政方略. Beijing: People's Publishing House北京:人民出版社. (pp. 107–108).

Shanghai Hotel to Huangmugang. Additionally, Shenzhen accelerated the construction of the initial four major parks.[37] Aside from Singapore, Hong Kong, which is also one of the Four Asian Tigers, served as another model for Shenzhen's urban and park construction in the early 1980s.[38]

### *4.3.2 The National Significance of SSEZ and Its Construction Performance*

The external experience that has been obtained will ultimately be utilized to practically develop this new domestic city. In contrast to the extensive chronicles of Chinese civilization, the SSEZ can be considered a relatively recent urban development. However, its urban and park construction is leading to the fervent pursuit and imagination of modernization among the Chinese people. Their fervent pursuit and resolute commitment to modernization is a form of resistance against Western modernity and may be traced back to the mid-nineteenth century when they were confronted with formidable ships and guns. Both the SSEZ and several developing countries and territories in East Asia have experienced a history of colonization and dominance. The catalyst for their economic ascent in the 1970s was a "nationalist project."[39] State power encompasses various interconnections, such as economics, culture, and space. This form of state power predominantly influenced the formation of the SSEZ. The urban construction was characterized by a strong desire to showcase their distinctive modernity, leading to the frequent utilization of abstract and formal ways of representation.

Gu Mu (穀牧, 1914–2009) played a proactive role in overseeing and leading urban planning and construction issues across the country during the initial stages of the establishment of the PRC. This distinguished political leader, who has devoted his life to the pursuit of national independence since his youth, has also actively participated in efforts related to the development of SSEZ.[40] His viewpoint is as follows:

[37] Duan, Y. B.段亞兵(2018). *Pioneering Chronicles of Shenzhen*深圳拓荒紀實. Beijing: People's Publishing House北京:人民出版社. (pp. 375); Shenzhen Special Economic Zone Research Association深圳經濟特區研究會(2010). *A Special History on the Reform and opening-up of Shenzhen Special Economic Zone*深圳經濟特區改革開放專題史. Shenzhen: Haitian Publishing House深圳:海天出版社. (pp. 375).

[38] According to the interview with Mr. Guo Binghao, 2023-02-16.

[39] Castells, M. (2009). *The rise of the network society (2nd ed.)*. Newark: John Wiley & Sons, Incorporated. (pp. 198).

[40] Gu, M. 穀牧(2009). *Memoirs of Gu Mu*穀牧回憶錄. Beijing: Central Literature Publishing House北京:中央文獻出版社. (pp. 359–370).

> The primary duty of the mayor is to execute urban planning... It is advisable for newly constructed buildings to prioritize the preservation of traditional Chinese traits in order to allow foreigners to easily recognize that they are in China.[41]

The above-mentioned perspectives were reproduced on the inaugural page of the "Shenzhen Special Zone Daily" on September 8, 1981. Zhou Ding,[42] like Gu Mu, also participated Anti-Japanese War.[43] In May 1986, he expressed his support for Gu Mu's viewpoints:

> Regarding the design and construction of urban public buildings, particularly those that serve as prominent landmarks in cities, we proceed with care and strive to develop a plan that successfully embodies the desired urban planning vision while also highlighting contemporary and national characteristics...[44]

The perspectives of Gu Mu and Zhou Ding should not be seen as subjective aesthetic preferences, but rather as manifestations of nation-state power, where architecture and urban environment serve as symbolic spaces. Given the context, it is evident why a significant number of initial prominent structures in SSEZ exhibit distinct "national characteristics," as depicted in Fig. 4.10.

The national policy of Reform and Opening Up can be interpreted as the linkage between the nation-state and the global level. Regarding this matter, the urban space of SSEZ, specifically the Shangbu-Luohu Area near the Luohu Port (罗湖口岸), requires a distinct and patriotic character that can immediately convey to outsiders that they are in China, as emphasized by Gu Mu. Despite its abstract, the formal processing approach is the most straightforward way to obtain recognition at all levels. Even in 1999, the invitation letter for the international consultation on the comprehensive planning of Luohu Port explicitly mentioned the spatial aim requirement of "City Gate-National Gate" to address its "political nature".[45] The correlation between the construction projects with national features and the consolidation of nation-state power requires additional clarification, regardless of whether it pertains to buildings or parks.

---

[41] Shenzhen Special Zone Daily深圳特區報. (1981, September 8). *Vice Premier Gu Mu and Wan Li Discuss Strengthening Urban Construction*萬裡、穀牧副總理談加強城市建設 [News article]. Shenzhen Special Zone Daily深圳特區報, 01.

[42] Zhou Ding (1927–2019), formerly served as the Deputy Mayor of Shenzhen and the First Deputy Director of the Shenzhen Urban Planning Committee.

[43] Shen, Z.沈祖(2019, August 10). *Comrade Zhou Ding's Life*周鼎同志生平. Shenzhen Special Zone Daily深圳特區報, A04.

[44] Zhou, D. 周鼎(1987). A Retrospective on Urban Planning and Construction in Shenzhen深圳城市規劃和建設的回顧. In Hai Tian Publishing House 海天出版社. *Overall Review of the General Plan of Shenzhen Special Economic Zone*深圳經濟特區總體規劃評論集. Shenzhen: Hai Tian Publishing House深圳:海天出版社. (pp. 19).

[45] Tan, Z., & Xiao, Y. B.譚崢,肖映博(2020). *Bridge and Door: The Evolutionary Public Space of the Luohu Port in the Perspective of "Borderscape"* 橋及閘——"關境"視角下的羅湖口岸公共空間演化史. New Architecture新建築, 2020(04), 11–17.

**Fig. 4.10** Public and tourist buildings constructed in the 1980s (Source from: *Annual Shenzhen Special Economic Zone Yearbook*(深圳經濟特區年鑒) of the 1980s). (**a**) Luohu Port Inspection Building 羅湖口岸邊檢大樓(1985). (**b**) Shenzhen Library 深圳圖書館 (1986). (**c**) Xili Lake Resort西麗湖度假村 (1979). (**d**) Xiangmi Lake Resort香蜜湖度假村 (1981)

## *4.3.3 National Forms in Architecture and Parks*

The discourse surrounding the concept of "national form" in the realm of built environment in China originated at the start of the previous century. Between 1927 and 1937, the Nanjing Kuomintang Government implemented the "Capital Plan" and initiated the discussion of "China's inherent forms (中國固有之形式)".[46] As a nationalist regime, the Kuomintang (KMT) during that period commissioned architects such as Henry Murphy, Lv Yanzhi (呂彥直, 1894–1929), Fan Wenzhao (范文照, 1893–1979), and Yang Tingbao (楊廷寶, 1901–1982) to create numerous public and monumental buildings. This led to the creation of many buildings that had distinct "national form" qualities.[47]

Shortly after the establishment of PRC, the architectural concept of "national form" received official recognition from the Communist Party of China, which shared a common commitment to nationalism. In the 1950s, there was widespread dissemination of the concept of combining "national form and socialist content"

[46] Capital Design and Technology Commissioner's Office國都設計技術專員辦事處(2006). *Capital Planning*首都計畫. Nanjing: Nanjing Publishing House南京:南京出版社. (pp. 60–63).

[47] Chen, C., & Wang, K. 陳晨,王柯(2019). *The Path of Architectural Creation about the National Form by the Architects of The Republic of China in Nanjing*民國建築師在南京的民族形式建築創作軌跡(1927—1937). Architect建築師, (No. 200), 107–114.

from Soviet architectural theory with traditional Chinese architectural forms throughout the architectural academia. Mr. Liang Sicheng (梁思成, 1901–1972), who was employed by the Beijing Urban Planning Commission at that time, played a prominent role in both the theoretical and practical aspects of this endeavor.[48] The classical architectural education he received at the University of Pennsylvania under the influence of the École des Beaux-Arts in Paris was primarily centered on historical styles. This influence, exemplified by him as a representative figure of the first generation of Chinese architectural historians, has continued to exert a profound and enduring impact on the academic framework of Chinese architecture.[49] Nevertheless, upon examining the roots of the discourse on architectural history pertaining to "style," it becomes evident that this type of architectural history writing, centered around "style," typically takes the form of national history, serving to consolidate the identity of the Western European nation-states and territorial authority that emerged after the eighteenth century.[50] Modern Chinese thinkers, like Liang Qichao (梁啟超) and Luo Zhenyu (羅振玉), have been affected by the Western European nation-state structure from the late nineteenth century. Hence, they assert that in order to oppose external forces, it is imperative to construct a distinct national narrative and thereafter establish sovereign nation-state.[51] In the context of an aspiration for the "revival of the nation (復興國家民族),[52]" Mr. Liang Sicheng embarked on the study of Chinese architectural history. In short, Chinese architectural historians, led by Mr. Liang Sicheng, aim to utilize discourse frameworks derived from Western Europe to establish a historically significant account of Chinese architecture with a national significance. Within the historical backdrop of the 1950s, this framework of discourse evolved within the process of learning from the Soviet Union, giving rise to the design principle of "national form with socialist content (民族形式、社会主义内容)."[53]

Although initially derived from architecture, with a primary emphasis on the materials, structure, and form of buildings, this principle extended its influence on the pertinent theories of urban park design throughout that period. Figure 4.11

---

[48] Xue, C. (2006). Building a revolution: Chinese architecture since 1980. Hong Kong: Hong Kong University Press.

[49] Zhao, C. 趙辰(2007). Nationalism and Classicism: The Contradictions and Tragedies in Liang Sicheng's Architectural Theoretical System民族主義與古典主義——梁思成建築理論體系的矛盾性與悲劇性. In *Misunderstanding of the "Facade": Architecture, Theory, History*"立面"的誤會:建築·理論·歷史. Beijing: Joint Publishing北京:生活·讀書·新知三聯書店. (pp. 39–40).

[50] Hisa C. J. 夏鑄九(2016). *On Heterotopias: Selected Essays*異質地方之營造:理論與歷史(I). Taipei: Tonsan Publications臺北: 唐山出版社. (pp. 185).

[51] Levenson, J. (1970). *Liang Chi-chao and the mind of modern China (2nd rev. ed.)*. Berkeley: University of California Press.

[52] Liang, S. C., & Lin, Z. 梁思成 著 林洙(Eds.) (2007). *Grand Simplicity and Beauty: Liang Sicheng's Most Beautiful Architectural Writings*大拙至美: 梁思成最美的文字建築. Beijing: China Youth Publishing House北京:中國青年出版社. (pp. 30–42).

[53] Wang, Y. S. 王元舜(2014). *Exploration on the Origin and Evolution of Soviet Architectural Theory "National in Form, Socialist in Content"*蘇聯"民族形式、社會主義內容"建築理論來源與演變之初探. New Architecture新建築, (04), 91–93.

illustrates the Huagang Guangyu Park (花港觀魚) in Hangzhou, which was constructed in the 1950s. This park served as a pioneering example of modern urban park design in China, showcasing the use of the design principle mentioned above.[54]

The designer, Mr. Sun Xiaoxiang (孫筱翔), aims to embody the people's aesthetic aspirations for the gardens of their homeland when outlining the primary functions of this park. In the handling of park topography, the Ming dynasty garden designer Ji Cheng's principle of "ingeniously borrowing from existing conditions" is employed as the primary guiding principle. The architectural layout and space zoning were influenced by the concept of "roads winding, spaces changing"

**Fig. 4.11** Master plan of Huagang Guanyu Park (Sun, X., & Hu, X. 孫筱祥,胡緒渭(1959). Planning and Design of Huagangguanyu Park in Hangzhou杭州花港觀魚公園規劃設計. *Architectural Journal*建築學報, (05), 19–24)

[54]Chen, X., & Zhao, J. J. 陳曦,趙紀軍(2020). A Comprehensive Study of Huagangguanyu Park: A Discussion on the Research of Modern Chinese Garden History花港觀魚公園研究面面觀——兼論中國現代園林史研究. In *Proceedings of the 2020 Annual Conference of the Chinese Society of Landscape Architecture*中國風景園林學會2020年會論文集(下冊) Vol. 2. (pp. 95–101).

prevalent in Chinese classical gardens, creating a layered and ever-changing space. All the buildings in the park incorporate national forms. Mr. Sun asserts that the Chinese classical garden is an exceptional paradigm for planting design. He envisions a perpetual spring garden, teeming with delightful avian melodies and exquisite blossoms. The park would showcase a wide array of vibrant and unusual flowers, as well as unique varieties of grasses.[55]

In Stan Fung's perspective,[56] the discourse surrounding "Chinese classical garden" grounded in the context of the nation-state abstracts both "Chinese" and "garden." One aspect involves the simplification of the diverse architectural and garden culture, which encompasses various ethnic groups and rich regional characteristics, into the architecture and garden culture represented by the Han cultural circle. However, various forms of gardens that existed in history have been simplified to a few gardens, with the existing Jiangnan gardens serving as the classical prototypes.

Likewise, urban parks built in the initial stages of SSEZ also serve as symbolic spaces. Architectural elements of park structures, such as pavilions, corridors, further demonstrate the significance of the nation-state. However, this is fundamentally no different from the afore-mentioned public buildings. What often overlooked is the national significance created by natural features such as landforms, water systems, and plants and how these were incorporated into urban process at that time.

### 4.3.4 The Nation-State Significance in the Structure of Terrain-Water System

The cultivated land in the Shangbu-Luohu Area, which used to represent the agricultural economy and way of life, underwent a symbolic transformation during the urbanization process of the SSEZ in the 1980s. Consequently, the farmland of the Caiwuwei Brigade and the Shuibei Brigade, along with the vegetable plots of the Cainong Brigade, underwent morphological changes and now resemble a terrain-water structure inspired by the design principles of traditional Chinese gardens, as Fig. 4.12.

> The arrangement and division of space are focused on bodies of water... Water is a defining characteristic of the southern region... The landscape is designed in accordance with the specific characteristics of the local environment... All of them are completed using established methodologies... During that period, the islands and embankments were excavated extensively from the fields and subsequently accumulated...[57]

[55] Sun, X., & Hu, X. 孫筱祥,胡緒渭(1959). Planning and Design of Huagangguanyu Park in Hangzhou杭州花港觀魚公園規劃設計. *Architectural Journal*建築學報, (05), 19–24.

[56] Fung, S. 馮仕達(2017). Longing and Belonging in Chinese Garden History中國園林史的期待與指歸. *Heritage Architecture*建築遺産, (006), 39–47.

[57] According to the interview with Mr. Shen Baojiu, 2023-02-12.

> …as I moved over the vast grassy field, I bent down to carefully study the landscape, identifying the raised sections and the lower ones. The soil was replenished following the precipitation and sedimentation events. The terrain remains exceptionally gorgeous…[58]
>
> …the elevated areas should be raised somewhat higher to construct islands, while the depressed areas should be dug slightly deepen to establish the water surface….[59]

The government acquires land to incorporate farmland into the urban process, which is controlled by state power. Additionally, planning and design utilize specific aesthetics to express the country's transformation of the terrain. The "national form" aesthetic extends beyond artificial architecture to encompass the natural terrain-water system. In the historical context, the power structure that produced physical spatial form and the spatial form itself, namely the space of representation and the representation of space, are highly unified here.

If we consider the planning and design of urban parks as the editing of topography, according to James Corner, this kind of "editing" is, in fact, a complex genealogy of plotting.[60] Within this genealogy, natural elements such as topography, water systems, and plants construct the urban process under the dominance of state power.

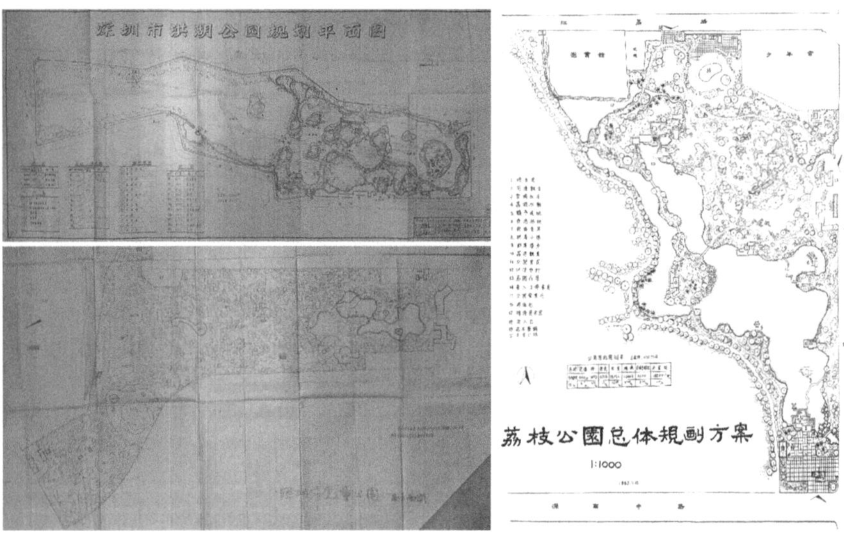


**Fig. 4.12** Master Plan of Honghu Park (1984), Children's Park (1985), and Lychee Park (1983) (The top-left image is the master plan of Honghu Park, provided by the Shenzhen Park Management Center. The bottom-left image is the master plan of the Children's Park, also provided by the Shenzhen Park Management Center. The right image is the master plan of Lychee Park, provided by the Shenzhen Lizhi Park Management Office)

---

[58] "*An Interview with Zhang Xinsi, Recipient of the Lifetime Achievement Award*張信思 終身成就獎獲得者專訪." Guangdong Landscape Architecture廣東園林, 5th Issue, 2017.

[59] According to the interview with Mr. Shen Baojiu, 2023-02-12.

[60] Corner, J. (2014). *The landscape imagination : collected essays of James Corner, 1990–2010*. Princeton Architectural Press. (pp. 101–104).

Those natural elements are all nature of cultural form, that is, nature that people "see" within specific social and cultural contexts. In the 1980s, the SSEZ in mainland China represented the intersection of globalization and the nation-state. This is why the SSEZ aimed to showcase its own modernity. The determination behind urban parks is intricate and multifaceted. Firstly, it involves examining the intricate and multifaceted determination found in urban parks. Firstly, it involves examining the architectural history of Western European nation-states in the eighteenth century and their dominant "style." Secondly, it encompasses the formalized nationalism and socialist design principles derived from the Soviet Union. Lastly, it encompasses the deep-rooted classical garden culture and natural aesthetic tradition in China. The inclusion of urban parks with "national characteristics" in the SSEZ is not solely a result of designers' individual choices or direct administrative mandates, but rather stems from a genealogical rationale.

> Not only Zhou Ding, but those of us who are involved in technical issues have consistently adhered to this approach…. After liberation, the education we received was basically based on these ideas…[61]

This kind of education can be found in Mr. Sun Xiaoxiang's book *Garden Art and Garden Design*(園林藝術及園林設計),[62] as mentioned above. In this renowned textbook, which enjoyed widespread usage from the 1960s to the 1980s, Mr. Sun posited that the garden's design should possess a discernible national aesthetic while also including socialist themes. "When designing a park, it is important to incorporate elements of traditional classical national heritage in order to create a park that reflects our own socialist values and national identity." When designing urban parks, it is important to consider the topography and transform the landforms in accordance with the specific conditions of the location and traditional Chinese landscaping techniques. The plant landscaping should aim to create an atmosphere reminiscent of the saying, "out of the shade of the willows came bright flowers and another village." We should take inspiration from the design of Chinese gardens, which often feature a central focus on the water surface, with the main roadways encircling it.

As a result of the landscape alterations, the SSEZ has seen the construction of four notable parks, including Lychee Park and People's Park. The meaning and cultural practices once linked to agricultural land have become obsolete, and the land itself has likewise vanished. Three decades after the formation of the Communist Party of China, a socialist-nationalist regime, a novel human-local dynamic arose in the agricultural sector of this nation, portraying a country that not only supplies its people with sustenance but also facilitates their reintegration into the global community, embodying the modernity that the SSEZ sought to emphasis during its globalization-focused endeavors in the early 1980s.

---

[61] According to the interview with Mr. Guo Binghao, 2023-02-16.

[62] Sun, X. X. 孫筱祥 (2011). Garden Art and Garden Design園林藝術及園林設計. Beijing: China Architecture & Building Press北京:中國建築工業出版社. (pp. 8–9, 181–194).

Undoubtedly, the park exerts a substantial influence on the city's development through the strategic utilization of natural elements such as the terrain and water systems. It is an emerging city built under the supervision of a powerful country. Emerging urban centers must demonstrate a combination of distinctive cultural characteristics and extensive international integration. The establishment of urban parks is positioned within a particular framework. They integrate environmental aspects that protect national identity by drawing inspiration from "Chinese classical gardens." Urban parks, on the other hand, are spatial constructions that were originally established in the nineteenth century in European and American countries. They exemplify the attributes of bourgeois advancement and advocate for universal ecological principles.

## 4.4 Marketing-Oriented Nature: The Transformation in Meaning of the Lychee Forest

### *4.4.1 Lychee from Two Perspectives*

The Lychee Park in the Shangbu-Luohu Area derives its name from the pre-existing lychee grove that was already on the location. The construction of Lychee Park stands out as one of the government's foremost priorities among the initial batch of parks. At that time, due to the large and thriving lychee orchards in these rural areas, the planners of the *Program for Urban Development of Special Economic Zone of Shenzhen Municipality* (1980) deliberately preserved them and decided to use lychee as the main landscape resources for the design of a park.[63] For this reason, they even adjusted Hongli Road (紅荔路) to the north to create a larger area for the park, indicating the significant attention given to the park in the urban planning of that time. The site selection for Lychee Park played a crucial role in inspiring the locations of other important public cultural facilities such as the Grand Theatre, Shenzhen Museums, Shenzhen Library, and Science Museum. As the municipal party committee and government were already located nearby, planners realized the opportunity to integrate the park with surrounding administrative functions and public cultural facilities, creating a low-density, modern urban image.[64]

Obviously, the lychee forest growing on the site before the establishment of the SSEZ laid the foundation for building an urban park here. It also determined, to a certain extent, the location of many large-scale cultural facilities in the surrounding area, thus creating a modern image of the Shangbu-Luohu Area as the city center at that time. The earliest origin of these lychee forests is unknown, but at least this lychee forest existed during the Anti-Japanese War. At that time, the invading Japanese Army even cut down these lychee forests to use as fuel, but fortunately, the

[63] According to the interview with Mr. Chen Shouya, 2022–12–17.

[64] According to the interview with Mr. Guo Binhao, 2023–02–16.

roots of these lychee trees began to sprout and grow new lychee trees. So far, this is the earliest account detailing the origin of the lychee grove at Lychee Park. When the construction of the park began, the lychee forest was a fruit farm for the Caiwuwei Brigade, and local villagers collected and sold the fruit every year for their livelihood.

So far, this is the earliest account detailing the origin of the lychee forest at Lychee Park. In 1980, Caiwuwei was the sole brigade out of the 16-production brigades in Fucheng Commune (附城公社) to cultivate lychees.[65] The 1980 Agricultural Production Statistical Annual Report of the brigade stated that it had a lychee cultivation area of 200 mu, and the yield for both 1980 and 1981 was 200 dan. The Latin name for lychee is *Litchi chinensis Sonn.* According to modern botanical taxonomy, it belongs to the genus Lychee within the *Sapindaceae* family. As stated in the *Flora of China* (中國植物志):

> It is an evergreen tree that typically reaches a height of no more than 10 meters, occasionally extending to 15 meters or beyond. It has grey-black bark and cylindrical branchlets that are brownish-red in color, adorned with numerous white lenticels. Its fruit has an oval to almost spherical shape, measuring 2–3.5 cm in length. It often appears dark red to bright red when it reaches maturity. The seeds are all enveloped in fleshy arils. The flowering phase occurs throughout the spring season, whereas the fruiting phase takes place in the summer…It is grown in the southwestern, southern, and southeastern regions of China, particularly in Guangdong and southern Fujian. The origin of it is widely acknowledged to be in the tropical and subtropical regions of southern China.

This is a conventional scientific description of lychee, employing precise terminology, concise language, an impartial tone, and avoiding of any emotional sentiment. From this standpoint, ancient Chinese writings regarding lychee exhibit notable dissimilarities:

> Viewed from afar, the hill's paved with brocades in piles(長安回望繡成堆);
> The palace doors on hilltops opened one by one(山頂千門次第開).
> A steed which raised red dust won the fair mistress' smile(一騎紅塵妃子笑);
> How many steeds which brought her fruit died on the run(無人知是荔枝來)!
> —Du Mu(杜牧) in Tang Dynasty, The Summer Palace(過華清宮絕句).[66]

> Beneath Mount Luofu, spring arrives in all four seasons, with new crops of oranges, loquats, and waxberries in succession. Eating three hundred lychees a day, one does not hesitate to become a Lingnan local. (羅浮山下四時春，盧橘楊梅次第新。日啖荔枝三百顆，不辭長作嶺南人。).
> Su Shi in Song Dynasty, Poems written in Huizhou.[67]

The portrayal of lychee in these poems serves to depict either the opulent life of the feudal emperor or the optimistic mindset of the exiled poet. All of them possess

[65] Compiled by the Commune Office公社辦公室 編. Basic Information of Fucheng Commune附城公社基本情況 [Unpublished document]. Archive Fond Number C1; Catalog Number A1.6; Volume Number 9; Item Number 1 (檔案全宗號C1;目錄號A1.6;案卷號9;件號1). Preserved at the Luohu Archives, Shenzhen.

[66] Xu, Y.C. 許淵沖 (Trans.) (2011). *300 Tang Poems: A Chinese-English Bilingual Edition* 唐詩三百首:漢英對照. Beijing: Wuzhou Communication Press北京:五洲傳播出版社. (pp. 163).

[67] Trans by the author.

a symbolic and humanistic hue, and Lychees are imbued with cultural and human qualities. Su Shi's poem not only enhanced the Chinese people's literary perception of lychees, but also bestowed a thousand-year-old historical association on the goods of the Lingnan region. Naturally, our research does not prioritize the examination of lychees' portrayal in ancient literature. We seek to understand the impact of the lychees and lychee forests in Lychee Park on the historical and social context of the SSEZ during the 1980s. However, it is undeniable that the two share clear resemblances—they are not the "scientific/neutral" lychees, but rather the "cultural/humanized" lychees.

### *4.4.2 Urban Marketing and Lychee*

Even before 1980, lychees were a significant commodity for international trade in Shenzhen. The lychee plantations owned by local fruit growers yield hundreds of metric tonnes of lychees each year, primarily for the purpose of export.[68] The *Reply of the State Council on the Planning of Foreign Trade Bases and Municipal Construction in Bao'an and Zhuhai Counties* (國務院關於寶安、珠海兩縣外貿基地和市政建設規劃設想的批復) highlighted the significance of drawing financial capital from compatriots of Hong Kong and Macao, as well as overseas Chinese. Similarly, the *Regulations on Special Economic Zone of Guangdong Province* (廣東省經濟特區條例) explicitly stated the encouragement of foreign nationals, overseas Chinese, Hong Kong, and Macao residents, and their respective companies and enterprises to invest and establish factories, either independently or through joint ventures. The *Preliminary Ideas of the Guangdong Provincial Committee of the Communist Party of China on the Trial Establishment of Shenzhen, Zhuhai, and Shantou Export Special Zones* (中共廣東省委關於試辦深圳、珠海、汕頭出口特區的初步設想) explicitly states that the special zones permit overseas Chinese, compatriots from Hong Kong and Macao, and selected foreign manufacturers to directly invest in establishing factories. The SSEZ, in its "Sixth Five-Year Plan (1981–1985)," prioritized the guiding concept of relying primarily on foreign investment as the main financial source for the development of the special zone. During the first stages of establishing the SSEZ, the government placed great importance on engaging in external activities. The development of Shenzhen's lychee industry and its associated forests played a significant role in this endeavor.

The inaugural Lychee Festival took place in Shenzhen from June 28 to July 7, 1988, with the main focus being the Lychee Park area. The event generated

[68] Li, D. 李定(2008). How Shenzhen Explored the Path of Market Economy深圳是如何探索市場經濟之路的. In Shenzhen Municipal Committee of the Chinese People's Political Consultative Conference, Cultural and Historical Committee, and Learning Committee深圳市政協文史和學習委員會(Eds.). *A Miracle of a City*一個城市的奇跡. Beijing: China Social Sciences Press北京:中國文史出版社. (pp. 166–167).

significant enthusiasm and energy during its ten-day duration. Mayor Li Hao's (李灏, 1927–) introductory statement articulated the official purpose of the Lychee Festival:

> This is… a significant event that combines economic, cultural, and social activities with distinct local features. The event serves as a platform for compatriots from Hong Kong and Macao, overseas Chinese, international acquaintances, and individuals from diverse backgrounds to engage in communication and commercial negotiations….The Lychee Festival was organized with the aim of fostering both domestic and international interactions in Shenzhen… Various industries in the city host events centered around Lychee, such as product fairs, investment opportunities, and trade negotiations. Additionally, there are theatre performances, large-scale flower shows, and economic and technological exhibitions….[69]

The first Lychee Festival met all expectations and received substantial coverage from Hong Kong's main publications, *Ta Kung Pao* (大公報)[70] as in Fig. 4.13 and *Overseas Chinese Daily* (華僑日報).[71] The sector of economy and commerce accomplished a trade agreement of 800 million yuan, with a modest expenditure of 280,000 yuan from the local government.[72]

Ding Bo (丁波, 1915–2001) was the one who initially suggested to the government the idea of establishing the Lychee Festival. The knowledgeable cultural worker, who frequently guided groups to engage in performances and visit different regions of the world, particularly in industrialized nations throughout the 1970s,[73] recollected:

> In June 1986, I received an invitation to travel to Shenzhen, coinciding with the optimal time for harvesting lychees. I authored a document titled "Shenzhen Culture and Arts Development (Recent) Vision" (深圳市文化藝術發展(近期)設想), in which I made reference to the Lychee Festival… Having travelled to over 40 countries across Europe, Asia, and the United States, I have encountered numerous cities that boast their own distinctive festivals. For instance, notable cultural celebrations are Japan's Cherry Blossom Festival, Bulgaria's Rose Festival, and Switzerland's Grape Harvest Festival. Shenzhen should establish its own festival to showcase its abundant lychee trees and highlight activities centered around lychee tasting. Additionally, the festival might feature product exhibitions, industrial introduction negotiations, technological exchanges, and trade negotiations.[74]

---

[69] Li, H. 李灝(1988, June 29). *Opening Speech for the First Lychee Festival in Shenzhen*深圳市首屆荔枝節開幕詞. Shenzhen Special Zone Daily深圳特區報, 01.

[70] Chen, L. F.陳來複(1988, June 29). *Opening of the First Lychee Festival in Shenzhen: Hong Kong Guests Attend Lychee Product Exhibition*深圳首屆荔枝節開幕 香港來賓出席品荔會. Ta Kung Pao大公報, 04.

[71] Xinhua News Agency新華社(1988, June 29). *Lychee Festival Held in Shenzhen*深圳舉辦荔枝節. Overseas Chinese Daily華僑日報, 2.

[72] Shenzhen Government Online深圳政府線上. (2023, January 24). *Shenzhen Lychee Festival*深圳荔枝節. Retrieved from http://www.sz.gov.cn/szstoryfr/202301/content/post_10391219.html

[73] Ding, B.丁波. *Recollections of Days at the China National Publications Import& Export CO., LTD.*回憶在中圖的日子. Retrieved from http://cnpeak.com/page/ztdw01.html

[74] Qiu, Y. 秋盈(1988, June 28). *Visiting Ding Bo Before the Litchi Festival*荔枝節前訪丁波. Shenzhen Special Zone Daily深圳特區報, 01.

六日　廿四小暑　第四版

遊行隊列長近三公里氣氛熱烈

深圳首屆荔枝節開幕

香港來賓出席品荔會

李灝談深圳成就感謝港澳台各界支持

【本報記者關來復深圳二十八日專電】成千上萬深圳市民今日走上街頭，觀看首屆荔枝節開幕式盛大遊行。

今日中午，深圳當局在銀湖賓館舉行荔枝品嘗會，三百多名中外來賓參加，共度一年一度的荔枝節。

深圳市長李灝在品荔會上表示，衷心感謝港澳台及海外各界人士對深圳建特區八年來的關心與支持。

李灝說，今年以來，特區經濟發展的勢頭更好，一至五月份，工業總產值二十七億九千萬元，比去年同期增加百分之四十四點四；社會商品零售額十八億五千萬元，增長百分之四十六點五；基本建設投資八億九千萬元，增長百分之六十五點一；出口貿易總額六億三千萬美元，增長百分之八十二點四；預算內財政收入達五億元，增長百分之三十九點三。一至五月，引進外資項目比去年增加了二點二倍，僅工業項目就已批准一百二十多項，投資總額增加了三點六倍。

李灝透露，根據中央領導人提出，政治體制改革，深圳也要走在全國的前面的要求，已成立了政治體制改革辦公室，正在組織力量，在調查研究基礎上，盡快制訂出特區政治體制改革的總體方案，並着手進行幹部制度改革和推行公務員制度的試點。

出席今天品荔會和荔枝節開幕式的有深圳市委、市府各級領導，以及國務院僑務辦公室主任廖暉、中國工商銀行行長張肖、政協常委丁家駒、中華全國醫學會副會長林佳楣、全國工商聯副主席張敬禮和廣東省有關部門負責人等。新華社香港分社副社長毛鈞年、張浚生及各部門負責人譚茀芸、楊舜、紀穎祥、林克平等亦出席。出席品荔會及開幕式的香港中資機構負責人有王光英、莊世平、袁庚、佟志廣、朱賓玉、張學堯、黃振聲等；香港知名人士有郭宜興、翁錦通、張永珍、李達生、吳多泰、余國春、楊光、伍淑清、賈桂蕃、黃祖芬、脫錦華、黃克立、蔡演雄、李以勵、倪少傑、唐驥千、張人龍、劉森淡、楊孫西、歐陽瑞、劉世仁、朱蓮芬、劉浩清、趙澤隆、曾敏之、鄒莘等。

今日下午四時開始的荔枝節開幕式遊行有儀仗隊、荔枝彩車隊，四十二軍軍樂團、彩旗隊、醒獅隊、廣告車隊以及由一百名大學女生組成的模特兒隊等，遊行隊列近三公里長。

從今天開始一連十天，深圳市內將有貿易洽談、香市、商品展覽、美食節、歌手大賽，以及書法、攝影、美術比賽等一系列活動，並有荔枝集市。

**Fig. 4.13** Ta Kung Pao of June 29, 1988 (Fourth section, partial)

The 1970s marked a significant period in which Ding Bo's overseas visits coincided with the fall of the Keynesian welfare state model in developed countries and the rise of neoliberalism. The process of urban deindustrialization and suburbanization gradually led to an economic downturn. Neoliberal urban government prioritizes local economic growth over the delivery of welfare. Urban festivals serve as a crucial strategy for urban marketing, contributing significantly to urban governance. Festivals are thought to bolster social cohesiveness and improve the local economy, hence sustaining local competitiveness, by integrating business, tourism, literature, and art.[75] The renowned Lingnan lychee culture, particularly the lychee grove located in Lychee Park, has evolved into an urban culture that Sharon Zukin describes as increasingly becoming the city's economic driver and a fundamental element in its ability to attract tourists.[76] Having visited numerous countries, Ding Bo introduced the concept of urban festival management to Shenzhen, resulting in the groundbreaking first Lychee Festival in the SSEZ.

The "Shenzhen Lychee Festival" was held from 1988 to 1997, spanning a total of ten occurrences. This event had a substantial impact on multiple fields of Shenzhen, such as attracting investments, expanding trade and tourism, facilitating scientific and technological exchanges, fostering cultural exchanges, promoting friendship activities with "Three Compatriots," and improving the image of the city.[77] Over the course of 10 years, Shenzhen's agreement resulted in a cumulative foreign investment of US$740 million, domestic investment of 2.26 billion yuan, and a commodities turnover of 8.45 billion yuan.[78]

### *4.4.3 Lychee Park in Lychee Festival*

The Shenzhen Lychee Festival takes place at various locations, including the Economic Building (經濟大廈), Shenzhen Gymnasium (深圳體育館), International Trade Centre (國貿中心), World Hotel (寰宇大酒店), Youth Activity Centre (青少年活動中心), Workers' Cultural Palace (工人文化宮) in Shangbu-Luohu Area, as well as Shekou, Nantou Lychee Garden, Nantou, and other places in the western part of the SSEZ. However, the Municipal Government area in Shangbu-Luohu Area serves as the central hub for this festival. Lychee Park, located in this central region, provides the location for a variety of lychee-centric events, including activities like lychee harvesting and lychee commerce. The quality of lychees in the Nanshan district of SSEZ is mostly characterized by types such as Guiwei (桂味) and Nuomizi (糯米滋), but the lychee varieties found in the Lychee Park, such as Heiye (黑葉) and Hezhi (荷枝), are typically less sensitive and less juicy compared to Guiwei and Nuomizi.[79] Nevertheless, being far from the administrative center at that time, Nanshan's endeavors were inadequate to establish it as the central focus

[75] Waitt, G. (2008). Urban festivals: Geographies of hype, helplessness, and hope. *Geography Compass*, 2(2), 513–537.

[76] Zukin S. (1995). The Cultures of Cities[M]. Oxford: Blackwell. Publishers. (pp. 1–8).

[77] Shenzhen Government Online深圳政府線上. (2023, January 24). *Shenzhen Lychee Festival深圳荔枝節*. Retrieved from http://www.sz.gov.cn/szstoryfr/202301/content/post_10391219.html

[78] Sunshine Daily晶報, 2018-07-03(A14).

[79] Shenzhen Nanshan District Chronicle Editorial Committee Office深圳市南山區區志編委會辦公室 (2012). *Nanshan District Chronicle of Shenzhen深圳市南山區志*. Beijing: Fangzhi Publishing House北京:方志出版社. (pp. 411); Yuchun雨純(1981, June 12). *Shenzhen Specialties深圳特產*. Shenzhen Special Zone Daily深圳特區報, 04.

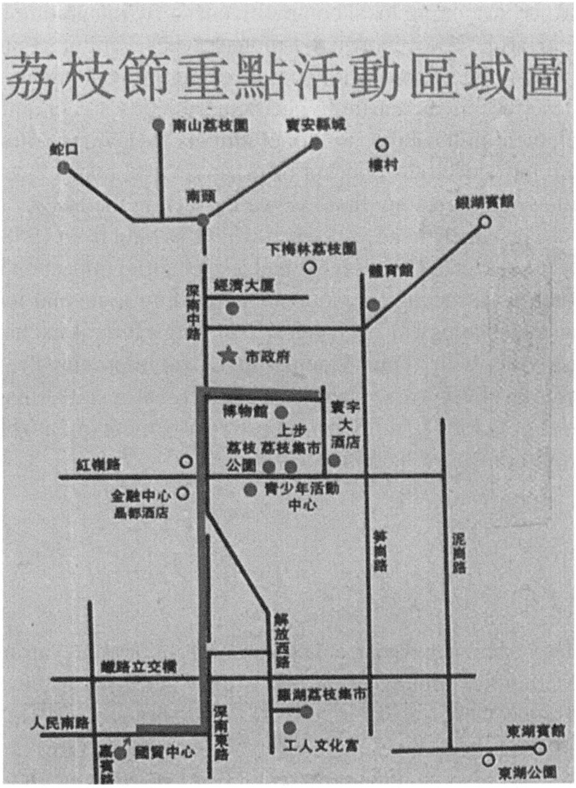


**Fig. 4.14** Map of key activity areas for the first Lychee Festival (in Traditional Chinese) (Litchi Festival Album Editorial Committee. 《荔枝節》畫冊編委會.*Litchi Festival Album*《荔枝節》畫冊. Preserved at the Local Literature Room of the Shenzhen Library)

of the Lychee Festival. After all, the lychee itself does not serve as the primary objective of the lychee festival. The primary objective is to conduct business discussions and showcase the city's image by use of lychees.

Figure 4.14 depicts the spatial configuration for the Lychee Festival. World Hotel, which started operating on the inaugural day of the Lychee Festival, is located on the eastern side of Lychee Park. This hotel was designated as the site for trade negotiations and product demonstrations during the Lychee Festival more than two months ago. The library, situated on the northern side of Lychee Park, functioned as a marketplace for the trade of both international and domestic books and audio recordings. The festival was initiated and sponsored by the city government, situated adjacent to Lychee Park. The Shenzhen Gymnasium, situated within a range of 1–2 km, served as the venue for the Shenzhen-Hong Kong sports festival. In addition, the International Trade Centre served as the venue for the opening ceremony and as a site for local officials to entertain international dignitaries. Lychee Park hosts various exhibitions, such as floral, avian, aquatic, and entomological showcases, along with competitions for plant bonsai. Additionally, it serves as a venue for musical and choreographic performances. Moreover, the northern and western roads adjacent to the park are embellished with colorful flags placed at regular intervals. Furthermore, it functions as a transient marketplace where local individuals can participate in the purchase and sale of freshly harvested lychees. Additionally, the Lychee Festival Organizing Committee is situated within the premises of the Lychee Park Management Office. Moreover, it functions as a venue where municipal authorities and officials from various government agencies in Hong Kong convene to consume lychees and engage in social activities.[80]

Lychee Park relies on a range of urban sites to support and stage the Lychee Festival. The festival's central location is determined by Lychee Park and its adjacent public facilities. The Lychee Festival at a metropolitan level is intended to promote investment. However, the significance of lychees as a symbol outweighs the actual quality of the fruit. However, it is undeniable that the existence of the lychee park provides a legitimate reason for organizing the lychee festival in Shangbu-Luohu Area.[81]

Furthermore, the Lychee Festival showcased the city's freshly urban image to a larger audience of both local and international visitors. This occurrence is not a mere coincidence, which was already taken into account during the construction of the Shangbu-Luohu Area in the early 1980s. At that time, the Lychee Park and many public cultural facilities were mostly concentrated in this area, largely because of the consideration of the urban image.[82]

### 4.4.4 Lychee Forest Prohibited from Reproduction

In 1998, the government halted the Lychee Festival and rebranded it as the "High-tech Fair(高新技術交易會)," revamping it into a more unique high-tech display.[83] This occurrence is not a result of chance or coincidence. In 1987, prior to the inaugural Lychee Festival, Shenzhen released the *Interim Regulations on Encouraging Scientific and Technological Personnel to Establish Private Technology Enterprises* (关于鼓励科技人员兴办民间科技企业的暂行规定). In 1988, the inaugural "Private Science and Technology Enterprise Product Exhibition(民间科技企业产品展览会)" coincided with the opening of the first Lychee Festival at the Shenzhen Science Museum.[84] However, the overwhelming enthusiasm for the Lychee festival overshadowed the significance of the exhibition. Upon the organization of the Eighth Lychee Festival in 1995, after eight years of development, the high-tech

[80] According to the *Shenzhen Special Zone Daily* during the Lychee Festival from June 28 to July 8, 1988.

[81] According to the interview with Mr. Ye Guidong, 2023-12-30.

[82] According to the interview with Mr. Guo Binghao, 2023-02-16.

[83] Li, Z. 李子彬. (2020). *My Time as the Mayor of Shenzhen*我在深圳當市長. CITIC Publishing Group中信出版社. (pp. 95–97).

[84] Zeng, J. Y.曾紀允(1988, July 7). *Opening of the Shenzhen Folk Technology Enterprise Product Exhibition*深圳民間科技企業產品展覽開幕. Shenzhen Special Zone Daily深圳特區報, 01.

industry was officially recognized as the major industry in Shenzhen.[85] The inaugural China International High-Tech Fair(中國國際高新技術成果交易會), often known as the Hi-Tech Fair (高交會), took place in Shenzhen in October 1999. It is at this juncture, Pony Ma (馬化騰), who had founded Tencent just one year prior, successfully secured the first venture capital investment of $2.2 million. Deng Guoshun (鄧國順) and Cheng Xiaohua (成曉華), representing Netac Technology, introduced the first flash drive to the global market, leading to the rapid displacement of magnetic discs and optical discs by the so-called "U disc".[86]

During the Lychee Festival, visitors can engage in the activity of collecting lychees and savor the flavor of freshly harvested lychee fruits in the lychee grove located within Lychee Park. Annually, during the period when lychees are maturing, there is a lively activity of picking lychees beneath the forest. Following the closure of the Lychee Festival, the lychees in the park became increasingly problematic for urban management, as depicted in Fig. 4.15. As early as the 1980s, in order to make this lychee forest get more fruit, the park management specially hired professional fruit farmers from Dalingshan Mountain in Dongguan to be responsible for fertilization, pruning and taking care of the fruits after ripening.[87]

Starting in 2002, Lychee Park implemented pesticide spraying to control fruit development. As the Lychee Festival has been closed for four years, the lychee trees at the Lychee Park no longer require the need to attract foreign guests and provide harvesting experiences. However, the abundance of fresh lychee fruits has brought forth numerous safety concerns because people may climb the lychee trees at any time to steal the fruits, and many of the three or four stores high lychee trees have repeatedly caused safety accidents. Ultimately, the park management's strategy was to spray the pesticide ethephon during the flowering period of lychee trees.[88]

Ethephon, also known as 2-chloroethylphosphonic acid, is mostly utilized as a substance that induces fruit ripening. At greater concentrations, it acts as a growth inhibitor for fruits. Given its extensive historical background, it is natural for the lychee tree to undergo the process of flowering and producing fruit. Nevertheless, within the particular circumstances of that period, the restriction on breeding and fruit control bestowed upon the lychee forest in Lychee Park has a novel urban characteristic: it becomes effortlessly manageable while retaining its natural semblance.

[85] Zhang, K. K. 張克科(2010). Witnessing the Advancement of Shenzhen's High-Tech Industries 見證深圳高新技術產業的前行步履. In *Shenzhen Municipal Political Consultative Conference History and Learning Committee*深圳市政協文史和學習委員會(Eds.). *The Rise of Shenzhen's Four Pillar Industries: High-Tech*深圳四大支柱產業的崛起 高新技術. Beijing: Chinese history and literature Publishing House.北京:中國文史出版社. (pp. 143).

[86] Lin, T.林濤(2018, November 14). *20 Years of China Hi-Tech Fair: Witnessing Shenzhen's Industrial Transformation and Innovation Development Dreams*高交會20年:見證深圳產業轉型與創新發展夢想. Southern News Network南方新聞網. Retrieved from https://www.sohu.com/a/275211650_222493

[87] According to the interview with Mr. Luo Xiaobing, 2023-04-02.

[88] According to the interview with Mr. Luo Xiaobing, 2023-04-02.

**Fig. 4.15** Lychee forest during and after the Lychee Festival (Source of the left image: *Scenery of Litchi Orchard*荔園風光, internal publication. Source of the right image: Zhang, Y. J. 張英俊 (2014, June 31). *Shenzhen Lychee Park Turned into "Orchard": Lychees Plundered by Tourists*深圳荔枝公園成"果園"荔枝遭遊客亂採摘. Retrieved from http://www.rmzxb.com.cn/c/2014-06-13/339145.shtml)

This shares the same logic as urban greening that undergoes manual pruning of branches and leaves; both are cultural forms of nature.

In contrast to the farmed land on the western side that was converted into a lake within the park, the lychee forest in Lychee Park has experienced minimal morphological alterations. Nevertheless, the lychee forest has undergone continuous reproduction throughout history. In the 1940s, it was ruthlessly destroyed by invaders, who used it as a source of fuel for military purposes. From the perspective of the indigenous population, lychees served as both sustenance and a potential source of income for farmers during the 1960s. As an expression of their beliefs, they bury their deceased ancestors beneath the lychee forest. During the construction of the park in the early 1980s, they relocated the tombs, transforming the lychee forest into a valuable landscape resource. In the late 1980s, lychees were strategically utilized to enhance foreign trade and stimulate economic growth during investment promotion. Fruit farmers with expertise were even entrusted with increasing lychee production. Following the rising of high-tech industries in the 1990s, lychees became a source of concern for urban management, and the growth of lychee fruits was subject to regulation. The lychee forest found at Lychee Park cannot be considered a permanent and unchanging reality within the vast expanse of history, as human social practices constantly shape its meaning. The process of change necessarily involves competition among multiple actors between invaders and native inhabitants, indigenous people and the government, citizens and administrative agencies, etc., all form their own aims and traditions centered around lychees and lychee forests.

In contrast to the farmed land on the western side that was converted into a lake within the park, the lychee forest in Lychee Park has experienced minimal morphological alterations. Nevertheless, the lychee forest has undergone continuous reproduction throughout history. In the 1940s, it was ruthlessly destroyed by invaders, who used it as a source of fuel for military purposes. From the perspective of the indigenous population, lychees served as both sustenance and a potential source of income for farmers during the 1960s. As an expression of their beliefs, they bury their deceased ancestors beneath the lychee forest. During the construction of the park in the early 1980s, they relocated the tombs, transforming the lychee forest into a valuable landscape resource. In the late 1980s, lychees were strategically utilized to enhance foreign trade and stimulate economic growth during investment promotion. Fruit farmers with expertise were even entrusted with increasing lychee production. Following the rising of high-tech industries in the 1990s, lychees became a source of concern for urban management, and the growth of lychee fruits was subject to regulation. The lychee forest found at Lychee Park cannot be considered a permanent and unchanging reality within the vast expanse of history, as human social practices constantly shape its meaning. The process of change necessarily involves competition among multiple actors between invaders and native inhabitants, indigenous people and the government, citizens and administrative agencies, etc., all form their own aims and traditions centered around lychees and lychee forests.

Subsequently, substitute the structure with lychee, as a means of rephrasing David Harvey's moving conclusion after his examination of the Sacre-Coeur in Paris: The lychee hides its secrets in sepulchral silence. Only the living, cognizant of this history, who understand the principles of those who struggled for and against the embellishment of that spot, can truly disinter the mysteries that lie entombed there and thereby rescue that rich experience from the deathly silence of the tomb and transform it into the noisy beginnings of the cradle.[89]

## 4.5 Governance-Oriented Nature: The Transformation of Honghu Park and Buji River

### *4.5.1 Sungang Flood Detention District in the City–Water Relationship*

The Buji River(布吉河) is a primary tributary of the Shenzhen River, it originates from Huangzhuli (黄竹瀝) in the central part of Shenzhen City and merges with the upstream tributaries of Shuijing (水徑) and Tangjing (塘徑) at Niulingxia(牛嶺嚇). The river then flows south through Buji Town (布吉鎮),[90] Caopu (草埔), and the Luohu commercial areas before eventually joining the Shenzhen River. Along the course of the Shenzhen River, there are secondary tributaries such as Dafenshui (大芬水), Lianhuashui (蓮花水), Qingshui River (清水河), and Bijiashan River (筆架

[89] Harvey, D. (2018). Monument and myth. In *Culture and Society*. Routledge. (pp. 159–178).

[90] In August 2004, Bujizhen was renamed Buji sub district. In April 2006, Buji sub district was divided into three sub-districts: Buji, Bantian, and Nanwan. For the sake of clarity, the term "Buji Town" has been used throughout the text, maintaining continuity with the name used since 1986.

山河),[91] as Fig. 4.16 illustrates. Buji Town, located outside Second Line, and Shangbu-Luohu Area, situated within Second Line, are both part of the Buji River drainage basin. However, in regard to their topography, the two regions exhibit dissimilar characteristics. The overall terrain of Buji Town is relatively high, dominated by hilly topography,[92] while the Shangbu-Luohu Area has a lower elevation and is primarily characterized by a coastal alluvial plain. The two areas are divided by a range of mountains and hills, including Yinhu Mountain (銀湖山), Weiling Hill (圍嶺山), and Qiushui Hill (求水山).

While Buji Town is a part of Shenzhen City, it does not fall within the jurisdiction of the SSEZ. Prior to 2004, Buji was predominantly rural until it was designated as Buji Subdistrict as a result of the urbanization occurring outside the Second Line.[93] In 1979, Buji Town initiated the implementation of the Reform and Opening-up policy, leading to the transition of the agricultural economic system from collective production to a household-based joint production contract responsibility system. During that period, various productive resources, including land, were allocated to each household.[94] If this type of land arrangement persists, Buji Town is likely to be similar to the extensive rural areas in China presently. The rapid growth of the industrial sector in the SSEZ has led to a substantial increase in the economic value of the land in Buji Town, surpassing its previous agricultural worth. In 1986, Buji Town transferred all land within its jurisdiction back to rural communal ownership to integrate the management of land.[95] This served as the basis for the swift growth of Buji Town's industrial economy, earning it the title of "Top Economically Dominant Town in Guangdong (廣東第一經濟強鎮)" in 2001.[96]

The era from 1983 to 1986 marked the early phase of Buji Town's industrialization, while the years 1987 to 1989, following the re-acquisition of property under community ownership, represented a significant leap forward in the development of Buji Town's industry. By 1989, the town had implemented 395 processing projects, which included foreign-funded firms. The total foreign investment in these projects

[91] Shenzhen Local Chronicles Compilation Committee深圳市地方誌編纂委員會 (2014). *Shenzhen Chronicles: Infrastructure Volume*深圳市志·基礎設施卷. Beijing: Fangzhi Publishing House北京:方志出版社. (pp. 201).

[92] Shenzhen Municipal History and Chronicles Office深圳市史志辦公室 (1996). *Brief Records of the Nineteen Towns in Shenzhen*深圳市十九鎮簡志. Shenzhen: Haitian Publishing House深圳:海天出版社. (pp. 107).

[93] Before the division into Bao'an District and Longgang District in 1992, the two areas were collectively referred to as Bao'an County.

[94] Bujizhen Chronicles Compilation Committee布吉鎮志編纂委員會(2009). *Buji Town Chronicle*布吉鎮志. Shenzhen: Bujizhen Chronicles Compilation Committee深圳:布吉鎮志編纂委員會. (pp. 76).

[95] Shenzhen Municipal History and Chronicles Office深圳市史志辦公室 (1996). *Brief Records of the Nineteen Towns in Shenzhen*深圳市十九鎮簡志. Shenzhen: Haitian Publishing House深圳:海天出版社. (pp. 109).

[96] Guan, G., Lin, R., & Li, B.關國林 林若飛 李博. (2001, December 20). *Buji: Building Guangdong's Premier Town This Way*布吉:這樣打造廣東第一鎮. Shenzhen Economic Daily深圳商報, A1.

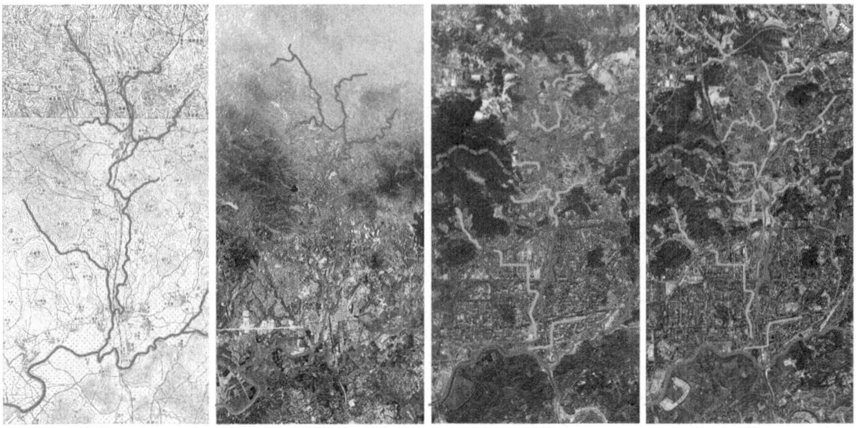


**Fig. 4.16** Changes in the Buji River Basin since the 1930s (dark green represents underground channels, blue surface rivers, and red the Special Zone boundary) (From left to right, source from the Central Research Institute (Taiwan) Digital Map Collection Integration and Query System中央研究院(臺灣)地圖數位典藏整合查詢系統. Data Identifier: map_loc_033061; USGS 1980 imagery, Document No.: D3C1216-300662F047(Sep. 1980); Google historical imagery; Liang, Q. Y. 梁巧茵(2022). Research on Joint Flood Control Scheduling of the Sungang Detention Area and Downstream River Channels.筍崗滯洪區及下游河道聯合防洪調度研究. *Water Resources Planning and Design*水利規劃與設計, (03), 76–79)

amounted to HK$1.6 billion, resulting in a cumulative foreign exchange income of HK$790 million. Additionally, these projects employed a workforce of 42,000 individuals. The town's industrial output value constituted 78% of the overall output value this year.[97]

Until 1980, the Buji River Basin was predominantly recognized for its mountainous terrain, dense forests, agricultural land, and indigenous villages. Prior to the implementation of extensive artificial construction in the middle and upper sections, the Shangbu-Luohu Area in the lower section was susceptible to regular flooding due to topographical factors. During the rainy season, the north-to-south flowing Buji River is prone to the influence of tidal surges from Shenzhen Bay, causing difficulties in the smooth flow of Buji River water into the Shenzhen River. Consequently, frequent occurrences of urban waterlogging often occur in the low-lying Shangbu-Luohu Area, as depicted in Fig. 4.17. However, the level of attention given to these water hazards at that time was far less than that observed after the establishment of the SSEZ in 1980. This transformation resulted in a sudden change in the political and economic significance of the Shangbu-Luohu Area, which shifted from being a border town in South China into a prominent location for

[97] Shenzhen Municipal History and Chronicles Office深圳市史志辦公室 (1996). *Brief Records of the Nineteen Towns in Shenzhen*深圳市十九鎮簡志. Shenzhen: Haitian Publishing House深圳:海天出版社. (pp. 111).

implementing Reform and Opening-up policy. Therefore, the concern of its image emerged as the main impetus in flood treatment.[98]

Moreover, the *Shenzhen Special Economic Zone Urban Development Outline* deliberated in November 1980 addressed flood management and drainage in the Shangbu, Luohu, and Huanggang (皇崗)areas, taking into account the following factors:

> Realign the curves in specific parts of the Shenzhen River and excavate them to improve the flow of water. Nevertheless, in matters concerning diplomatic issues, it is advisable to seek the evaluation of higher-ranking authorities. During periods of intense precipitation, the larger tributaries will collect more rainfall and should channel it into artificial lake situated in verdant areas. The lake functions as reservoir, temporarily storing water until the floods in the Shenzhen River recede, thereby averting flooding in the city.[99]

Moreover, the *Shenzhen Special Economic Zone Urban Development Outline* deliberated in November 1980 addressed flood management and drainage in the Shangbu, Luohu, and Huanggang areas, taking into account the following factors:

Furthermore, as of October 1981, the Luohu Commercial District had attracted almost 1.915 billion Hong Kong dollars in international investment, signed 22

**Fig. 4.17** Floods in the Luohu urban area, summer 1979 (Shenzhen Luohu District Chronicle Editorial Committee Office深圳市羅湖區志編委會辦公室 (2012). *Nanshan District Chronicle of Shenzhen*深圳市羅湖區志. Beijing: Fangzhi Publishing House北京:方志出版社. (pp. 164))

[98]Zhang, M. J. 張明傑(2010). The Initial Development of Jiang Zemin and China's Economic Special Zones江澤民與我國經濟特區的初創. *Literature of Chinese Communist Party*黨的文獻, (137), 23–28+120. Chen, H. (Eds.) 陳宏 編. (2011). Genealogy of Shenzhen Indigenous People深圳原住民家譜*Shenzhen: Shenzhen Press Group Publishing House*深圳:深圳報業集團出版社. (pp. 253).

[99]Program for Urban Development of Special Economic Zone of Shenzhen Municipality (1980), preserved at the archive office of Shenzhen Urban Planning and Natural Resources Bureau.

contracts with foreign entrepreneurs, and seen significant progress in many projects. At that time, multiple construction teams, totaling nearly 2000 individuals, were involved in the construction of the Luohu Commercial District.[100] The occurrence of floods during this period would result in significant damage to urban industries with high value-added as well as to commercial and residential properties. Moreover, it could also have an adverse impact on the national plan for reform and opening-up policy. The flood issue required immediate attention, and the approach involved creating "artificial lake" that would hold the excess water. In January 1982, the government formed the Sungang Flood Detention District(筍崗滯洪區) Project Headquarters with the purpose of filling 1.3 million cubic meters of stone and pouring 18,000 cubic meters of reinforced concrete. The Sungang Flood Detention District was finished in November 1983, with a design water level of 11.3 m and a design flood detention capacity of 2.5 million $m^3$.[101] The flood detention district was integrated with the development of municipal roads, and the Sungang Flood Release Gate(筍崗洩洪閘) was constructed by combining embankment roads and bridge gates. Both the east and west embankment tops have planned heights of 13.0 m. The eastern embankment is integrated with roads, and the breadth of the embankment's top is 20 meters, and the length of the structure is 2.1 kilometers. The width of the top of the western embankment is 4 meters, and the embankment itself has a length of 2.0 kilometers.[102] Figure 4.18a illustrates the Sungang Flood Detention District, which used to be agricultural land located in the middle part of the Buji River. The ownership of this property was held by the Shuibei and Tianbei villages. However, the flooding of the Buji River was frequent, leading to highly unstable agricultural output in this area.[103]

The conversion of cultivated land on the alluvial plain into a flood detention district serves to mitigate urban waterlogging, thereby altering the dynamic between humans and the land. This is the urbanization of land. Subsequently, the flood detention district will undergo a conversion into an urban park, necessitating an understanding of its significance within a new historical framework.

---

[100] Zeng, J. T. 曾錦棠(2009). *Old News of Shenzhen*深圳舊聞. Shenzhen: Shenzhen Press Group Publishing House深圳:深圳報業集團出版社. (pp. 133–134).

[101] Lei, S., & Li, J. 雷聲,李峻(1998). Design and Practice of Sediment Discharge Channel in Sungang Detention District筍崗滯洪區排沙道工程設計實踐. *Guangdong Water Resources and Hydropower*廣東水利水電, 1998(S1), 22–24; Resolving Pollution, Beautifying the Environment, Benefiting the People: The Sand Discharge Channel Project in Sungang Detention Area in Shenzhen Won the Third Prize for Excellent Design in 1995解決污染,美化環境,造福人民——深圳市筍崗滯洪區排沙道工程榮獲95年深圳優秀設計三等獎. *World Architecture Review*世界建築導報, 1996(03), 74.

[102] Ye, L. Y. 葉林宜(1995). The Flood Control Effect of the Shenzhen Reservoir Combined with Sungang Floodgate on the Shenzhen R iver深圳水庫結合筍崗洩洪閘對深圳河的防洪作用. *Pearl River*人民珠江, (06), 6–8+14.

[103] According to the interview with Mr. Guo Binghao, 2023–02–16.

### 4.5.2 *Honghu Park: Sediment, Pollution, and Flood detention district*

Starting in approximately 1980, with the goal of saving construction funds, several reservoirs constructed in the 1950s and 1960s, including Shiyan Reservoir (石岩水庫, 1960), Xili Reservoir (西麗水庫,1960), and Bijiashan Reservoir (筆架山水庫, 1954), were repurposed as tourist attractions due to their favorable scenic resources. Similarly, the Sungang Flood Detention District was converted into an urban park during this period.[104] Although the urban park was not expected to immediately increase foreign exchange, it was considered to significantly enhance the city's investment environment.

Following the 1983 visit to Singapore, authorities reached a consensus on the significance of greening and beautifying the urban environment. Therefore, the construction of urban parks has experienced tremendous acceleration. The topography of the entire flood detention district was defined by the fact that the highest point of the east and west dikes reached an elevation of 13.0 m. The topography within the

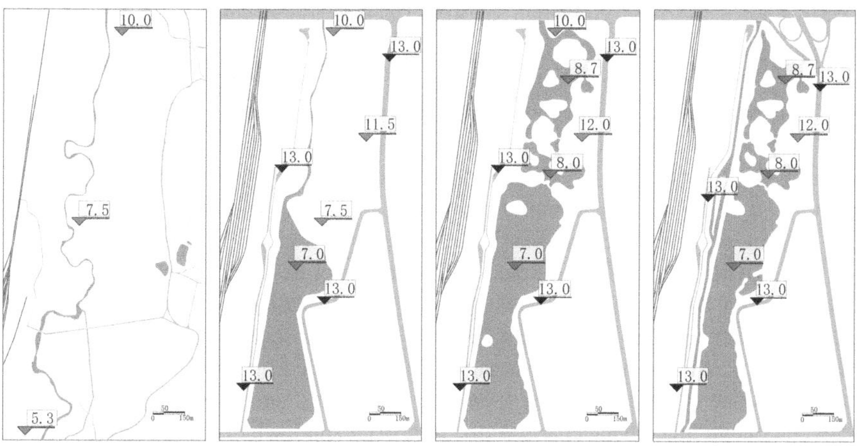


**Fig. 4.18** Morphology of water system in Honghu Park at different periods
((**a**) Based on the USGS 1980 imagery and Topographic map of Honghu Park in 1979 (Survey institution unknown), the original map is preserved in the Archive Room of Honghu Park Management Office.; (**b**) According to the topographic map surveyed in 1979 and relevant literature data on the flood detention district; (**c**) Redrawn based on Topographic Map of Shangbu-Luohu Area, Shenzhen City by Shenzhen Survey and Mapping Company in 1985, Original map is preserved at the Archive office of Shenzhen Park Management Center.; (**d**) Redrawn based on Topographic Map of Honghu Park, Shenzhen City by Shenzhen Survey and Mapping Company in 1994, original map is preserved at the Archive office of Shenzhen Park Management Center). (**a**) Buji River and farmland. (**b**) Sungang Flood Detention District. (**c**) Honghu Park. (**d**) Honghu Park and the sediment channel

[104] Wang, R. B. (Eds.) 王若兵 編. (1990). *Water Resources Chronicle of Shenzhen*深圳市水利志. Guangzhou: Guangdong Science and Technology Press廣州:廣東科技出版社. (pp. 123).

area exhibited an elevation gradient, with the northern part being higher and the southern part being lower. The northern elevation measured approximately 10.0 m, whereas the southern elevation measured around 5.3 m. The water level design of the Sungang Flood Detention District specified a maximum water level of 11.3 m and a minimum water level of 7.0 m. During the non-flood season, a significant portion of the lake bed in the northern section of the flood detention district, with an elevation above 7.0 m, would become visible above the water surface. This visibility was visually unappealing, as depicted in Fig. 4.18b. In the *Program for Urban Development of Special Economic Zone of Shenzhen Municipality* (1980), this area was initially considered as a flood detention district. However, it was later designated as a park for two main reasons:[105] Firstly, it had both water bodies and land, meeting the basic requirements for park construction; Secondly, it was conveniently located in the planned greenbelt area between Luohu and Shangbu areas, providing an opportunity to introduce greenery into the city along the Buji River. At that time, the site was mainly abandoned farmland with minimal artificial construction. In terms of vegetation, there were no particularly large trees; instead, there were some small trees, shrubs, and weeds. The key premise for constructing the park at that time was to retain its water detention function.[106]

To achieve the interconnected spatial effect of topography and water system commonly found in traditional gardens, it was essential for the design to include a specific quantity of water in the northern part of the park. Therefore, the park was designed to partition the entire water area into three distinct levels: Level 1 at elevation 7.0 m, Level 2 at 8.0 m, and Level 3 at 8.7 m,[107] as depicted in Fig. 4.18c. These levels were separated by movable water barriers and cascading waterfalls.

While it might seem reasonable to construct an urban park atop the flood detention district, it nevertheless resulted in complications. Initially, the flood section of the Buji River had a sudden expansion, resulting in a sharp decrease in the flow velocity of the river and a significant reduction in the sediment-carrying capacity of the water. Consequently, the sediment carried by the river water piled easily in Honghu Park. Furthermore, the situation was worsened by the construction of the industrial zone in Buji Town. In 1986, the collective management of land in Buji Town facilitated the advancement of the industrial zone, which make upper sections of the Buji River exhibited widespread visibility of uncovered loess and significant soil erosion. A significant portion of this soil erosion was deposited in Honghu Park, as illustrated in Fig. 4.19. In June 1990, estimations indicated that the storage capacity above an elevation of 6 m had decreased by 132,300 $m^3$.[108] Furthermore, the

---

[105] According to the interview with Mr. Guo Binghao, 2023–02–16.

[106] According to the interview with Mr. Shen Baojiu, 2023–02–12.

[107] Shen, B. J., Shen, T. X., & Shen, T. P. 沈葆久,沈天翔,沈天鹏(1994). *Shenzhen New Gardens: Abstract Gardens*深圳新園林——抽象式園林. Haitian Publishing House深圳:海天出版社. (pp. 129–132).

[108] Lei, S., & Li, J. 雷聲,李峻(1998). Design and Practice of Sediment Discharge Channel in Sungang Detention Area筍崗滯洪區排沙道工程設計實踐. *Guangdong Water Resources and Hydropower*廣東水利水電, (S1), 22–24.

sewage released from the Buji Town factory directly entered Honghu Park via the Buji River, resulting in severe eutrophication of the park's water body, which became visibly dark and malodorous.

Buji Town belongs to a hilly terrain, and once the developed land is subjected to rainfall, surface runoff will carry a large amount of loose sediment downstream, resulting in noticeable soil erosion. The main strategy to address the problem is to enhance the management of soil and water conservation upstream and cease the discharge of sewage. Nevertheless, it is evident that throughout the 1980s, when the economy heavily depended on industrial development, SSEZ were unable to address these challenges. After the large-scale dredging of Honghu Park in 1990, the water conservancy department constructed a sediment drainage channel in 1991, as depicted in Fig. 4.18d. The guiding concepts for designing the sediment drainage channel involve increase the sediment-carrying capacity of water flow, narrow the flood section to flush away sediment, and separate clean water from turbid water:

> Construct a sediment drainage channel along the western embankment of the flood detention district…to enhance the sediment-carrying capacity of water flow. Between the sediment drainage channel and the flood detention district, tow lateral weirs are constructed on both the upstream and downstream sides. The upstream side weir serves to segregate the sewage discharged from the upstream river during the dry season from the water body in the flood detention district… the downstream side weir serves the purpose of regulating the water level in the flood discharge channel, ensuring a consistent landscape water level even during the dry season. During the flood, adjusting the gates in various ways can effectively increase the velocity of water flow, facilitating the removal of sand.[109]

In fact, the problem of soil erosion in the Buji River Basin appeared before 1980, as documented in the *Buji Town Chronicle(布吉鎮志)*:

> The Buji River initially had a riverbed made of stones and appeared deep and transparent. Historically, there were wooden rafts that traversed between Buji and Shenzhen. Subsequently, because of soil erosion in the upstream areas, the river gradually transformed

**Fig. 4.19** Sediment and pollution in Honghu Park (1990) (Left image from Shenzhen Municipal Chronicles Compilation Committee深圳市地方誌編纂委員會(2009). *Chronicles of Shenzhen: Volume on Infrastructure Construction*深圳市志·基礎建設卷. Beijing: Fangzhi Publishing House 北京:方志出版社. (pp. 887); The right image from *Beautiful Lotus in Shenzhen*荷美深圳, an internal publication)

[109] Lei, S., & Li, J. 雷聲,李峻 (1998). Design and Practice of Sediment Discharge Channel in Sungang Detention Area筍崗滯洪區排沙道工程設計實踐. *Guangdong Water Resources and Hydropower*廣東水利水電, (S1), 22–24.

into a riverbed with a sandy bottom and shallow depths, resulting in the disappearance of rafts....[110]

The decision made by Buji Town in 1986 to reclaim land for collective use to stimulate industrial development clearly exacerbated the issue of soil erosion within a brief timeframe. Based on a survey conducted in 1995, the extent of soil erosion in the Buji River Basin was measured to be 10.12 square kilometers, making it the primary tributary basin to experience this problem.[111] It is evident that the essence of environmental issues is, in fact, a matter of production methods, simultaneously dictating the institutional arrangements made by the farmers in Buji regarding land resources. Honghu Park, situated on the alluvial plain, is one of the rare urban areas in the Buji River Basin that has managed to preserve its original natural features. Before the construction of the sediment drainage channel in 1991, it could only humbly look up to Buji Town as a topographic and economic highland, passively accepting the sediment and pollution brought to it by urbanization.

### *4.5.3 Lotus: From Pollution Control to Social Administration*

From 1980 to 1991, when the sediment drainage channel of the Sungang Flood Detention District was constructed, untreated industrial sewage from Buji Town was frequently released into the Buji River. As a result, the levels of COD, BOD, nitrogen, and total phosphorus in the lower reaches of the river exceeded the standards set for Class V water bodies. The sewage seeping into Honghu Park has caused considerable eutrophication of the park's water quality.[112]

In 1988, Honghu Park implemented the introduction of aquatic plants, including lotus and water lily, sourced from Wuhan (武漢), Guangchang (廣昌), and other locations. This initiative aimed to effectively address the issues of black, malodorous water pollution and the proliferation of mosquitoes.[113] The lotus, scientifically known as *Nelumbo nucifera*, is a perennial aquatic herb belonging to the

[110] Bujizhen Chronicles Compilation Committee布吉鎮志編纂委員會 (2009). *Buji Town Chronicle*布吉鎮志. Shenzhen: Bujizhen Chronicles Compilation Committee深圳:布吉鎮志編纂委員會. (pp. 194–197).

[111] Wang, R. B.王若兵(Eds.). (2001). *Water Affairs Chronicle of Shenzhen*深圳市水務志. Shenzhen: Haitian Publishing House深圳:海天出版社. (pp. 179).

[112] Bujizhen Chronicles Compilation Committee布吉鎮志編纂委員會(2009). *Buji Town Chronicle*布吉鎮志. Shenzhen: Bujizhen Chronicles Compilation Committee深圳:布吉鎮志編纂委員會. (pp. 131); Shenzhen Municipal Chronicles Compilation Committee深圳市地方誌編纂委員會(2009). *Chronicles of Shenzhen: Volume on Infrastructure Construction*深圳市志·基礎建設卷. Beijing: Fangzhi Publishing House北京: 方志出版社. (pp. 886).

[113] Shenzhen Urban Management and Comprehensive Law Enforcement Bureau深圳市城市管理和綜合執法局 (2021). *Shenzhen in the Parks*公園裡的深圳. Shenzhen: Haitian Publishing House 深圳:海天出版社. (pp. 117).

Nymphaeaceae family.[114] Research in the natural sciences, including wetland research, plant research, and environmental protection, has consistently demonstrated that lotus plants possess a remarkable capacity to absorb nitrogen and phosphorus nutrients across a wide range of concentrations. As a result, lotus plants are highly effective in mitigating water eutrophication and improving water quality.[115] In the summer of 1989, vast expanses of lotus blossoms flourished in the Lianxiang Lake (莲香湖) and Xiaoyao Lake (逍遥湖) regions of the park, covering hundreds of square meters. In that particular year, Honghu Park seized the chance to organize the first Lotus Festival, showcasing about 600 containers of lotus plants from over 70 different species. This event successfully drew the attention of over 10,000 visitors. The Lotus Festival has been successfully hosted for 34 consecutive years as of 2022, with the exception of 1991 and 1993 when it was suspended due to lake dredging and urban flooding,[116] as of 2022, the Lotus Festival has been successfully held for 34 consecutive sessions.[117]

The exceptional significance of the lotus in the spiritual realm of the Chinese attributes to the success. In the early twenthieth century, Chinese intellectuals began incorporating Western natural disciplines, which subsequently influenced and transformed their own approaches and methodologies for problem-solving. Nevertheless, within the context of ancient Chinese culture, the lotus holds a distinct significance that diverges from its interpretation in modern botany. Furthermore, it possesses a more extensive historical legacy and profound humanistic implications compared to the previously described lychee.

> Uphill stands mulberry(山有扶蘇);
> And lotus in the pool(隰有荷華).
> The handsome I don't see(不見子都);
> Instead I see a fool(乃見狂且).
> Uphill stands a pine-tree(山有橋松);
> And in the pool leaves red(隰有游龍).
> The pretty I don't see(不見子充);
> I see the sly instead (乃見狡童).
> ——Songs Collected in Zheng, Modern Henan·A Joke(詩經·鄭風·山有扶蘇)[118]

---

[114] Chinese Academy of Sciences, Editorial Committee of Flora of China中國科學院中國植物志編輯委員會(1979). *Flora of China (Volume 27)*中國植物志(第27卷). Beijing: Science Press北京:科學出版社. (pp. 032).

[115] Thongtha, S., Teamkao, P., Boonapatcharoen, N., Tripetchkul, S., Techkarnjararuk, S., & Thiravetyan, P. (2014). Phosphorus removal from domestic wastewater by Nelumbo nucifera Gaertn. and Cyperus alternifolius L. *Journal of environmental management*, 137, 54–60.

[116] Shenzhen Municipal Chronicles Compilation Committee深圳市地方誌編纂委員會. (2009). *Chronicles of Shenzhen: Volume on Infrastructure Construction*深圳市志·基礎建設卷. Beijing: Fangzhi Publishing House北京: 方志出版社. (pp. 733).

[117] Shenzhen Urban Management and Comprehensive Law Enforcement Bureau深圳市城市管理和綜合執法局(2021). *Shenzhen in the Parks*公園裡的深圳. Shenzhen: Haitian Publishing House深圳:海天出版社. (pp. 117).

[118] Xu, Y. C. (Trans.) 許淵沖 譯. (2021). *Book of Poetry Book of Songs: Feng (Chinese-English Bilingual Edition)* 許淵沖譯詩經·風:漢英對照. Beijing: China Translation and Publishing Corporation北京:中譯出版社.

The uncommon West Lake in the midst of sixth moon(畢竟西湖六月中);
Displays a scenery to other months unknown(風光不與四時同).
Green lotus leaves outspread as far as boundless sky(接天蓮葉無窮碧);
Pink lotus blossoms take from sunshine a new dye(映日荷花別樣紅).
——Yang Wanli(楊萬里) in Song Dynasty·The Lakeside Temple at Dawn (曉出淨慈寺送林子方)[119]

…but my favourite is the water-lily. How stainless it rises from its slimy bed! How modestly it reposes on the clear pool—an emblem of purity and truth! Symmetrically perfect, its subtle perfume is wafted far and wide, while there it rests in spotless state, something to be regarded reverently from a distance, and not be profaned by familiar approach.

予獨愛蓮之出淤泥而不染，濯清漣而不妖，中通外直，不蔓不枝，香遠益清，亭亭淨植，可遠觀而不可褻玩焉。

——Zhou Dunyi(周敦頤) in Song Dynasty·The Language of Flowers(爱莲说)[120]

This is but a little glimpse into the vast number of poems dedicated to the lotus in ancient China. Out of all of them, Zhou Dunyi's "Love of the Lotus" is the most extensively disseminated. This poem constructs a noble sentiment for lotus flowers: they are unwilling to be tainted by association with "dirty mud," a meaning evidently incomprehensible to a positivistic worldview that only takes observed facts as the basis.[121] Just as in Chinese tradition, the color red is associated with joy and luck, while from a positivist perspective, red merely corresponds to visible light with a wavelength ranging from 380 to 780 nanometers. Another illustration is the Pavilion of Surging Waves, located in Suzhou. The Chinese perceive the hermit emotions and life philosophies of the past, whereas positivists can only interpret it as a square structure situated on a rock formation. Chinese individuals possess a remarkable ability to attribute a humanistic significance to the realm of human experience, and this is equally applicable to the lotus blossoms seen at Honghu Park. Initially, Honghu Park had no association with lotus blossoms throughout its creation:

> Regarding plant design, the northern region is primarily characterized by the prevalence of southern palm trees, the central region is dominated by southern broadleaf trees, and the southern region by coniferous and aquatic plants… At that time, the design did not include a lotus theme, which might have been a later addition by the park's administration….[122]

The initial intention of introducing them in 1988 was to address water pollution and enhance the aquatic landscape. However, throughout the following operation and maintenance of Honghu Park, lotus blossoms acquired additional significance within the historical and social backdrop of that period. Like the industrial metropolises of New York and Chicago throughout the nineteenth century, Shenzhen in the 1980s and 1990s experienced issues with inadequate public security, disorderly

[119] Xu, Y. C. Xu, M. (Trans.) 許淵沖 許明 譯. (2021). *Goden Treasure of Quatrains and Octaves (Chinese-English Bilingual Edition)* 許淵沖譯千家詩:漢英對照. Beijing: China Translation and Publishing Corporation北京:中譯出版社.

[120] Trans. by the author.

[121] Ou, L. T. 歐力同(1987). *Kong De and His Positivism*孔德及其實證主義. Shanghai: Shanghai Academy of Social Sciences Press上海:上海社會科學院出版社. (pp. 28).

[122] According to the interview with Mr. Shen Baojiu, 2023–02–12.

ethics, frequent criminal activities, and the widespread presence of pornography, gambling, and narcotics.[123] The government's public security and law enforcement departments have frequently undertaken initiatives such as "crackdowns (嚴打)" and "anti-gang operations (反黑清幫). "[124] Thus, the "Seventh Five-Year Plan (七五計畫)" (1986–1990) released by the Shenzhen Municipal Government in August 1986, expressly included the development of spiritual civilization as a key objective:

> ...Specifically, we need to enhance intellectual and political efforts, resolutely resist the influence of decaying capitalist ideas and lifestyles, and employ a vivid and entertaining approach to promote patriotism and collectivist education... the special zone is situated inside an intricate and multifaceted environment... The frequent influx of foreign exchanges and the infiltration of corrupt capitalist ideas have led to the decision that the special zone must prioritize the development of spiritual civilization. Efforts will be made to promote our country's outstanding national art, while simultaneously combating and prohibiting pornographic and other detrimental cultural influences.[125]

In addition, the "Seventh Five-Year Plan" explicitly highlighted the correlation between urban greening and the construction of spiritual civilization:

> Enhancing environmental conservation and implementing urban greening initiatives is not solely an issue of urban planning, but also a crucial component in fostering spiritual civilization and enhancing the cultural well-being of individuals.[126]

Urban parks have long been recognized by the government to promote spiritual civilization and improve the social environment. The lotus blossom, with its anthropomorphic aspect of emerging from the dirt unstained, has the potential to influence the civic conscience of this immigrant metropolis, drawing from a rich historical legacy. The individual accountable for the inaugural Lotus Festival recalled:

> The lotus characterizes being untainted while emerging from muddy surroundings... The social impact is undeniably substantial, as it plays a crucial role in promoting transparent governance and undoubtedly contributes significantly to public education. The primary objective is to cleanse water and offer landscape functionalities. Over time, people in Shenzhen increasingly acknowledges the roles of Honghu Park in promoting good governance and education... The lotus is often associated with incorruptibility and cleanliness due to the popular statement "out of mud but not stained."[127]

---

[123] Shenzhen Local Chronicles Compilation Committee深圳市地方誌編纂委員會(2006). *Chronicles of Shenzhen: Volume on Legal System and Administrative Affairs*深圳市志·法制政務卷. Beijing: Fang Zhi Publishing House北京: 方志出版社. (pp. 123–136).

[124] Shenzhen Local Chronicles Compilation Committee深圳市地方誌編纂委員會(2006). *Chronicles of Shenzhen: Volume on Legal System and Administrative Affairs*深圳市志·法制政務卷. Beijing: Fang Zhi Publishing House北京: 方志出版社. (pp. 137–141).

[125] Shenzhen Development and Reform Commission深圳市發展和改革委員會 (2017). *Shenzhen's Five-Year Plans*深圳歷次五年規劃. Shenzhen: Haitian Publishing House深圳:海天出版社. (pp. 29; 81–82).

[126] Shenzhen Development and Reform Commission深圳市發展和改革委員會 (2017). *Shenzhen's Five-Year Plans*深圳歷次五年規劃. Shenzhen: Haitian Publishing House深圳:海天出版社. (pp. 29; 81–82).

[127] According to the interview with Mr. Huang Chongyuan, 2023–06–15.

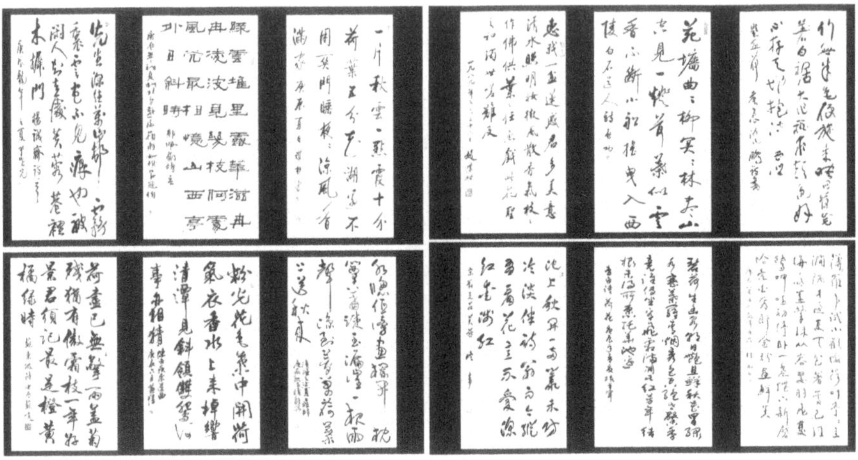

**Fig. 4.20** Many renowned calligraphers inscribed poems about lotus on the walls of the stele gallery

Crime can be seen as a symbolic representation of social contamination, while lotus is associated with the contrasting concept of involvement or collaboration. After it, a sequence of construction endeavors centered around the motif of lotus took place in Honghu Park. The government aims to instill law-abiding behavior among citizens by including the idea of lotus in the park through the naming of scenic sites, poetry, plaques, and couplets. As illustrated in Fig. 4.20, the construction of the stele gallery for chanting the lotus is scheduled to take place in 1994. It has been meticulously crafted by numerous renowned contemporary calligraphers across the country, including Zhao Puchu (趙朴初), Qi Gong (啟功), Shen Peng (沈鵬). By extolling the lotus, it evokes individuals' sentiments of integrity and self-driven determination, inciting clarity amidst confusion and fostering purity, with the park management's aim of nurturing the spirit and fostering societal harmony.[128]

Since the 1990s, Lotus symbols have been extensively utilized in park features, including paving, landscape walls, and bridges. Furthermore, there are aquatic landscapes symbolizing the act of cleansing one's conscience to atone for previous errors, as well as sculptures depicting lotus flowers that symbolize a refuse to embrace the spring breeze. The term "lian(廉)," which is a homophone of "lotus," has also been inscribed on stones and has become a point of attraction in the park to advocate for the integrity of officials. The Chinese homophony of the term "Lotus" has also been expanded to encompass officially recognized meanings such as "harmony," "cooperation," and "core," which are also depicted in sculptures. These meanings indirectly evoke the significance that the lotus holds for citizens.

[128] Shenzhen Honghu Park Management Office深圳市洪湖公園管理處 (2009). *Lotus*荷花. Beijing: China Science and Education Publishing House北京:中國科教出版社. (pp. 98).

The cultural value of the lotus is continuously being examined and enhanced, not just for the public but also for government authorities. In 2004, a bridge pavilion in the park was adorned with couplets and renamed the "Qingyi (清漪)" pavilion to commemorate the integrity of government officials:

> Integrity is a quality that will benefit the people and lead to prosperity in all aspects. The water emanating from the newly bloomed lotus will possess a pristine quality, permeating every body of water it encounters. The pristine water in the marsh exudes an abundance of aesthetic appeal for the visitors. You will be able to enjoy the pristine clarity of Honghu Lake's water.

Honghu Park organized an exhibition in September 2017 showcasing calligraphy and painting works centered around the theme "Clean Breeze and Lotus Charm, Utilizing Lotus as a Representation of Integrity."

May party members, cadres, community residents, and tourists be imbued with the ethos of ethical governance while enjoying themselves and appreciating the beauty of lotus blossoms during their leisure activities.

Furthermore, multiple collaborative endeavors were undertaken between the government's public security law enforcement department and Honghu Park. In December 2015, the Shenzhen Municipal Public Security Bureau inaugurated an "Anti-Drug Warning Education Base (禁毒警示教育基地)" within Honghu Park. In August 2018, it inaugurated an "Anti-Cult Warning Education Base (反邪教警示教育基地)." The "Luohu District New Era Civilization Practice Centre (羅湖區新時代文明實踐中心)" opened in June 2022 at the Dutch Exhibition Hall located within the park.

Michael Foucault's concept of "gouvernement" pertains to the fundamental regulation of life itself, rather than solely focusing on the body, since the ruler embodying royal authority has vanished. This concept is strongly intertwined with the notion of "police," resulting in politics being transformed into a form of "art de gouverner." Since the 1990s, Honghu Park has served as a flood detention district to manage Buji River floods, and the presence of lotus plants in the park has been intended to address water pollution in the Buji River. Both of them actively participated in social governance by utilizing particular significance derived from traditional culture, thereby embodying the concept of "art de gouverner." It did not involve physical punishment, but rather a more cunning method of disciplining incoming immigrants from all over China, including government officials.

## 4.6 Summary

Through the study of landscape history in the Shangbu-Luohu area, it becomes evident that it possesses a rich history before reform and opening up, rather than presenting itself as a "blank slate." Over the course of centuries, a small-scale commodity economy and corresponding social networks have emerged within this agricultural plain landscape. However, the influx of Hong Kong capital following the reform and opening up reshaped this agricultural plain landscape. Within the global industrial division of labor, the special economic zone seized opportunities to integrate with Hong Kong's industrial transfer and foster the development of the "Three-Processing and One Compensation" manufacturing industry. This, in turn, propelled the construction of numerous residential areas and public service facilities. Over a decade, all of these urban constructions depleted the land resources of the agricultural plain, transforming it into a modern urban landscape. This transformation is not only evident in physical form but also in terms of land ownership. Under China's dual urban-rural land system, vast expanses of land previously collectively owned by rural villages transitioned into state-owned land under local government control. Whether for the construction of industrial zones, residential areas, transportation infrastructure, or urban parks, all acquisitions of land resources occurred within such institutional frameworks. In the 1980s, the land resources utilized for constructing the four major parks in the Shangbu-Luohu area primarily originated from the collective land of several production brigades, including Caiwuwei, Hubei, Cainong, and Shuibei and Tianbei. The swift change in ownership subsequently brought about morphological transformations. Through specific aesthetic principles in urban park planning and design, these farmlands, vegetable plots, orchards, and fishponds were repurposed as landscape resources within urban parks, thereby imbuing them with a sense of urbanized aesthetics. Throughout this transition, the land retained a facade of "natural" appearance, yet its cultural form underwent significant changes. This marks a shift in significance and power relations.

In the early 1980s, the Shangbu-Luohu area constructed four parks atop its plain topography, serving the citizens and bolstering the confidence of external investors. These urban parks, drawing inspiration from traditional garden-making principles, transformed former agricultural land into urban parks with traditional garden features, showcasing corresponding national characteristics. The concept of creating urban parks with national characteristics originates from three aspects: (1) nation-state system and the discourse of stylistic classification in architectural history; (2) formalized nationalism and socialist design principles derived from the former Soviet Union; (3) the profound classical garden culture and natural aesthetic tradition of the country. In fact, including China, many Asian countries and regions experienced economic take-off around the 1970s, largely stemming from the nation-state power system. Behind the national style evident in the urban construction of the Shangbu-Luohu area in the early 1980s lies the emphasis on national identity during China's initial integration with globalization. As understood by Crang, within the context of the nation-state system, the state will establish national images through symbols, power, and meaning will also be imbued into human–land relationships. In urban parks, this national identity is manifested not only through architecture but also through the terrain-water system. However, before the construction of the parks, the natural elements such as plants, topography, and water systems in this area were entirely associated with agricultural production and did not possess any national characteristics, which signifies a meaningful transformation of nature.

For Lychee Park, this transformation also reflects the corresponding economic significance. The 1980s marked a period when Shenzhen integrated with neoliberal globalization. In neoliberal urban governance, economic growth takes absolute political priority. While the relationship between China's reform and opening up and neoliberalism is complex, urban festivals as a strategy for promoting economic growth under neoliberal urban governance were indeed adopted in the SSEZ. In the late 1980s, the Lychee Festival held in Lychee Park became a crucial pivot for foreign trade and economic work, making significant contributions to investment promotion, urban marketing, and other aspects. During this period, lychee trees and the lychee forest in Lychee Park cannot be viewed solely through the lens of modern botany; instead, they should be understood as a cultural form of nature intertwined within the social and historical context of the special economic zone at that time. During the Lychee Festival, the lychee forest in Lychee Park was managed by professional fruit farmers to increase yield. However, following the high-tech transformation of urban industries, the Lychee Festival was discontinued, and the lychee forest was sprayed with chemical pesticides for fruit sterilization due to management reasons. At this point, the lychee forest was neither natural nor unnatural but rather represented a cultural form of nature.

While Lychee Park exemplifies nature's positive role in economic development, the case of Honghu Park presents a stark contrast. Situated in the middle reaches of the Buji River, Honghu Park was originally the farmland of Shuibei and Tianbei villages. Due to frequent flooding downstream in the Luohu area, it was designated as a flood detention district and subsequently transformed into an urban park to improve the city's investment environment and provide leisure spaces for citizens. However, upstream along the Buji River, Buji Town, known as the "Top Economically Dominant Town in Guangdong," began vigorous industrial development in the late 1980s. The wastewater discharged, along with sedimentation, accumulated in Honghu Park, severely damaging the park environment and illustrating the negative impact of economic development on nature. The construction of a sediment drainage channel addressed the park's pollution issues, while previously, Honghu Park attempted to introduce lotus flowers to mitigate water pollution and enhance the lake's aesthetics. The literary imagery of lotus flowers in traditional Chinese culture, symbolizing purity despite emerging from mud, was employed in the park's construction and management. A series of attractions and designs revolving around this literary imagery aimed to cultivate civic consciousness in this emerging immigrant city amidst the chaotic public security environment of Shenzhen in the 1980s and 1990s. Honghu Park thus became an exemplary "art de gouverner": not only treating pollution but also governing society.

Whether it be the terrain-water system structure that presents a "national gate image" with ethnic features, the gently sloping lands in Lychee Park planted with lychee groves that engage in urban marketing, the terrain and economic highlands of Buji Town that serve as the source of silt for Honghu Park, or the park waters in Honghu Park that can hold a vast area of lotuses and "governing society," all are closely tied to the topographical conditions of the Shangbu-Luohu plains. In the 1980s and 1990s, the numerous urban parks constructed on this plain either facilitated urban development or suffered negative impacts from it, reflecting the richness and complexity of the relationship between nature and urban social life.

# Chapter 5
# Meilin-Gangxia Park Group: The Humanization of Hills

## Contents



**Abstract** This chapter examines the urban parks developed in the hilly terrain of the Meilin-Gangxia area around the turn of the millennium. Through an analysis of historical processes such as population resettlement and forest landscape transformation related to these parks, the chapter demonstrates how these developments were connected to Shenzhen's pursuit of becoming an "international city" and its deeper integration into the global market at the time.

**Keywords** Meilin-Gangxia · Hilly area · International city · Globalization



X. Liu, *Fabricating Nature*, The Urban Book Series,
https://doi.org/10.1007/978-981-95-8208-2_5

# 5.1 The Landscape History of Meilin-Gangxia Area

## *5.1.1 Before Reform and Opening Up*

The Shangbu-Luohu area, constructed in the 1980s, is widely recognized as the inaugural urban center in the Shenzhen Special Economic Zone (SSEZ). The land and space resources throughout the entire special economic zone have been consumed rapidly due to the process of urbanization and considerable landscape transformation. By 1990, given the same circumstances, the Shangbu-Luohu area had almost exhausted all its accessible acreage for development. The city's growth into the Meilin-Gangxia area in the west was inevitable as a result of geographical limitations. The topographical features of mountains and hills in the northern and eastern directions hindered urban sprawl.

The north of the Meilin-Gangxia area is the Tanglang Mountain (塘朗山)-Meilin Mountain (梅林山)-Yinhu Mountain range (銀湖山系), a consisting segment of the Second Line Mountain System. The south of the area is the coastal alluvial plain in the special economic zone. With the exception of Lianhuashan Hill (蓮花山) and Bijiashan Hill (筆架山), the majority of the area is flat.[1] Figure 5.1 illustrates the

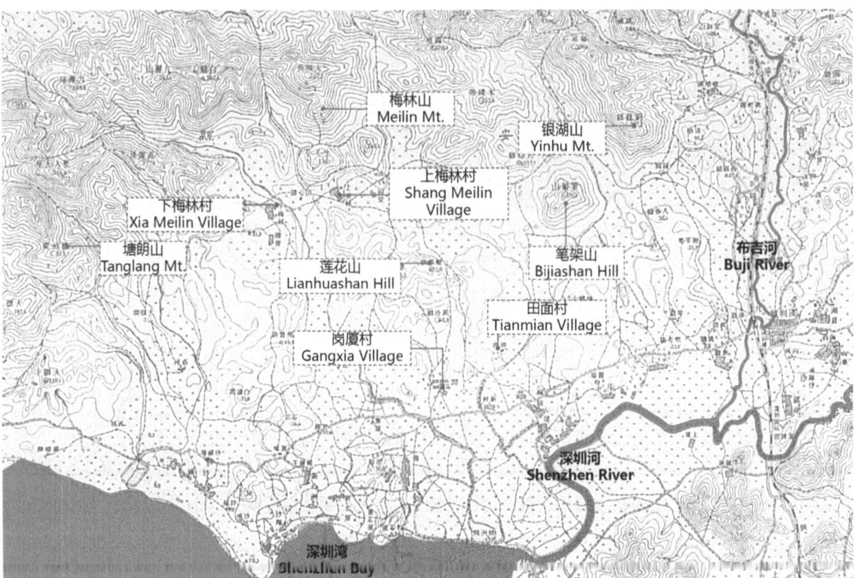


**Fig. 5.1** Meilin-Gangxia Area in 1930 [Source: Taipei Academia Sinica Digital Map System, Data Code: map_moi1_01802]

[1]Compilation Committee of Local Records of Futian District 深圳市福田區地方誌編纂委員會. Shenzhen City. *Futian District Annals 1979–2003*深圳市福田區志1979–2003. Beijing: Fangzhi Publishing House北京:方志出版社. (pp. 32).

geographical layout of the northern near-mountain region, which comprises four primary villages: Shangmeilin, Xiameilin, Gangxia, and Tianmian. Expansive forests and agricultural fields surround these villages in the pre-mountainous areas. To the south of the region, the Shenzhen River courses its way toward the Shenzhen Bay. Additionally, there are settlements situated here on the alluvial plains, including Xiasha, Shangsha, Huanggang, and Futian.

The establishment of Xiameilin Village, currently inhabited by the Zheng clan, can be traced back to the early Ming Dynasty at the latest, approximately 600 years ago.[2] Aside from Xiameilin, the Meilin area is also home to Shangmeilin, also a large community located in the northeastern of Xiameilin, which was inhabited by the Huang clan almost 800 years ago.[3] The two predominant clans in the Meilin area established their settlements with a time gap of over two centuries; however, they encountered nearly identical natural and topographical circumstances. The Meilin area exhibits a topography characterized by mountainous and hilly terrain, with a generally flat terrain except for the presence of the mountain range in the northern part. The flat terrain of the Meilin area is oriented east-west, with the northeastern region at a higher elevation and the southwestern region at a lower elevation. There are southern mountains and hills, such as Lianhuashan Hill, and northern mountains, including Meilin Mountain and Bijiashan Hill.[4] For ancestors in the past, the region was located at a considerable distance from the sea, which posed challenges in terms of exploiting marine resources for fishing purposes. Nevertheless, the mountains were teeming with Chinese bayberries (Yangmei), hence the appellation "Meilin." The topography of hills and mountains rendered it favorable for successive generations of the Zheng and Huang lineages, who primarily pursued agricultural activities. The Gangxia Village, located south of Lianhuashan Hill, has residents engaged in both agricultural and fishing activities, which was a little bit different from Zheng and Huang lineages. Gangsha Village is situated to the south of the coastal alluvial plain. During the early Ming Dynasty, namely in 1374, the Wen clan laid the foundation in Gangxia. This historical event is documented in the *Gangxia Wen Clan Genealogy (崗廈文氏族譜):*

*Our ancestor initially located at Shibu, subsequently moved to Zhoutou… and ultimately reached the footing of a mountain. Observing a location encompassed by water on three sides and mountains on the remaining side, featuring rolling peaks that overlook the sea, which was beneficial for both agricultural and fishing activities… the village was given the name "Gangxia Village."*[5]

---

[2] Shenzhen Futian District Historical Records Office 深圳市福田區史志辦公室 (2012). *Futian Tracing the Roots* 福田溯源. Shenzhen: Haitian Publishing House 深圳:海天出版社. (pp. 95).

[3] Shenzhen Futian District Historical Records Office深圳市福田區史志辦公室(2012). *Futian Tracing the Roots* 福田溯源. Shenzhen: Haitian Publishing House 深圳:海天出版社. (pp. 95).

[4] Compilation Committee of *Shangmeilin Village Chronicle*《上梅林村志》編纂委員會(2007). *Shangmeilin Village Chronicle*上梅林村志. Haikou: Southern Publishing House海口: 南方出版社. (pp. 8).

[5] Gangxia Wen's Clan Association Limited崗廈文氏宗親會有限公司, Shenzhen Gangxia Industrial Co., Ltd. 深圳市崗廈實業股份有限公司.*Gangxia Wen's Genealogy*崗廈文氏族譜,

Furthermore, there is the Tianmian Village where people with various surnames, such as Lin, Li, Zhang, coexist. It was established during the Kangxi period of the Qing Dynasty and is the smallest indigenous village in Futian District. Before the reform and opening up, it functioned as a production team in Gangxia, and later, it gained independence and became an independent village.[6] After the PRC was founded, Meigang Township[7] was collectively referred to as Gangxia Village, along with Shangmeilin and Xiameilin villages. According to the statistics of Futian Commune in October 1979, the then state of cultivated land in the main villages, specifically Shangmeilin, Xiameilin, Gangxia, and the southern villages, is as shown in Table 5.1:

Prior to the reform and opening up, the local predecessors gradually enhanced the flat terrain in the Meilin-Ganglia area, transforming it into arable land. Due to the prosperity of future generations and the scarcity of available fields, the local predecessors deliberately transformed the surrounding hillside property into a plantation for cultivating various fruit trees, including bayberry, lychee, and others.[8]

### *5.1.2 Three-Processing and One Compensation Period*

During the initial stages of the reform and opening up, the Meilin-Gangxia area possessed a substantial amount of arable land. Prior to the government's initiation of land acquisition, each village in the Meilin-Gangxia area possessed several fishponds,

**Table 5.1** Land cultivation status of various brigades in Futian Commune[a]

| Brigade name | Amount of families | Population | Area of arable land(mu) | |
|---|---|---|---|---|
| | | | Water field | Dry field |
| Shangmeilin | 142 | 604 | 1276 | 160 |
| Xiameilin | 200 | 809 | 1920 | 210 |
| Gangxia | 291 | 1063 | 2095 | 710 |
| Shangbu | 314 | 1188 | 2029 | 808 |
| Futian | 302 | 1126 | 2200 | 1160 |
| Huanggang | 467 | 1944 | 3145 | 1210 |
| Shangsha | 236 | 901 | 1916 | 548 |
| Xiasha | 325 | 1264 | 1497 | 981 |

[a]Shenzhen Futian District Historical Records Office 深圳市福田區史志辦公室(2012). *Futian Tracing the Roots* 福田溯源. Shenzhen: Haitian Publishing House 深圳:海天出版社. (pp. 9).

preserved at the Genealogy Reading Room of Bao'an District Library, Shenzhen藏於深圳市寶安區圖書館族譜閱覽室.

[6]Wu, J. X.吳錦雄(Eds.). (2016). *Village Histroy and Intangible Cultural Heritage of Futian District*深圳市福田區村史與非遺文化.Shenzhen: Shenzhen Huifengtang Cultural Communication Co., Ltd. 深圳: 深圳市惠風堂文化傳播有限公司. (pp. 023).

[7]Shenzhen Futian District Historical Records Office 深圳市福田區史志辦公室(2012). *Futian Tracing the Roots* 福田溯源. Shenzhen: Haitian Publishing House 深圳:海天出版社. (pp. 54–55).

[8]Shenzhen Futian District Historical Records Office 深圳市福田區史志辦公室(2012). *Futian Tracing the Roots* 福田溯源. Shenzhen: Haitian Publishing House 深圳:海天出版社. (pp. 150).

orchards, and vast expanses of mountains and forests,[9] in addition to the cultivated land recorded in Table 5.1. In an agrarian society, the symbiotic relationship between people and land was crucial, and for centuries, their existence supported the economic and social functioning of the agricultural settlements in the Meilin-Gangxia area. However, soon they were engulfed in the rapid urbanization process of the SSEZ.

As shown in Table 5.2, in 1981, the agricultural land area in Shangmeilin declined by 108 mu,[10] which was the third year since the founding of Shenzhen as a city and the second year since the establishment of the special economic zone. With the development of the "Three-Processing and One Compensation" industries in the Meilin area, these agricultural lands are rapidly being converted for urban construction, primarily into various types of industrial factories.[11] By the end of 1990, the Shenzhen Municipal Government had acquired all the land in Shangmeilin Village, reserving only 345 mu of land for the village's own use.[12]

As early as 1986, the *Master Plan of Shenzhen Special Economic Zone (1986)* had already outlined the inclusion of 15 industrial zones within the special economic zone. Four of these were situated in Futian District: Meilin Industrial Zone (108 ha.), Bagualing Industrial Zone (85 ha.), Shangbu Industrial Zone (89 ha.), and Huanggang Industrial Zone (236 ha.). Among them, Meilin Industrial Zone spanned both sides of Beihuan Road, with Lianhuashan Hill to the south. During that time, the developed area of Meilin Industrial Zone accounted for 14.6%, approximately 15.77 ha. As shown in Fig. 5.2, the surrounding area was also planned as urban parks, public buildings, and various types of residential land, much of which was located on previously cultivated and forested land.

**Table 5.2** Land statistics of Shangmeilin Village in 1981[a]

| Farmland (mu) | | | | | | |
|---|---|---|---|---|---|---|
| | Year-ending | | | Fruit and vegetable(mu) | | |
| Year-beginning | Water field | Dry field | Forestry(mu) | Vegetable | Banana | Lychee |
| 1136 | 1028 | | 12,050 | 535 | 20 | 200 |
| | 993 | 35 | | | | |

[a]Compilation Committee of *Shangmeilin Village Chronicle*《上梅林村志》編纂委員會(2007). *Shangmeilin Village Chronicle* 上梅林村志. Haikou: Southern Publishing House海口: 南方出版社. (pp. 120–133).

[9]Compilation Committee of *Shangmeilin Village Chronicle*《上梅林村志》編纂委員會(2007). *Shangmeilin Village Chronicle* 上梅林村志. Haikou: Southern Publishing House海口: 南方出版社. (pp. 8).

[10]Compilation Committee of *Shangmeilin Village Chronicle*《上梅林村志》編纂委員會(2007). *Shangmeilin Village Chronicle* 上梅林村志. Haikou: Southern Publishing House海口: 南方出版社. (pp. 121).

[11]Compilation Committee of *Shangmeilin Village Chronicle*《上梅林村志》編纂委員會(2007). *Shangmeilin Village Chronicle* 上梅林村志. Haikou: Southern Publishing House海口: 南方出版社. (pp. 139–141).

[12]Compilation Committee of *Shangmeilin Village Chronicle*《上梅林村志》編纂委員會(2007). *Shangmeilin Village Chronicle* 上梅林村志. Haikou: Southern Publishing House海口: 南方出版社. (pp. 8).

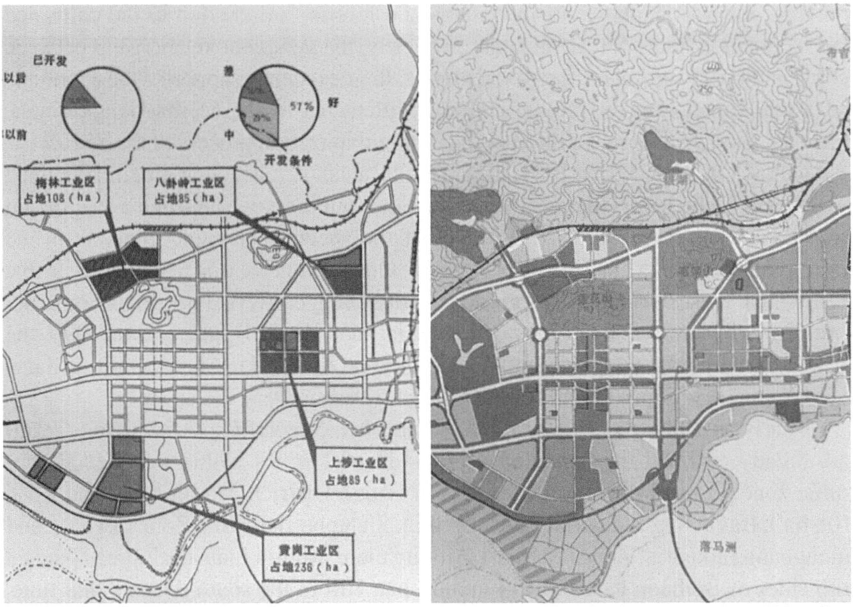


a) Industrial Land in " Master Plan of Shenzhen Special Economic Zone (1986)"

b) Land Use Planning in " Master Plan of Shenzhen Special Economic Zone (1986)"

**Fig. 5.2** Meilin Industrial Zone and surroundings in the 1986 edition of Urban Master Plan. (**a**) Industrial land in "Master Plan of Shenzhen Special Economic Zone (1986)". (**b**) Land use planning in "Master Plan of Shenzhen Special Economic Zone (1986)"

### *5.1.3 Initial Phase of Establishing an International City*

In the early 1990s, the shift of the world economic center toward the Asia-Pacific region had already become a noticeable trend. The Shenzhen Municipal Government has acknowledged the limitations of previous growth efforts that prioritized low-end industries and has recognized the need for industrial restructuring.[13] Consequently, throughout the early 1990s, the goal and consensus among various government departments were to develop Shenzhen into an international city. Li Hao, the ex-mayor of Shenzhen, expressed:

*An international metropolis is characterized by its multifaceted urban functions, akin to places such as Hong Kong and Singapore, not just industrial and tourism*

[13] Shenzhen Municipal Political Consultative Conference Cultural深圳市政協文化文史和學習委員會.Historical, and Learning Committee (2010). *The Rise of Shenzhen's Four Pillar Industries: High-Tech深圳四大支柱產業的崛起·高新技術*. Beijing: China WenShi Publishing House北京: 中國文史出版社. (pp. 44–54).

*aspects. Primarily, it functions as a transportation hub, a commerce center, and an international financial hub.*[14]

During that period, the development of an international city involved multiple facets. Shenzhen's role as the region's financial center was of utmost importance since it served as a catalyst for the globalization of different urban functions. Furthermore, it became imperative to enhance the advancement of high-tech industries, attract talented individuals, and enhance urban governance and infrastructure to meet global benchmarks, as part of the process of becoming an international metropolis.[15] Such concept of an "international city" aligns closely with the urban theorist Saskia Sassen's concept of a "global city," emphasizing the importance of finance, high technology, markets, and the city's position within the world economy.[16] Given these principles, it is important to analyze how the transformation in urban landscape represented the government's perception of a global metropolis.

By 1990, the Luohu area had exhausted its available space for urban development, but the Meilin-Gangxia area to its west still contained extensive farmland that could be converted into urban construction with Futian district to the west. As shown in Fig. 5.3, the entire Futian district, including the Meilin-Gangxia area, was still predominantly agriculture. Apart from some localized industrial zones, large-scale urban construction had not yet commenced, providing the spatial requirements for the new industry. As early as 1989, the Futian district government allocated funds for land acquisition, with the intention of requisitioning these farmlands from rural collectives. In 1990, a combined area of 80 ha. of rural land was acquired, with 766.2 ha. of land from multiple villages were designated for the development of the new urban center.[17] The total area of both farmland and forested land in Futian district had a dramatic reduction from 2333 ha. in 1987 to 1301 ha. by 2003.[18] All these changes in the landscape were undertaken to align with the grand vision of the government in developing the financial industry and constructing an international city.

Furthermore, in the development of high-tech industries, several industrial zones planned and constructed in the 1980s in Futian are also aligning themselves with the

---

[14] Li H.李灝(1993). Taking practical steps and actively and steadily developing Shenzhen into an international city將深圳建成綜合性國際城市(代序) In Lin, Z. J. 林祖基 編(Eds.), *Essays on Shenzhen - An International City*深圳-國際性城市論文集. Beijing: China Economic Publishing House北京:中國經濟出版社. (pp. 4–11).

[15] Li H.李灝 (1993). Taking practical steps and actively and steadily developing Shenzhen into an international city. In Lin, Z.林祖基(Eds.). *Essays on Shenzhen - An International City*深圳-國際性城市論文集. Beijing: China Economic Publishing House 北京:中國經濟出版社. (pp. 4–11).

[16] Sassen, S. (1991). *The global city: New York, London, Tokyo (second ed.)*. Princeton University Press. (pp. 1–3).

[17] Chen, Y. X.陳一新 (2015). *A Thirty-Year Historical Study on Urban Planning and Construction of Shenzhen Futian Central Business District (CBD)* 深圳福田中心區(CBD)城市規劃建設三十年歷史研究. Nanjing: Southeast University Press南京:東南大學出版社. (pp. 41).

[18] Shenzhen Futian District Local Records Compilation Committee.深圳市福田區地方誌編纂委員會 (2012). *Futian District Annals, Shenzhen 1979–2003*深圳市福田區志1979–2003. Beijing: Fangzhi Publishing House. 北京:方志出版社. (pp. 503–506).

**Fig. 5.3** A comparison between Meilin-Gangxia Area and Shangbu-Luohu Area (1990) [Chen, Y. X.陳一新 (2015). *A Thirty-Year Historical Study on Urban Planning and Construction of Shenzhen Futian Central Business District (CBD)* 深圳福田中心區(CBD)城市規劃建設三十年歷史研究. Nanjing: Southeast University Press南京:東南大學出版社. (pp. 3;16).]

transformation. *Urban Master Plan of Shenzhen Municipality (1996–2010)* aimed at gradually moving out industrial zones from the central area of Futian District, thus providing space for the development of relevant functions (Table 5.3).[19]

Subsequently, the government detailed these requirements in the specifically formulated *Futian District Zoning Plan in Shenzhen City* (深圳市福田區分區規劃) (1998–2010) emphasizing the preservation of the existing functional layout of the Meilin Industrial Zone, with a focus on industrial and residential purposes. Among them, the industrial sector needs to undergo industrial restructuring, moving toward high-tech industries. The residential component requires structural organization and supplementary public facilities to form a concentrated and complete urban living area of medium to high quality.[20]

In the 1990s, the industrial zones developed in the 1980s in Futian District underwent a progressive transformation: (1) Aligning with the industries and functions of the Futian Central Area, including finance, commerce, and administration. (2) Shifting conventional low-end industries toward advanced manufacturing and high-tech industries. (3) These two aspects, in turn, drove the surrounding urban spaces to provide living facilities for the newly arrived personnel and institutions.

Several urban parks such as Meilin Park, Lianhuashan Park, Bijiashan Park, Huanggang Park, and Central Park were all completed and made accessible around the year 2000 in relation to urban management and municipal construction. The "Rectification Action (梳理行動)" aimed at clearing illegal constructions across the city also unfolded swiftly in 2003, which will be detailed in the subsequent text.

[19] Shenzhen Municipal People's Government 深圳市人民政府(1997). *Urban Master Plan of Shenzhen Municipality (1996–2010)*. Preserved at Futian Library, Shenzhen,藏於深圳市福田圖書館. (pp. 13).

[20] *Futian District Zoning Plan in Shenzhen City*(深圳市福田區分區規劃) (1998–2010), the official website of the Urban Planning and Natural Resources Bureau of Shenzhen, http://pnr.sz.gov.cn/ywzy/ghzs/fqgh/ftfjgh/5.htm

**Table 5.3** Retained and improved without expanding its scale, focusing on the electronics and communication industries[a]

| Industrial zone | Planning and adjustment of industrial land |
|---|---|
| Bagualing Industrial Zone | Partial retention of industrial land in the near term, with gradual transformation into a comprehensive zone in the long term. |
| Shangbu Industrial Zone | Partial retention for high-tech enterprises, with other areas transformed into a comprehensive zone with residential, commercial, financial, and office functions. |
| Meilin Industrial Zone | Retained and improved without expanding its scale, focusing on the electronics and communication industries. |

[a]Shenzhen Municipal People's Government 深圳市人民政府(1997). *Urban Master Plan of Shenzhen Municipality (1996–2010)*. Preserved at Futian Library, Shenzhen,藏於深圳市福田圖書館. (pp. 6–7).

During the period spanning from the 1980s to the 1990s, there was a discernible shift in the government's stance and anticipated outcomes for the Meilin-Gangxia area. During the 1980s, Meilin, Shangbu, and Huanggang were predominantly committed to industrial growth. However, in the 1990s, the government gradually restricted industrial development, steering toward finance and high-tech industries and concurrently developing corresponding residential functions and supporting facilities. Additionally, there has been a large-scale rectification in urban management, specifically targeting the removal of illegal constructions. Three mountain parks in the Meilin-Gangxia area, namely, Meilin Park, Bijiashan Park, and Lianhuashan Park, completed their construction and opened to the public. These parks primarily exhibit undulating topography, which is notably different from the four main parks built on the flat landscape of the Shangbu-Luohu region. There are notable differences in the internal arrangements and spatial forms of the parks in the two areas. Additionally, the ownership changes related to these parks are also varied.

## 5.2 Alternations in Ownership and Morphology

### *5.2.1 The Evolutionary Process of Forestry and Forest Ownership*

Prior to the Republican era, forest ownership in Guangdong was categorized into two types: official-owned and privately owned. These were, respectively, known as official mountains and forests and private mountains and forests. During the first year of the Republic of China in 1912, the Beiyang Government established the Ministry of Agriculture and Forestry (農林部), which subsequently classified forest ownership into two categories: state owned and privately owned. In November of 1914, the Beiyang Government enacted the first Forest Law, which classified forests into three categories: state-owned, publicly owned, and privately owned. By the end of the Republic of China and just before the foundation of the PRC, forest ownership in Guangdong had mostly transitioned to clan and private ownership. Following the foundation of the PRC in 1949, the government authorities in Guangdong province systematically assumed control over state-owned and publicly owned forests that had been inherited from the previous Republican era, the majority of which were subsequently nationalized. The execution of the registration and validation of forest rights accompanied the implementation of land reform in 1951. In April 1953, the Guangdong Provincial Government proposed four principles to immediately establish forest rights, which included:

*The lands and forests owned by landlords that should be confiscated, as well as public lands and forests, excluding large forests and desolate mountains far from villages where the population is sparse, are all to be nationalized. The rest were uniformly allocated to the local mountain-dwelling farmers, resulting in either individual private ownership for those near the mountains or community ownership for the entire hamlet.*[21]

Subsequently, following the periods of primary and advanced agricultural cooperatives, by the latter half of 1956, various forms of forest ownership in rural regions had effectively transformed into consolidated community ownership. The urgent demand for timber during the Great Leap Forward and socialist industrialization in 1958 caused a disruption in forest ownership. The onset of the Cultural Revolution in 1966 exacerbated the disorder in forest ownership in Guangdong Province. The Communist Party of China Guangdong Provincial Committee and the provincial government initiated the "Three Determinations in Forestry (林業三定)"[22] project over the entire province in June 1981, concluding it by the fall of 1982 with the end of the Cultural Revolution. The province forestry department submitted the *Report on Relaxing Forestry Policy Issues* (關於放寬林業政策有關問題的報告) to the People's Government of Guangdong Province in March 1984. Consequently, there was an improvement[23] in the disorganized administration of forest land and forest rights in Guangdong Province, and similar progress was seen at the Bao'an County level.[24]

Regarding forest resources in Guangdong province, collectively owned forest land spans more than 42.1 million ha., representing 78.8% of the total area, while state-owned forest land spans over 11.3 million ha., accounting for 21.2%[25]. The

[21] Guangdong Provincial Local Chronicles Compilation Committee廣東省地方市志編纂委員會 (1998). *Guangdong Annals: Forestry Annals*廣東省志·林業志. Guangzhou: Guangdong People's Publishing House廣州: 廣東人民出版社.(pp. 248–249).

[22] Stabilize mountain ownership, delineate reserved mountains, and establish the responsibility system for forestry production.

[23] Guangdong Provincial Local Chronicles Compilation Committee 廣東省地方市志編纂委員會 (1998). *Guangdong Annals: Forestry Annals* 廣東省志·林業志. Guangzhou: Guangdong People's Publishing House 廣州: 廣東人民出版社.(pp. 248–249).

[24] Bao'an County Local Chronicles Compilation Committee寶安縣地方誌編纂委員會(1997). *Chronicles of Bao'an County*寶安縣誌. Guangzhou: Guangdong People's Publishing House 廣州: 廣東省人民出版社.(pp. 226–232).

[25] Editorial Committee of "Guangdong Forest." 《廣東森林》編委會 (1990). *Guangdong Forest* 廣東森林. Guangzhou: Guangdong Science and Technology Press 廣州:廣東科技出版社, China

specified areas at the Bao'an County level that adhere to the "Three Determinations" work are as follows: The state-owned forest lands cover 20,455 mu, the production team forest areas cover 45,229 mu, the public welfare areas cover 380,066 mu, the responsibility areas cover 361,855 mu, the self-retained areas [26] cover 163,089 mu, the responsible fruit areas cover 46,500 mu, and the self-retained fruit areas have 81,458 trees.[27] It is evident that, whether at the provincial level or the Bao'an County level, collective ownership of rural mountain and forest rights is substantial.

### *5.2.2 The Land Source of the Three Mountain Parks in Gangxia-Meilin Area*

In the context of the chaotic ownership of mountain and forest rights throughout the province during the 1950s and 1960s, there were nuanced alterations in the particular circumstances within Futian District. The primary focus of Futian District, particularly the Meilin-Gangxia area, is the cultivation of economically valuable products, including fruit trees and bamboo. From 1963 to 1966, the district actively encouraged the practice of scientific afforestation and the cultivation of fast-growing tree species.[28] The rationale behind this emphasis was multifaceted. In terms of topography, the area was characterized by a series of small, hilly mountains and was located to the south of the Tanglangshan-Meilinshan-Yinhushan mountain range, making use of the advantageous landscape. In aspect of labor, Futian District possessed multiple sizable agricultural villages with ample labor resources, facilitating the cultivation of a significant quantity of cash crops. Historically, during the period of socialist industrialization, there were economic difficulties and a demand for timber.

From 1967 to 1977, communes and townships established forest farms to promote afforestation through collaboration between the government and communal entities. Between 1978 and 1990, there was a campaign to encourage the use of aerial seeding afforestation combined with voluntary tree planting. From 1982 to 1984, the Bao'an County government actively promoted the production of lychee in

Forestry Publishing House中國林業出版社. (pp. 48).

[26] Public welfare mountains(公益山), established with the purpose of improving the ecology, are owned by the state. Responsibility mountains(責任山) and self-retained mountains(自留山) are collectively owned.

[27] Bao'an County Local Chronicles Compilation Committee 寶安縣地方誌編纂委員會(1997). *Chronicles of Bao'an County*寶安縣誌. Guangzhou: Guangdong People's Publishing House 廣州: 廣東省人民出版社 (pp. 227).

[28] Shenzhen Futian District Local Chronicles Compilation Committee 深圳市福田區地方誌編纂委員會 (2012). *Futian District Annals, Shenzhen 1979–2003*深圳市福田區志1979–2003. Beijing: Fangzhi Publishing House北京: 方志出版社. (pp. 507).

villages located near the mountains.[29] Commencing in 1988, Futian District initiated the afforestation policy of *Integrating Afforestation with Fruit Tree Cultivation* (造林種果相結合): (1) Villages and sub-districts without appropriate slopes for fruit trees, predominantly cultivated *Taiwan acacia*; (2) Six villages,[30] specifically Shangmeilin, Xiameilin, Gangxia, Tianmian, Huanggang, and Xiasha, with favorable slopes, prioritized the cultivation of fruit trees. Building upon this, when we integrate the land ownership structure of the three mountain parks mentioned in the table, we can comprehend the historical origins of the vegetation on the site before the construction of the parks.

Meilin Park is primarily characterized by mountainous terrain. Before the constuction, the Guli Zone has numerous lychee trees, and both the Tingtao Zone and Fangxiang Zone also feature a significant number of lychee trees, while the Caimei Zone has relatively fewer. Among the various zones of Meilin Park, in addition to lychee trees, there are also many originally introduced fast-growing tree species such as *Eucalyptus*, along with a small quantity of *Acacia mangium Willd.*[31] South of the main peak of Lianhuashan Park, a substantial number of lychee trees[32] are also planted in a terraced manner. Before the establishment of Bijiashan Park, its plant composition differed from the first two parks. It mainly featured *Pinus massoniana*, with a small quantity of eucalyptus, and very few other tree species.[33]

In Table 5.4, the land for these three mountain parks, except for the portion of land in Bijiashan Park that was acquired from Shenzhen Forest Farm (深圳林場), was sourced from various villages involved in the *Integrating Afforestation with Fruit Tree Cultivation* program. In simpler terms, the presence of lychee trees on the sites of Meilin Park and Lianhuashan Park before their construction was not a natural fact; rather, it was a result of the land used for constructing these parks being sourced from villages such as Shangmeilin, Xiameilin, and Gangsha, where the government encouraged the cultivation of fruit trees. Furthermore, the *Pinus massoniana* in Bijiashan Park were planted in 1958 when Bijiashan Hill served as the Shenzhen Forest Farm, and the *Eucalyptus* trees were also products of afforestation efforts in Guangdong Province around the same period.[34] Although these mountain parks began their journey as urban parks only from the day of their establishment, the presence of vegetation in these parks can be traced back to an earlier period.

[29] Shenzhen Futian District Local Chronicles Compilation Committee深圳市福田區地方誌編纂委員 (2012). *Futian District Annals, Shenzhen 1979–2003*深圳市福田區志1979–2003. Beijing: Fangzhi Publishing House北京:方志出版社. (pp. 151).

[30] Shenzhen Futian District Local Chronicles Compilation Committee 深圳市福田區地方誌編纂委員會 (2012). *Futian District Annals, Shenzhen 1979–2003*深圳市福田區志1979–2003. Beijing: Fangzhi Publishing House 北京:方志出版社. (pp. 507).

[31] According to the interview with Mr. Wang Xiaoming, 2023-06-06.

[32] According to the interview with Mr. Wang Xiaoming, 2023-06-06.

[33] According to the interview with Mr. Huang Chongyuan, 2023-06-15 and Mr. Wu Qinxiang, 2023-08-01.

[34] According to the interview with Mr. Huang Chongyuan, 2023-06-05 and Mr. Wu Qinxiang, 2023-08-01.

**Table 5.4** Composition of land sources for Gangxia-Meilin Park Group[a]

| Park name | Meilin Park | | Lianhuashan Park | | | Bijiashan Park | | |
|---|---|---|---|---|---|---|---|---|
| Construction time | 2002–02 | | 1992–10 | | | 1989–10 | | |
| Opening time | 2004–07 | | 1997–06 | | | 1998–11 | | |
| Total park construction area | 115.44[b] | | 180.6[c] | | | 128.13[d] | | |
| Land ownership before the construction of the park | Xia meilin | Shangmeilin | Shangmeilin | Xiameilin | Gang-xia | Shenzhen Forest Farm | Gangxia | Tianmian |
| The corresponding land area | 85.43 | 30.01 | 39.6 | 39.0 | 102.0 | 97.93 | 18.6 | 11.6 |
| Percentage of the total park area | 74.0% | 26.0% | 21.9% | 21.6% | 56.5% | 76.4% | 14.5% | 9.1% |

[a]Source: Compilation based on relevant documents on land acquisition and construction during the park establishment period, and oral accounts from eye-witnesses.
[b]Based on the land parcel maps compiled by the Shenzhen Planning and Land Resources Bureau in July 2002 and March 2010, internal data.
[c]Based on the land parcel maps (宗地圖) compiled by the Shenzhen Planning and Land Resources Bureau in Dec. 2005, internal data.
[d]Based on the land parcel maps (宗地圖) compiled by the Shenzhen Planning and Land Resources Bureau in Feb. 1992, internal data.

They were planted in these areas within specific social and historical contexts, making it part of their historicity.

Amidst this vegetation, it is evident that these collectively owned mountain woods formerly sustained the livelihood and productivity of the indigenous population. In 1978, the overall income of the forestry sector in Futian People's Commune amounted to 0.48 million yuan. In 1984, the creation of several forest and fruit-breeding farms led to a significant growth in the forestry total income of Shangbu Administrative District, reaching 3.025 million yuan.[35] The primary source of revenue from forestry was derived from the rugged topography in the northern section of the area, as seen in Table 5.5:

The four villages, specifically Shangmeilin, Xiameilin, Gangxia, and Tianmian, have traditionally depended on the natural surroundings of the Meilin-Gangxia area for their sustenance, participating in agricultural, forestry, animal husbandry, and fishing industries. Table 5.6 displays the revenue generated by agricultural-related sectors in the years 1980 and 1981.

### *5.2.3 Nationalization of Collective Mountains and Forests*

The prerequisite for establishing the three mountain parks is the nationalization of these common mountain forests. In June 1992, with the commencement of the collective land acquisition in the special economic zone, the collective land within the special zone underwent a slow process of nationalization, becoming designated as reserve land for urban development. By applying Meilin as a case, the Bureau of Planning, Land, and Resources of Shenzhen Municipality (深圳市規劃國土局), operating on behalf of the government, entered into a *Land Requisition Agreement*[36] (統征土地協議書) with Shenzhen Xiameilin Industrial Co., Ltd.[37] (深圳市下梅林實業股份有限公司) in April 1996. The deal consolidated the acquisition of all collective land owned by Xiameilin Village, amounting to 9149.35 mu (609.96 ha.).

**Table 5.5** 1978–1985 forestry income table for Futian District[a]

| Year | 1978 | 1979 | 1980 | 1981 | 1982 | 1983 | 1984 | 1985 |
|---|---|---|---|---|---|---|---|---|
| Forestry income (in ten thousand yuan) | 0.48 | 1.29 | 5.39 | 1.06 | 0.57 | 0.79 | 30.25 | 14.03 |

[a]Shenzhen Futian District Local Chronicles Compilation Committee深圳市福田區地方誌編纂委員會 (2012). *Futian District Annals, Shenzhen 1979–2003*深圳市福田區志1979–2003. Beijing: Fangzhi Publishing House 北京:方志出版社. (pp. 507).

[35] Shenzhen Futian District Local Chronicles Compilation Committee深圳市福田區地方誌編纂委員會. (2012). *Futian District Annals, Shenzhen 1979–2003*深圳市福田區志1979–2003. Beijing: Fangzhi Publishing House 北京:方志出版社. (pp. 507).

[36] *Land Acquisition Agreement* 統征土地協議書, preserved at the Archives office of Shenzhen Park Management Center, with governmental file number: 深規土征字[1996]04號.

[37] Shangmeilin Industrial Co., Ltd. is a joint-stock enterprise formed by the conversion of collective assets, including land and properties, from four production teams in the original Shangmeilin Brigade. This is a result of rural urbanization in Shenzhen in 1992.

**Table 5.6** Agricultural industry income for four villages in 1980 and 1981[a]

| | 1980 | | | | 1981 | | | |
|---|---|---|---|---|---|---|---|---|
| Village | Forestry | Agriculture | Fishery | Animal husbandry | Forestry | Agriculture | Fishery | Animal husbandry |
| Shang-meilin | 35,888 | 722,276 | 0 | 0 | 929 | 78,871 | 179 | 5580 |
| Xiameilin | – | – | – | – | 0 | 190,878 | 1500 | 0 |
| Gangxia | 11,068 | – | – | – | 200 | 98,569 | 11,017 | 4701 |
| Tianmian | – | – | – | – | – | – | – | – |

[a]*Summary of the agricultural production team income distribution for the major brigades of Futian Commune* in 1980 and 1981福田公社各大隊1980及1981年度農業生產隊收益分配匯總表, document number: 全宗號FL6;目錄號A12.1;案卷號5;件號6, 全宗號FL6;目錄號A12.1;案卷號6;件號1, preserved at the Futian District Archives.

The agreement explicitly states that the second party had completely acquired all collective land within and outside the special economic zone, as mandated by the first party, upon the signing of this agreement. If land is rediscovered or omitted in the future, both parties will be required to provide compensation or refunds in accordance with the compensation standards outlined in this agreement. The remuneration standards outlined in the agreement are as follows in Table 5.7:

The present-day Caimei Zone and Guli Zone in Meilin Park were established right through this land acquisition from Xiameilin Village in 1996.[38] In 2002, following the completion of acquisition, the government consolidated the property obtained from the acquisition of Shangmeilin Village (known as the Tingtao Zone and Fangxiang Zone) and designated it collectively for the building of Meilin Park. The process of acquiring land for Bijiashan Park commenced earlier. In December 1992, the Bijiashan Park Construction Office (筆架山公園籌建辦公室) entered into a land acquisition agreement with the state-owned Shenzhen Forestry Farm. As both entities were responsible for managing state-owned land, the land acquisition agreement did not provide compensation for the mountainous landscape; it solely addressed compensation regarding trees and structures (Table 5.8).

## 5.3 Sheltering Nature: Population Resettlement

### *5.3.1 Population and Housing Issues*

The *Interim Measures for the Implementation of the Labor Contract System in Shenzhen* (深圳市實行勞動合同制暫行辦法) were announced by the Shenzhen Municipal Government in August 1983. Shenzhen pioneered the statewide

**Table 5.7** Compensation table for Meilin Park land acquisition[a]

| Compensation item | | Area (mu) | Compensation fee (mu) | Amount of money |
|---|---|---|---|---|
| Compensation for land | Water field | 463 | 5900 | 2,731,700 |
| | Vegetable field | 258 | 6600 | 1,702,800 |
| | Dry field | 350 | 3000 | 1,050,000 |
| | Forests and hills | 8078.35 | 920 | 7,432,082 |
| Compensation for objects on the ground | | – | – | 16,893,120 |
| Total | | 9149.35 | – | 29,809,702 |

[a]*Land Acquisition Agreement* 統征土地協議書, preserved at the Archives office of Shenzhen Park Management Center, with governmental file number: 深規土征字[1996]04號.

[38] *Notice on the Handover of Meilin Park Land* 關於做好梅林公園用地交接工作的通知, preserved at the archives office of Shenzhen Park Management Center.

**Table 5.8** Land acquisition compensation table for Bijia Mountain Park[a]

| Compensation object | Quantity | Unit price | Total amount |
|---|---|---|---|
| *Pinus massoniana* | 116,662 | 12.0 | 1399944.0 |
| *Cunninghamia lanceolata (Lamb.)Hook.* | 19,862 | 27.0 | 536274.0 |
| *Pinus elliottii Engelmann* | 4114 | 13.0 | 53482.0 |
| *Eucalyptus* | 1995 | 15.0 | 29925.0 |
| Bamboo | 765 | 30.0 | 22950.0 |
| *Pyrus ussuriensis maxim* | 242 | 500.0 | 121000.0 |
| Seedling in nursery | 13,496 | – | 116433.0 |
| House ($m^2$) | 714.116 | 160.0 | 114258.56 |
| Total amount | | | 2394266.56 |

[a]*Land Acquisition Agreement of Bijiashan Park, Shenzhen* 深圳市筆架山公園征地協議書, Document Number: 全宗號GY20;年度1992;件號16.preserved at the Archives office of Shenzhen Park Management Center.

implementation of a complete labor contract system for all employees in early 1993.[39] The increasing relaxation of talent policies in Shenzhen, together with the abundance of job opportunities resulting from the construction of the special economic zone, attracted individuals from many regions of the country, including cadres, laborer, technical personnel, and farmers. As a result, the permanent resident population of Shenzhen[40] increased rapidly (Tables 5.9 and 5.10).

The above table indicates that throughout the 1990s, the yearly increase in the city's permanent resident population was considerably greater compared to the 1980s. While there was some population growth in locations outside the special economic zone, the population within the special economic zone experienced significant and rapid growth. Moreover, the trend indicates a noticeable concentration of the population in the Futian District in the 1990s. By the year 2000, the population in Futian District had reached 784,300, a 7.42% increase from the previous year, making it the largest in terms of population among the four administrative districts within the special economic zone.[41] Indeed, during the whole decade of the 1990s, Futian District boasted the highest population density per square kilometer among all administrative districts within the special economic zone.[42]

[39] Cui, H., Pan, M., & Wu, S. 崔宏軼,潘夢啟,吳帥 (2020). Changes and Countermeasure Suggestions of Science and Technology Talents Policies in China's Economic Special Zones: A Case Study of Shenzhen我國經濟特區科技人才政策變遷及對策建議——以深圳為例. *Jianghuai Tribune*江淮論壇, 303(05), 30–36.

[40] Including Registered and Temporary Resident Populations.

[41] Shenzhen Statistical Information Bureau 深圳市統計資訊局(2001). *Shenzhen Statistical Yearbook 2001*深圳統計資訊年鑒2001. Beijing: China Statistical Publishing House 北京:中國統計出版社.(pp. 95).

[42] Editorial Committee of "Shenzhen Real Estate Yearbook." 《深圳房地產年鑒》編委會 (2001). *Shenzhen Real Estat Yearbook 2001*深圳房地產年鑒2001. Shenzhen: Haitian Publishing House 深圳:海天出版社.(pp. 8).

**Table 5.9** Annual population figures of permanent residents inside and outside the special economic zone in Shenzhen City [Shenzhen Statistical Information Bureau 深圳市統計資訊局 (1999). *Shenzhen Statistical Yearbook 1999*深圳統計資訊年鑒1999. Beijing: China Statistical Publishing House 北京:中國統計出版社. (pp. 109–121); Shenzhen Statistical Information Bureau (2001). *Shenzhen Statistical Yearbook 2001*. Beijing: China Statistical Publishing House. (pp. 95–106); Shenzhen Statistical Information Bureau (2003). *Shenzhen Statistical Yearbook 2003*. Beijing: China Statistical Publishing House. (pp. 63–73); Shenzhen Statistical Information Bureau (2005). *Shenzhen Statistical Yearbook 2005*. Beijing: China Statistical Publishing House (pp. 57–71); Shenzhen Statistical Information Bureau (2011). *Shenzhen Statistical Yearbook 2011*. Beijing: China Statistical Publishing House. (pp. 47–62).]

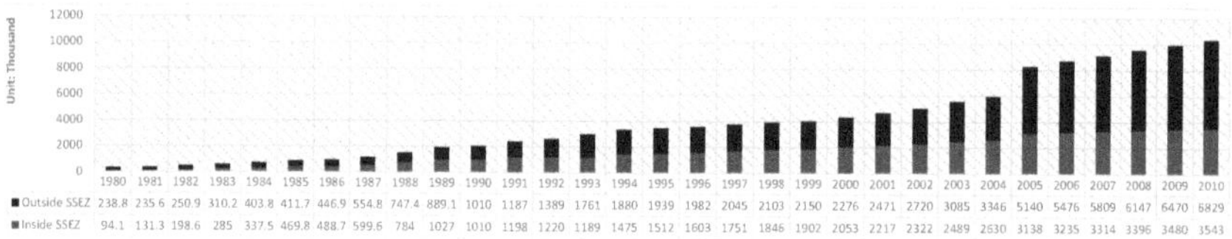


**Table 5.10** Annual growth in the permanent resident population of Shenzhen City [Ibid.]

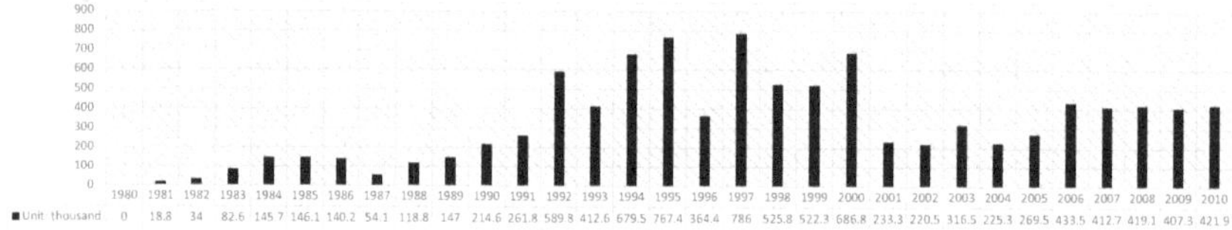


The substantial increase in population posed a major challenge to the development of the Special economic zone, particularly in terms of housing. In 1990, the estimated housing shortfall for personnels of party and government organizations and public service units was 2.762 million square meters.[43] By 1994, while the housing issues faced by personnels of these institutions had been largely handled, the general housing conditions in the city still exhibited major disparities due to variations in individual economic circumstances and enterprise conditions. The discrepancy was most pronounced for temporary residents[44] without Shenzhen household registration, possessing low earnings, and originating from different regions of the country.[45] The government indeed recognized and valued the economic

[43] Editorial Committee of "Shenzhen Real Estate Yearbook." 《深圳房地產年鑒》編委會 (2001). *Shenzhen Real Estat Yearbook 2001*深圳房地產年鑒2001. Shenzhen: Haitian Publishing House 深圳:海天出版社.(pp. 110).

[44] In government statistics, non-local registered population is commonly referred to as temporary residents, requiring a temporary residence permit according to the regulations at the time. Individuals without a temporary residence permit, without stable employment, and fixed residence are termed "Three No's Personnel" by the government. This study, however, does not distinguish "Three No's Personnel" based on the possession of a temporary residence permit and collectively refers to them as temporary residents.

[45] Editorial Committee of "Shenzhen Real Estate Yearbook." 《深圳房地產年鑒》編委會 (2001). *Shenzhen Real Estat Yearbook 2001*深圳房地產年鑒2001. Shenzhen: Haitian Publishing House 深圳:海天出版社.(pp. 89).

contributions generated by the temporary residents. However, many breeders and individual vendors have long resided in makeshift sheds along the hillsides, roadsides, and pond banks, adding to the pressure on the government in terms of urban management. In September 1990, the Shenzhen Municipal Government issued the *Trial Measures for the Construction and Management of Settlement sites for Temporary Residents in Shenzhen Special economic zone*[46](深圳特區暫住人員安置區建設管理試行方案), aiming to implement housing resettlement for temporary residents in a relative reasonable manner. This regulation outlined five principles for the development of temporary resident settlement locations, three of which involve aspects of construction:

1. *The ideal location for the settlement site should be situated away from the bustling downtown and residential areas, equipped with sewage facilities, and conveniently accessible by transportation. Its purpose is to provide services for temporary residents in Shenzhen who meet the residency criteria, operate on a low-profit basis, and its management processes also need to be simplified.*
2. *The settlement site will adhere to a cohesive plan encompassing unified design, construction, and management. The Bureau of Planning, Land, and Resources will uniformly plan the land, allocating a blue-line land arrangement to each district for a five-year term. Every district will draft design plans, submit drawings and layouts, and acquire approval from the Urban Management Office.*
3. *The planning and design must adhere to the criterion of utilizing semi-permanent two-story structures for a duration of three to five years, or maybe longer. Living facilities should have adequate infrastructure to handle sewage and waste disposal, together with essential public services and fencing.*

From 1992 to 1995, three primary temporary settlement sites were established in the densely populated Futian District (Fig. 5.4). These places offered affordable accommodation to more than 100,000 temporary residents in Shenzhen during that period. By 1995, the Lianhuashan Hill Settlement site (蓮花山安置區) was equipped with various amenities, including a savings bank, post and telecommunications office, medical and health center, market, stores, restaurants, kindergarten, and parking spaces. Also, greenery was arranged both in front and behind the housing, as well as on both sides of the road.[47] Many staff from prominent companies in Shenzhen have relocated to these affordable housing units, guaranteeing the fundamental stability of these enterprise personnel, and fostering business growth.

[46] Shenzhen Municipal People's Government深圳市人民政府.. (1990, September 30). Notice on Issuing the *Trial Implementation Plan for the Construction and Management of Settlement Areas for Temporary Residents in Shenzhen Special Economic Zone*. 深圳市人民政府關於印發《深圳特區暫住人員安置區建設管理試行方案》的通知 [EB/OL]. Retrieved from www.sz.gov.cn/zfgb/1900/gb9/content/post_10036113.html

[47] Shenzhen Special Economic Zone Yearbook Editorial Committee 深圳經濟特區年鑒編輯委員會 (1996). *Shenzhen Special Economic Zone Yearbook 1996*深圳經濟特區年鑒1996. Shenzhen: Shenzhen Special Zone Yearbook Press深圳:深圳特區年鑒社.(pp. 456).

Simultaneously, numerous newly hired civil officers and clerks in government institutions would temporarily rent similar temporary accommodations (Table 5.11).[48]

The temporary residents in these provisional settlement zones demonstrate a comparatively elevated degree of mobility, which introduces specific characteristics of instability in relation to public security. Nevertheless, the government has specifically issued documents to encourage construction and has established systematic management departments. Notably, the Lianhuashan Hill Settlement site was awarded the prestigious designation of "Safe and Civilized Community (安全文明社區)" by the Shenzhen Municipal Government in 1995, making it the sole temporary settlement site to attain this distinction.[49] Before the year 2000, what truly concerned the government were not these temporary settlement sites with relatively good management levels, but rather the self-formed habitations that sprawled across the mountains on the edges of the economic zones, constructed by temporary residents themselves.

*The tin cottages in Meilin Park were predominantly situated in the western section of the Caimei Zone. Adjacent to the rear mountain of Meishan Middle School*

**Table 5.11** Specific information table of the three major resettlement areas

| | Donggualin Settlement Site[a] | Huangmugang Settlement Site[b] | LianhuashanSettlement Site[c] |
|---|---|---|---|
| Start construction year | 1992 | 1990–05 | Before 1995 |
| Demolition time | 1998-09-04 | 2004-05-28 | 2001-08-11 |
| Area | 15 ha. | 9 ha. | 15 ha. |
| Settlement capacity | 9000 | 25,000 | 12,000 |
| Amount of buildings | 116 | 129 | 149 |
| Location | Caitian first Village in the northern part of Lianhuashan Hill | The southern part of current Bijiashan Park | The southeast and northeast parts of current Lianhuashan Park |

[a]*Shenzhen Economic Daily*深圳商報(September 1, 1998); *Shenzhen Special Zone Daily*深圳特區報(September 5, 1998).

[b]*Shenzhen Economic Dail*深圳商報 (May 29, 2004); *Shenzhen Special Zone Daily*深圳特區報 (May 26, 2004); *Shenzhen Special Zone Daily* 深圳特區報(September 7, 1998); *Daily Sunshine*晶報 (June 10, 2017).

[c]*Shenzhen Special Zone Daily*深圳特區報 (June 23, 2010); *Shenzhen Economic Daily*深圳商報 (August 11, 2001); *Shenzhen Economic Daily*深圳商報(August 1, 2001).

[48] Mang, J.莽疾 (1998). Resettlement Areas: An Important Form of Providing Affordable Housing for Low-Income Families by the Government 安置區: 國家為低收入家庭提供廉租房的一種重要形式. *China and Foreign Real Estate Guide*中外房地產導報, (10). (pp. 38–39).

[49] Shenzhen Special Economic Zone Yearbook Editorial Committee 深圳經濟特區年鑒編輯委員會 (1996). *Shenzhen Special Economic Zone Yearbook 1996*深圳經濟特區年鑒1996. Shenzhen: Shenzhen Special Zone Yearbook Press 深圳:深圳特區年鑒社. (pp. 456).

**Fig. 5.4** Locations of the three settlement sites [Based on historical satellite imagery from Google Earth in February 2000]

(梅山中學), *there was a proliferation of illegal constructions in this vicinity...in contrast, the Guli Zone and the Tingtao Zone had a lower number of such constructions. The Tingtao Zone was characterized by the presence of local farmers' residences and temporary constructions. The Guli Zone was quite simple, with minimal complexity. The primary concentration of tin dwellings was in the Caimei Zone, where a multitude of terraced vegetable fields span the entire slope. These people were all "Three without Individuals* (三無人員)": *without a fixed residence, without regular employment, and without proper identification...these people make a living through vegetable cultivation, chicken farming, and pig breeding. Notably, there were more than one hundred pigs located beside Meishan Middle School. During that period, the Urban Management Bureau undertook a comprehensive citywide "Rectification Action* (梳理行動)," *demolishing all the improvised tin dwellings constructed on the slopes...those people were quite aggressive...dealing with these situations required caution and a gentle approach, avoiding excessive actions....*[50]

*The illegal constructions in Lianhuashan Park are mainly associated with two groups of people. The southern part was associated with Huatai Company* (华泰公司), *which could be considered organized by the government. The second group, situated at the back of the mountain, consisted of individuals engaged in lychee cultivation, pig farming, vegetable farming, and even dog slaughter. How this situation developed remains unclear...at that time, there were at least five hundred people residing behind the mountain...the presence of "three without individuals" posed a significant wildfire risk to the forest. Handling the situation with Huatai was relatively easy as it was organized by the government. However, dealing with the "three without individuals" behind the mountain was more challenging. It wasn't until December 1997 that we organized a group of several hundred people*

[50] According to the interview with Mr. Wang Xiaoming, 2023-06-06.

*to dismantle the constructions. Subsequently, individuals involved in dog slaughter, some of whom had knives on their belts, approached me…after a lengthy conversation, we managed to handle the situation amicably…their location was around the current overpass coming from the Upperhills. It was indeed quite frightening….*[51]

*(Huangmugang Settlement Site) It was approved by the Futian District government as a settlement site. In the late 1980s and early 1990s, there were tens of thousands of people living here. In the late 1980s and early 1990s, there were tens of thousands of people living here. Although the government approved the southern part of Huangmugang, in reality, shanties were built all over the hill. It was lively, with numerous restaurants, at least one or two dozen on the side near Sungang Road. There were also kindergartens, primary schools, markets, and a temple where the incense offerings were abundant. It was challenging to dismantle due to its vibrant community.*[52]

From 1980 to 2000, based on the government's statistics on various criminal cases in the city, temporary residents constituted an overwhelming majority in terms of proportion.[53] Therefore, the government directed its attention toward temporary residents as the primary target of its public security management at that period. These persons who lacked appropriate legal documents, genuine job, and a fixed place of residence[54] were commonly known as "Three without individuals." Occasionally, they were pejoratively labeled as "blind wanderers (盲流)." Within the framework of urban–rural dynamics during that period, the majority of them relocated from rural areas to cities due to factors such as evading famine, finding shelter, or pursuing job opportunities. In 1985, Shenzhen pioneered the "Temporary Residence Permit (暫住證)" system, and the public security department began implementing various management systems for the floating population, regularly conducting clearances of temporary residents. Between 1985 and 2000, public security officials at all levels in Shenzhen conducted a total of 2.73 million[55] clearances and removals of temporary residents, as indicated by statistical data. During that period, a significant number of these temporary residents sought refuge in the underdeveloped hilly regions around the urban development of the special economic zone, including Meilin Mountain, Lianhuashan Hill, and Bijiashan Hill.

[51] According to the interview with Mr. Wang Xiaoming, 2023-06-06.

[52] According to the interview with Mr. Huang Chongyuan, 2023-06-15.

[53] Shenzhen Municipal Local Chronicles Compilation Committee 深圳市地方誌編纂委員(2014). *Chronicles of Shenzhen: Volume on Infrastructure Construction* 深圳市志·基礎建設卷. Beijing: Fangzhi Publishing House 北京: 方志出版社. (pp. 480).

[54] Shenzhen Municipal Local Chronicles Compilation Committee 深圳市地方誌編纂委員 (2014). *Chronicles of Shenzhen: Volume on Legal System and Administrative Affairs*深圳市志·法制政務卷. Beijing: Anzhi Publishing House 北京: 方志出版社. (pp. 114).

[55] Shenzhen Municipal Local Chronicles Compilation Committee 深圳市地方誌編纂委員 (2014). *Chronicles of Shenzhen: Volume on Legal System and Administrative Affairs*深圳市志·法制政務卷. Beijing: Fangzhi Publishing House 北京: 方志出版社. (pp. 114).

### 5.3.2 Population Settlement in Mountainous Hills

The Chinese characters 嶺(ling) in Donggualing, 崗(gang) in Huangmugang, and 山(shan) in Meilinshan, Lianhuashan, and Bijieshan all describe hilly topography. So, why did the government choose these locations to build temporary settlement sites? And why did the temporary residents from self-formed habitations choose to live in these mountainous areas on the outskirts of the special economic zone?

Let's start by examining the government-led construction of temporary settlement sites. By integrating the construction ideas outlined in the *Trial Measures for the Construction and Management of Settlement sites for Temporary Residents in Shenzhen Special economic zone,* we may endeavor to examine the process of selecting suitable locations for the three primary temporary settlement sites.

(I) Temporary settlement sites demonstrate considerably reduced population stability in comparison to residential areas. Frequent population mobility can easily lead to public security issues. Therefore, to protect the safety of local residential areas, the selection of temporary settlement sites must be "away from the bustling downtown and residential areas."

(II) Temporary residents in the settlement sites also need "conveniently accessible by transportation" near their place of work. Consequently, the choice of locations can only be somewhat prejudiced and should not be excessively distant.

(III) From 1992 to 1995, as the temporary settlement sites were being constructed, there was a sweeping surge in urban construction in Futian District. Temporary settlement sites should not encroach onto property earmarked for future development in terms of land use. Hence, the locations should be chosen in areas where there are no plans for future high-density urban expansion.

(IV) Although the foothill areas are sloping, the gradient is notably less pronounced when contrasted with the middle and upper sections of the hill. The difficulty and cost of development and construction in these areas are relatively low. Therefore, for the government, temporarily utilizing these spaces planned for urban parks and restoring the greenery after resolving the settlement issue is evidently the most cost-effective approach.

The mountainous terrain in the Meilin-Gangxia area, especially Meilin Mountain, Lianhuashan Hill, and Bijiashan Hill, and their surroundings, is difficult to develop and has become a later-developed area in urban construction. Within these later-developed areas, there are still some foothill areas with relatively flat terrain. These areas are frequently designated for the construction of urban parks in future planning. Therefore, selecting temporary settlement sites in these foothill areas aligns well with the afore-mentioned analysis.

The foothills of Meilin Mountain were not developed for settlement sites for two main reasons. Firstly, the Meilin area is situated at the northernmost tip of Futian District, which made it excessively isolated and hindered transportation, rendering it inconvenient for the government at that time. Furthermore, during that period, the government did not designate the foothills of Meilin as park construction land;

instead, they were considered peripheral green belts of the city. The implementation of a strategy to establish temporary settlement sites and then restore vegetation on these forests would encounter substantial legal impediments. This correlation with the terrain is not limited to these temporary settlement sites but also extends to self-formed habitations by temporary residents (Table 5.12).

Firstly, the temporary residents in self-formed habitations mostly come from rural areas across the country. They have lower educational levels and haven't received systematic vocational training. Despite their willingness to work hard, it is challenging for them to find suitable job opportunities in the factories and enterprises of the special economic zone. However, the mountainous terrain can provide them with a natural environment close to rural settings. Vegetable and fruit cultivation, poultry and livestock farming, and other economic activities dependent on the natural environment can be carried out here and sold to the citizens of the special economic zone. These mountainous areas serve as productive nature, offering them basic economic livelihood support. Sparse population and relatively flat terrain are two major reasons for conducting these types of economic activities.[56]

Furthermore, in contrast to the busy urban downtown, the temporary residents in hilly environment are exposed to a reduced number of arbitrary identification inspections by the authorities. The government incurs significantly higher administrative expenses when conducting random inspections on them. According to the *Measures for the Detention and Repatriation of Urban Vagrants and Beggars* (城市流浪乞討人員收容遣送辦法), issued by the State Council in May 1982, there are three categories of people who need to be detained and repatriated to their registered residence after entering the city: (1) those who beg in the city after migrating from rural areas; (2) city residents begging on the streets; (3) others who live without a

**Table 5.12** General overview of temporary population settlements in the three mountain parks

| Occupancy situation in Meilin Park[a] | | Occupancy situation in Lianhuashan Park[b] | | Occupancy situation in Bijiashan Park[c] | |
|---|---|---|---|---|---|
| Tin houses | More than 100 ones | Canvas shelters | More than 4000 ones, about140,000 m$^2$ | Canvas shelters | 200–300 ones |
| Canvas shelters | More than 30 ones | | | | |
| Vegetable field | Nearly 10,000 m$^2$ | | | | |

[a]*Explanation Regarding Issues Related to the Construction of Meilin Park* 關於建設梅林公園有關問題的說明, preserved at Archives office of Shenzhen Park Management Center深圳市公園管理中心檔案室, Document Code: GY13; Year 2001; File Number 3.
[b]*Investigation Report on Unauthorized Constructions within the Scope of Lotus Mountain Park* 關於蓮花山公園範圍內違章建築情況的調查報告, preserved at Archives office of Shenzhen Park Management Center深圳市公園管理中心檔案室, Document Code: GY10; Year 1993; File Number 5.
[c]According to the interview with Mr. Huang Chongyuan, 2023-06-15.

[56] According to the interview with Mr. Fu Keqin, 2023-02-07.

fixed residence. In June 2003, due to legal controversies, the *Detention and Repatriation of Urban Vagrants and Beggars* was ultimately abolished at the national level.[57] In the newly announced *Measures for the Assistance and Management of Homeless Vagrants and Beggars in Urban Areas* (城市生活無著的流浪乞討人員救助管理辦法), the government's administrative inclination shifted from "repatriation" to "assistance." However, in Shenzhen in the 1990s, once individuals were randomly checked by the government's public security department and couldn't provide relevant identification, they would be repatriated to detention centers in places like Zhangmutou, Dongguan (東莞樟木頭).[58]

In summary, in the mountainous hills of the Meilin-Gangxia area, both the temporary settlement sites constructed by the government at the foothills and the self-formed habitations formed by temporary residents on the slopes and hilltops are closely related to the topography that is interwoven into the social context of the rapid population expansion in the 1990s special economic zone. The hilly topography of the Meilin-Gangxia area, which was developed later for urban construction in the special economic zone, provided a sanctuary for these temporary residents during a period of substantial population growth. They are not only productive nature but also serve as protective nature.

### 5.3.3 Rights to the City: Regreening

These self-formed habitations and temporary settlement sites were doomed to be dismantled at a later time. During the 1990s, the Shenzhen Municipal Government at different levels reached an agreement on the construction of an international city. In the official view, attracting multinational companies to settle in Shenzhen became the government's critical task[59] in foreign economic and trade work, and having a clean environment, smooth transportation, and social tranquility were deemed "essential conditions for an international city."[60] The objective is to bring Shenzhen in line with international norms and ensure its influence extends globally. The former Mayor, Li Hongzhong, held the belief:

---

[57] Liu, W. W. 劉偉偉 (2015). Multiple Streams Analysis of Policy Termination:Empirical Study of the Custody andRepatriation System 政策終結的多源流分析——基於收容遣送制度的經驗研究. *Journal of Public Management*公共管理學報, 12(04), 21–38 + 154.

[58] Zhang, W., & Chen, Y. 章文,陳勇 (2001). Repatriation: When Will It Emerge from Sorrow? .收容遣送,何時走出悲辛? *South Reviews*南風窗, 14, 36–38.

[59] Zhang, X. W., & Rao, J. 張興文,饒潔 (2002, January 14). *Ye Minhui, Director of the Municipal Foreign Trade and Economic Cooperation Bureau, Discusses the Train of Thought for Foreign Trade and Economic Work This Year* 市外經貿局局長葉民輝談今年外經貿工作思路. Shenzhen Special Zone Daily. (01).

[60] Publicity Department of the Shenzhen Municipal Committee of the Communist Party of China, Shenzhen Press Group中共深圳市委宣傳部,深圳報業集團(2001). *Clear and Smooth—The Record of the Initial Battle in 2003* 淨暢寧——2003初戰實錄. Shenzhen: Haitian Publishing House 深圳:海天出版社. (pp. 5).

*An international city must first and foremost be a healthy, clean, and orderly city; otherwise, international is out of the question. For Shenzhen to achieve international recognition as a significant metropolis, it must priorities its urban environment, which is a "compulsory course."*[61]

(1) The Regreening of Self-formed Habitations.

The Shenzhen Municipal Government faced considerable pressure when developing an international city,[62] which led to the necessary action of removing temporary residents' makeshift huts and tin shacks. The Shenzhen Municipal Government held a mobilization meeting on March 17, 2003, to initiate the "Rectification Action" aimed at addressing a range of urban environmental management challenges throughout the city.[63] While the official commencement of the Rectification Action occurred at this time, the government had already initiated preliminary actions and expressed concern about the urban environment. These acts were bound to generate controversies about public opinion,[64] legal principles, and human rights. While this study primarily examines how urban parks, as a cultural form of nature, formerly concealed these conflicting and competitive values.

In the 1980s, the Meilin area saw a transition from being exclusively an industrial zone to a combination of high-tech businesses and affluent residential areas. This, together with the overpopulation in the Luohu district, resulted in the migration of multiple government institutions to the Futian district. Consequently, the government had to address the issue of providing nearby housing for government officials. In 1996, the Shenzhen Housing Bureau organized the planning and design of Meilin first Village (梅林一村), a large residential area known as the "First Village in Asia" at that time. Initially, this residential area was mainly intended to provide housing for government officials. It covered an area of 384,000 square meters, with a total construction area of approximately 800,000 square meters, comprising nearly 7000 housing units. In 1998, the first group of residents moved in.[65] By 2001, the population of Meilin first Village had exceeded 30,000 people, and the total population of the Meilin community radiating from it had surpassed

---

[61] Wu, Z. J. 吳子俊(Eds.) (2005). Shenzhen City Sorting Action Chronicle: Adjusting the Direction of Urbanization with a Scientific Development Perspective 深圳市梳理行動紀實:用科學發展觀調校城市化發展方向. Shenzhen: Haitian Publishing House 深圳:海天出版社. (pp.5).

[62] Jin C., Chen S.Z. 金城,陳善哲(2004, August 9). Shenzhen's "Rectification Action": An Exemplar of Rapid Urbanization in China 深圳"梳理行動":急速城市化的中國標本. *twenty-first Century Business Herald* 21世紀經濟報導. (001).

[63] Shenzhen Yearbook Compilation Committee深圳年鑒編纂委員會. *Shenzhen Yearbook 2005*深圳年鑒2005. Shenzhen: Shenzhen Yearbook Press深圳: 深圳年鑒社. (pp.48).

[64] Dang, G.Y. 党國英(2009, September 14). *Shenzhen's "Rectification Action": Struggle between the City and Outsiders*深圳"梳理行動"、城市與外來者的角力. Southern Metropolis Daily 南方都市報; Liu, X.D. 劉顯東 (2004, September 15). *Sorting Action: Who is Struggling with Whom*梳理行動:誰與誰角力. Daily Sunshine晶報.

[65] Meilin first Village: A Beautiful "Urban Business Card" in Shenzhen 梅林一村: 深圳一張靚麗的"城市名片" (2009). *China Property Management*中國物業管理, 70(01), 41–42.

300,000. However, at this time, there was still no urban park in the entire Meilin community.[66]

In this context, the construction of Meilin Park was brought onto the government's agenda. The park was divided from west to east into the Caimei Zone, Guli Zone, Tingtao Zone, and Fangxiang zone. Initially, they were four foothills of Meilin Mountain that extended into the city. According to the original plan, they were considered as part of the city's surrounding forests. However, with the transformation of the urban functions in the Meilin area, it has evolved into the backyard of Futian Central Business District. Consequently, these four foothills have transformed into urban parks.[67]

The westernmost part of Meilin Park, known as the Caimei Zone, experienced a significant change in this context. As per an article in the *Shenzhen Economic Daily* (深圳商報), prior to the construction of the park, this area was characterized as "a landscape of destruction."

*...The hillside is covered with terraced patches of green vegetables. Outside a two-story shack, a middle-aged couple sits, having breakfast. The man introduces himself as Mr. He, originally from Maoming. He has been growing vegetables in the back hills of Meilin for 12 years, currently earning thousands of yuan per month. He sends the money back home to support his children's education. He said that he is very familiar with the locals, and before Meilin first Village was built, the locals allowed him to cultivate several acres of land on the mountain. He grew vegetables on the mountain and sold them at the foot of the mountain, earning an annual income of thirty to fifty thousand yuan. Over the past decade, his total income has reached three to four hundred thousand yuan. Other families who were cultivating vegetables at the same time used to earn around 100,000 yuan per year due to the abundance of people and land. However, it's no longer possible now as there is no available land... He mentioned that he won't be cultivating vegetables here for much longer. Once the construction of Meilin Park project[68] begins, he plans to return to his hometown...in front of the temple, migrants raise flocks of chickens, ducks, and geese...many city residents come here to buy vegetables, chickens, and ducks from these temporary residents. The "three poultry" here are slaughtered and sold on the spot....*[69]

As shown in Fig. 5.5, the large vegetable fields that once covered the hillside became part of Meilin Park in the years following the initiation of construction in 2001. In the spring of 2003, the park management planted more than 18,900 trees on these former vegetable fields, comprising approximately 56 tree species, covering an area of over 100,000 square meters. In addition, grass was planted on the

---

[66] *Regarding the proposal for the construction of Meilin Park*關於建設梅林公園的請示, preserved at Archives office of Shenzhen Park Management Center 深圳市公園管理中心檔案室, Document Code: GY13; Year 2001; File Number 10.

[67] According to the interview with Mr. Wang Xiaoming, 2023-06-06.

[68] Refers to Meilin Park near the Meilin Reservoir.

[69] Sun, Z.H. & Lou, L. 孫中海,婁荔 (2002, January 13). *Meilin Back Mountain in Devastating Condition* 梅林後山滿目瘡痍. Shenzhen Economic Daily 深圳商報. (01).

**Fig. 5.5** Temporary residents at the foot of Meilin Mountain and their lifestyle [Sun, Z.H. & Lou, L. 孫中海,婁荔 (2002, January 13). *Meilin Back Mountain in Devastating Condition* 梅林後山滿目瘡痍. Shenzhen Economic Daily 深圳商報.(01).] Above: Temporary residents cultivating large plots of vegetables in Meilin Park. Lower left: Temporary residents selling freshly slaughtered chickens and ducks; Lower right: Temporary residents constructing makeshift shelters

cement slope at the southeast corner of the park,[70] where it meets Meilin first Village. After certain topographic adjustments, the farmers who once relied on these vegetable fields gradually left, and the shanties and tin houses they built were also demolished. In the historical images of the Caimei Zone (Fig. 5.6), the white mulch cloth physically covers the newly laid turf, and the new turf, under the dominant government power, conceals the underprivileged nature and the life forms attached to it.

(2) The Regreening of Governmental Settlement Sites.

[70] Meilin Park Management Office. (2002, December 12). *Summary of Work in 2003 and Work Ideas for 2004* 梅林公園管理處2003年工作總結及2004年工作思路, preserved at archives office of Shenzhen Park Management Center 深圳市公園管理中心檔案室, Document Code: GY13; Year 2003; File Number 5.

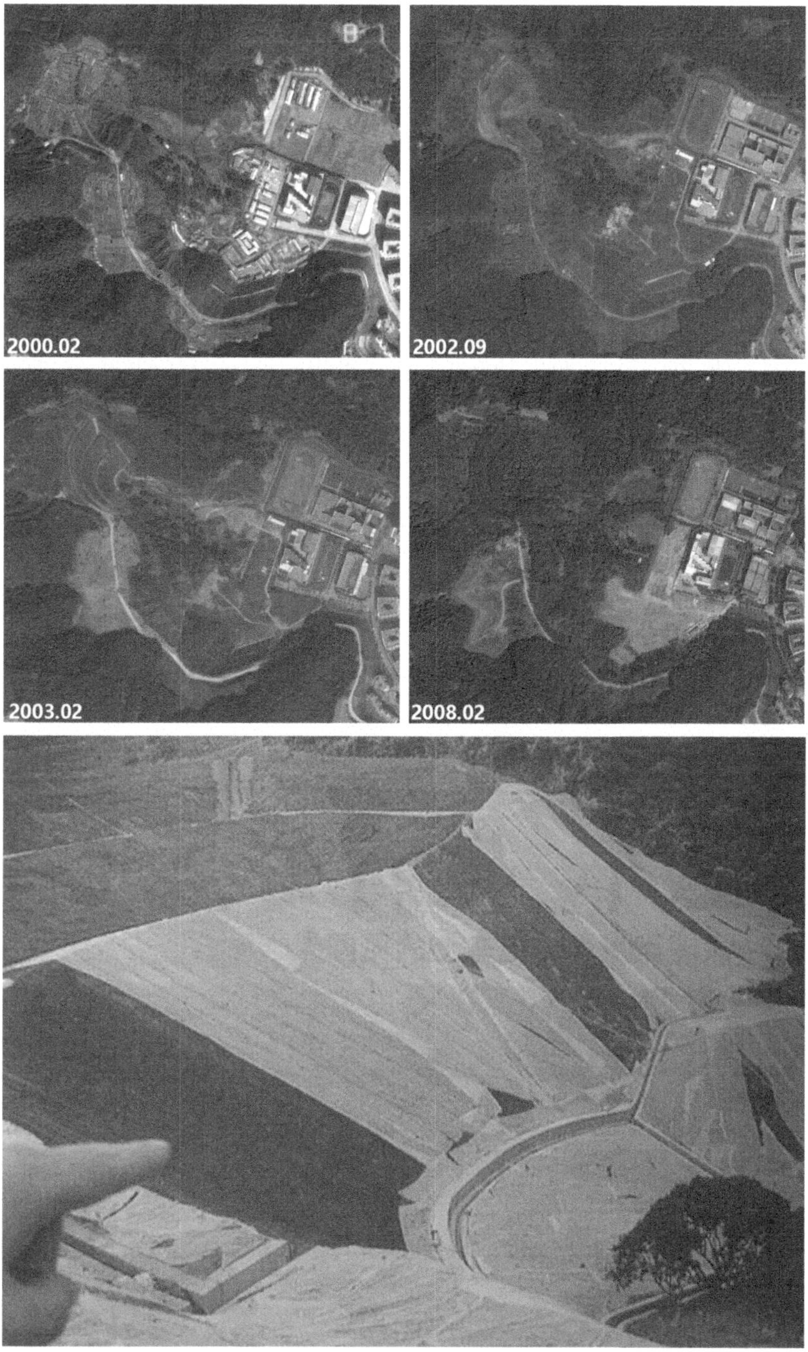


**Fig. 5.6** The transformation of Caimei Zone and the regreened slope inside it (December 2022) [Source from Google historical imagery and Shenzhen Urban Park Management Center with Document number: 全宗號GY13; 年度2002; 件號63.]

The construction of permanent settlement sites, such as Meilin first Village and Meishan Yuan Community, marks the successful achievement of the objectives set for the three major temporary settlement sites established in the early 1990s. In terms of usage duration, each of the three settlement sites has exceeded the 3–5 years usage period specified in the *Trial Measures for the Construction and Management of Settlement sites for Temporary Residents in Shenzhen Special economic zone.*[71] In May 1998, the Shenzhen Municipal Government issued *Notice on Further Cleaning and Demolishing Various Types of illegal constructions* (關於進一步清理和拆除各類違法違章建築實施方案的通知). In this notice, Donggualing Settlement Site was considered an "illegal constructions" by the government and was later demolished in August–September, 1998: "Donggualing Settlement Site in Futian District, allocated to the Municipal Housing Bureau for the construction of welfare housing."[72] It was redeveloped into the present-day Caitian first Village.

The Huangmugang Settlement Site and Lianhuashan Settlement Site, which were situated close to Futian Central Business District, also had a brief duration of existence (Fig. 5.7). Because of their influence on the city's image,[73] they were demolished in May 2004 and August 2001, respectively, and transformed into green spaces, once again becoming integrated into the urban parks. The demolition of the

**Fig. 5.7** The demolition of the Lianhuashan Settlement Site was being leveled in preparation for greening [Left image: Wang, L. J. 王蘭君 (2001, September 17). *Lianhuashan Settlement Site Adorns Green Attire to Welcome National Day* 蓮花山安置區改換綠裝迎國慶. Shenzhen Economic Daily 深圳商報.(A6); right image: Ma, J. Y. 馬驥遠 (2017, June 10). *Transitioning a generation of Shenzhen residents in resettlement areas.*安置區，過渡了一代深圳人. Daily Sunshine晶報. (A04).]

[71] Shenzhen Municipal People's Government深圳市人民政府. (1990, September 30). Notice on Issuing the *Trial Implementation Plan for the Construction and Management of Settlement Areas for Temporary Residents in Shenzhen Special Economic Zone*. 深圳市人民政府關於印發《深圳特區暫住人員安置區建設管理試行方案》的通知. Retrieved from www.sz.gov.cn/zfgb/1900/gb9/content/post_10036113.html

[72] Shenzhen Municipal People's Government深圳市人民政府. (1998, June 5). *Notice on Further Implementation Plan for Cleaning up and Dismantling Various Illegal Buildings*关于进一步清理和拆除各类违法违章建筑实施方案的通知. Retrieved from https://www.sz.gov.cn/zfgb/1998/gb143/content/post_10123163.html

[73] According to the interview with Mr. Yang Yibiao, 2023-06-18.

Huangmugang Settlement Site was regarded as an integral component of the "beautify the Bijiashan Hill." Following its demolition, a parking lot was built on the western side to mitigate parking problems within the park. For the eastern side, in the process of greening, sporting amenities such as tennis courts and basketball courts were constructed.[74] The regreening plan for the Lianhuashan Settlement Site was determined well before the demolition, employing over 100 plant species with the aim of achieving "continuous blooming throughout the seasons and ever-changing scenery[75] (時時花不斷，季季景不同)." Moreover, the Lianhuashan Settlement Site also preserved some original memory. As the preservation of buildings was not allowed, the banyan trees flanking the main road in the original settlement site were preserved as a tree-lined avenue, and the shared well, after being inscribed by Mr. Meng Zhaozhen (孟兆禎, 1932–2022), transformed into a water feature named "Si Yuan[76] (思源, Remember the Origin)."

During the process of demolition and regreening of the three settlement sites, there were no serious incidents of resistance or obstruction.[77] However, this does not imply that conflicts and competition regarding space meaning did not occur. On the one hand, government officials and journalists responsible for demolition work are facing personal threats,[78] on the other hand, the *Southern Metropolis Daily* (南方都市報) raised three questions from the perspective of relocated residents: (1) Rent around the relocated area has increased; (2) Should they return to their hometown or continue to stay? (3) What should be done about the over 1500 students?[79] Of course, the majority of the news reports still unfolded the news narrative from the government's standpoint, emphasizing the rationale behind "returning the green to the people," "building an international garden city," "constructing a green home,"

[74] Peng, S. 彭森 (May 21, 2004)0.*800-meter Green Belt to be Fully Displayed 800米綠化帶將完整展現*. Shenzhen Evening News 深圳晚報. (A7); Sun L. (May 21, 2004). *Temporary Settlement Sites in Huangmugang completely demolished on the 28th* 黄木崗臨時安置區28日全拆除. Southern Daily南方日報.(001).

[75] Xu, S.徐升 (2001, August 1). *The finalized greenery plan for the demolition site of the Lotus Mountain Park resettlement area* 蓮花山公園安置區拆除場地綠化方案敲定. Shenzhen Economic Daily深圳商報.(A5).

[76] According to the interview with Mr. Yang Yibiao, 2023-06-18.

[77] Huang, Y. L., & Li, H. 黄楊略,李紅 (2001, September 27). *Futian: Making Shenzhen's Face More Beautiful* 福田: 讓深圳的臉面更靚麗. Shenzhen Economic Daily深圳商報.(A4); Li, H., & Guan, Y. 李紅,管亞東 (2004, May 29). *How did the smooth progress of the citywide "cleanup" proceed successfully: Owners and residents consider the overall situation* 全市"梳理"之最何以進展順利:業主居民顧全大局. Shenzhen Economic Daily 深圳商報. (A3);姜東南 (1998, September 11). *No "Nail" That Can't Be Pulled Out - Lessons from the Demolition of Illegal Buildings in Donggualing* 沒有拔不掉的"釘子"——冬瓜嶺違章建築群拆除的啟示. Shenzhen Economic Daily深圳商報.(1).

[78] Jiang, D.N.姜東南(1998, September 11). *No "Nail" That Can't Be Pulled Out - Lessons from the Demolition of Illegal Buildings in Donggualing* 沒有拔不掉的"釘子"——冬瓜嶺違章建築群拆除的啟示. Shenzhen Economic Daily 深圳商報.(1).

[79] Feng, L., & Ya, N. 豐雷,亞牛 (2004, May 21). *Have You Found the House?* 你找到房子沒?. Southern Metropolis Daily 南方都市報. (A39).

and the legitimacy of the "Rectification Actions."[80] Subsequently, there were also some nostalgic reports that, after describing the spatial characteristics of these settlement sites as "cramped, very simple, and very crowded," often shifted the narrative toward themes like "youth," "memory," and "the lifestyle of the younger generation."[81]

As of August 30, 2004, citywide, the "Rectification Action" resulted in a total demolition area of 3833.00 ha., equivalent to the size of a medium-sized city. Out of these, a total of 575.98 ha. were successfully transformed into green spaces. More precisely, in Futian District, a total of 228.00 ha. of land were demolished, with 54.40 ha. being transformed into green spaces.[82]

After examining the origins of urban parks, John B. Jackson shifted his focus away from the expert-driven, urbanized parks that aimed for a "contact with nature." Instead, he saw rough nature, such as small groves, that existed outside the power of 19th-century liberal reformers, transcending class distinctions.[83] The three mountain parks in the Meilin-Gangxia area previously exhibited a rough quality, but they have been transformed according to the government's plan to create an international city, resulting in a more sophisticated and polished design. The verdant appearance of natural landscapes easily conceals the past and is more prone to adopting aesthetic language to mask the previously contested meanings about the subject, which is a conflict of power structures. If the ultimate challenge of historical writing is to reveal the roots of power,[84] this revelation cannot be confined to the local government level. Instead, it should delve into the brutality and complexity within the world economy: globalization involves not only international finance, multinational corporations, and professionals but also gatekeepers, domestic workers, industrial workers, and the temporary residents once sheltered by the hills of the Meilin-Gangxia area.[85] Their right to enter the city is equally governed by this brutality and

[80] Li, H., & Guan, Y. 李紅,管亞東 (2004, May 29). *How did the smooth progress of the citywide "cleanup" proceed successfully: Owners and residents consider the overall situation* 全市"梳理"之最何以進展順利:業主居民顧全大局. Shenzhen Economic Daily 深圳商報. (A3); Li, H. & Guan, Y. D.李紅，管亞東 (2004, May 29). *Yesterday's Demolition of Huangmugang Resettlement Area* 黃木崗安置區昨日開拆. Shenzhen Business Daily 深圳商報. (A3).

[81] Li, Z. Q. 李振岐 (2019, January 4). *Dongguling, Have You Ever Grown Winter Melons* 冬瓜嶺，種過冬瓜嗎? Daily Sunshine晶報(16); Long L.玲瓏(2010, March 1st).*Lianhuashan Hill: The Settlement Site is my first home* 蓮花山:安置區有我第一個家. Shenzhen Evening News 深圳晚報; Qiu, H. 邱禾(2010, August 2). *Lianhua Mountain and Donggualing Resettlement Areas: Carrying the Passionate Years of Entrepreneurship* 蓮花山、冬瓜嶺安置區:承載創業時的激情歲月. Shenzhen Evening News 深圳晚報.(24).

[82] Wu, Z. J. 吳子俊 (Eds.). (2005). *Shenzhen City Sorting Action Chronicle: Adjusting the Direction of Urbanization with a Scientific Development Perspective* 深圳市梳理行動紀實:用科學發展觀調校城市化發展方向. Shenzhen: Haitian Publishing House 深圳:海天出版社. (pp.59).

[83] Jackson J. B. (1984). *Discovering the Vernacular Landscape*. Yale University Press. (pp. 125–130).

[84] Trouillot, M. R. (2015). *Silencing the past: Power and the production of history*. Beacon Press (p. xvi).

[85] Sassen, S. (2014). *Expulsions: Brutality and complexity in the global economy*. Harvard University Press. (pp1–11).

complexity, closely related to what Aljun Appadurai calls "grass-root globalization."[86] From collective ownership in rural areas to government ownership, from temporary occupation by temporary residents to being redeveloped as urban parks, these hills are constantly being reproduced, and the cultural forms of nature are continually switched. Behind this lies Shenzhen's imagination of globalization as a city.

## 5.4 Spectacular Nature: The Plants in History

### *5.4.1 Historicity of Plants Communities*

From the 14th to the eighteenth century, the population growth in the Lingnan region led to deforestation primarily caused by agricultural development.[87] By the beginning of the twentieth century, the majority of the forests in the Lingnan region had vanished due to conflicts, corruption, and financial collapse. Following 1949, despite limited afforestation initiatives in Guangdong Province, they proved unable to satisfy the timber requirements during the succeeding period of socialist industrialization.[88] In reference to the antecedent of Shenzhen, Bao'an County, circa 1960, it was depicted as a locale where "desolation" aptly reflected its condition, characterized by scarce vegetation and bare soil.[89] From the 1960s onward, Bao'an County has consistently achieved an average annual artificial afforestation of over 10,000 mu and forest closure of over 20,000 mu.[90] By the conclusion of the 1970s, there was a noticeable improvement in contrast to the circumstances preceding the establishment of the PRC, as evidenced in Table 5.13. Nevertheless, the deterioration of the natural environment caused by excessive harvesting and insufficient afforestation has resulted a low survival rate for afforestation in this region.[91]

---

[86] Appadurai, A. *Grassroots globalization and the research imagination* In Appadurai, A. (Eds.). (2001). Globalization (Vol. 2). Duke University Press.(pp.1-21).

[87] Marks, R. (1998). *Tigers, rice, silk, and silt: Environment and economy in late imperial South China*. Cambridge University Press (pp. 35–46; 317–327).

[88] Zaizhi Z. and Chokkalingam U. Success and sustainability: Lessons from Guangdong Province In Chokkalingam, U. et al. (Eds.) *Learning lessons from China's forest rehabilitation efforts: national level review and special focus on Guangdong Province*. CIFOR. (pp. 69–134).

[89] He, F. 何發. Counting the Trees on Which Hill Even with Closed Eyes閉著眼睛都能數出哪個山頭種了什麼樹. In Shenzhen Urban Planning and Natural Resources Bureau, Shenzhen Forestry Bureau深圳市規劃和自然資源局 深圳市林業局(Eds.), *Memory Forest: Oral History of Shenzhen Forestry*記憶森林 深圳林業口述史 (1980-2020). Unpublished internal document.

[90] Bao'an County Local Chronicles Compilation Committee 寶安縣地方誌編纂委員會(1997). *Chronicles of Bao'an County*寶安縣誌.Guangzhou: Guangdong People's Publishing House 廣州: 廣東人民出版社. (pp. 115).

[91] Guangzhou Institute of Geography廣州地理研究所(1985). *Atlas of Natural Resources and Economic Development in Shenzhen*深圳市自然資源與經濟開發圖集. Beijing: Science Press北京: 科學出版社. (pp. 109).

**Table 5.13** Forest area estimation in Guangdong Province[a]

| Year | Total land area (10,000 $km^2$) | Forest area (10,000 $km^2$) | Percentage |
|---|---|---|---|
| BC 2700 | 22 | 20 | 90.9% |
| AD 1700 | 22 | 13 | 54.5% |
| AD 1937 | 22 | 2.23 | 10.0% |
| AD 1977 | 22 | 7.49 | 33.4% |

[a]Ling, D. X. 淩大燮 (1983). Changes in China's Forest Resources我國森林資源的變遷. *Agricultural History of China*中國農史, 1983(02), 26–36.

Guangdong is situated in a subtropical area characterized by elevated temperatures and ample precipitation, so the native vegetation consists of tall, dense evergreen broad-leaved tree communities. By the conclusion of the twentieth century, the native forest types—tropical seasonal rainforest, subtropical evergreen seasonal rainforest, and subtropical evergreen broad-leaved forest—had become small in area and were mostly located in remote and less disturbed areas with inconvenient transportation. By the end of the twentieth century, the regions that were formerly occupied by native plants had mostly transitioned into secondary ecosystems consisting of trees, shrubs, and herbaceous plants.[92] For example, in Shenzhen, the evergreen broad-leaved tree communities on Wutong Mountain (梧桐山), Qiniang Mountain (七娘山), and the north slope of Yangtai Mountain (羊臺山) have mostly succeeded into secondary mixed forests dominated by *Pinus massoniana*, with dominant tree species being *Pinus massoniana* (馬尾松), *Schima spp* (荷木), and *Itea oblonga Hand.-Mazz* (華鼠刺).[93]

During evolution, the habitat of the vegetation has also shifted from being humid to semi-humid. When human activities lead to a reduction in the coverage of native vegetation, the high temperatures, heavy rainfall, and monsoon climate conditions in Guangdong contribute to significant soil erosion and water runoff. The fertility, water retention, and permeability of the soil continuously decrease under these conditions, causing the habitat to gradually shift toward a drier direction. This change in habitat causes secondary vegetation communities to evolve toward semi-humid or dry conditions, making it challenging to restore the original tall, dense evergreen broad-leaved tree communities. Furthermore, as the community becomes more secondary, the habitat conditions become more arid. At this point, native tree species are less likely to thrive, so when afforestation is carried out in sparsely vegetated mountainous areas, Guangdong Province often uses drought-resistant species such as *Acacia confuse Merr* (臺灣相思), *Eucalyptus spp.* (桉樹), *Casuarina spp.*(木麻黃), and *Pinus massoniana* (馬尾松).[94] From 1963 to 1979, aerial seeding afforestation covered an area of 2.18 million ha., with an effective area of 1.138 million ha. *Pinus massoniana*

[92]Guangdong Institute of Botany廣東省植物研究所 (1976). *Vegetation of Guangdong* 廣東植被. Beijing: Science Press 北京: 科學出版社.(pp. 39-40).

[93]Bao'an County Local Chronicles Compilation Committee 寶安縣地方誌編纂委員會(1997). *Chronicles of Bao'an County*寶安縣誌. Guangzhou: Guangdong People's Publishing House 廣州:廣東人民出版社. (pp. 218).

[94]Guangdong Institute of Botany 廣東省植物研究所 (1976). *Vegetation of Guangdong* 廣東植被. Beijing: Science Press 北京: 科學出版社. (pp. 39–40).

and *Acacia confuse Merr,*[95] became the main tree species during this period. Corresponding to the situation in Guangdong Province, it can be observed that before the reform and opening up, the forestry land in Bao'an County and its subordinate Fucheng Commune (附城公社) mainly consisted of these tree species.[96]

Table 5.14 shows that whether it is in the forestry land of Bao'an County at that time or in the affiliated Fucheng Commune,[97] *Pinus massoniana* dominated the proportion of forestry land designated for timber production, with other species primarily including *Casuarina* and *Eucalyptus*, known for their drought resistance. The

**Table 5.14** Inventory of forest resources in Bao'an County and Fucheng Commune in 1975[a]

| Forest types/Tree species/Tree communities | | | | Bao'an County (now the Shenzhen City) | | Fucheng Commune (including current Futian District) | |
|---|---|---|---|---|---|---|---|
| | | | | Area (mu) | Percentage | Area (mu) | Percentage |
| Forestry land | Dense forest land | Timberland forest | Fir | 10,204 | 1.1% | 414 | 0.9% |
| | | | *Pinus massoniana* | 748,951 | 77.4% | 38,815 | 84.9% |
| | | | Mixed wood | 43,371 | 4.5% | 1036 | 2.3% |
| | | | Mixed pine and fir | 2395 | 0.2% | – | – |
| | | Protective forest | *Casuarina spp.* | 1624 | 0.2% | – | – |
| | | | *Eucalyptus* | 12,598 | 1.3% | 1070 | 2.3% |
| | | | Other | 453 | 0.0% | – | – |
| | | Economic forest | Bamboo/tea/fruit trees | 31,255 | 3.2% | 3570 | 7.8% |
| | | Sub-total | | 850,851 | 87.9% | 44,905 | 98.3% |
| | Sparse Forest Land | *Pinus massoniana* | | 116,070 | 12.0% | 362 | 0.8% |
| | | Mixed pine and fir | | 613 | 0.1% | 431 | 0.9% |
| | | Others | | 28 | 0.0% | – | 0.0% |
| | | Sub-total | | 116,711 | 12.1% | 793 | 1.7% |
| | Total | | | 967,562 | 100.0% | 45,698 | 100.0% |

[a]Bao'an County Local Chronicles Compilation Committee 寶安縣地方誌編纂委員會(1997). *Chronicles of Bao'an County*寶安縣誌. Guangzhou: Guangdong People's Publishing House 廣州:廣東人民出版社. (pp. 228–229).

[95] Bao'an County Local Chronicles Compilation Committee 寶安縣地方誌編纂委員會(1997). *Chronicles of Bao'an County*寶安縣誌. Guangzhou: Guangdong People's Publishing House 廣州:廣東人民出版社. (pp.338–360).

[96] Bao'an County Local Chronicles Compilation Committee 寶安縣地方誌編纂委員會(1997). *Chronicles of Bao'an County*寶安縣誌. Guangzhou: Guangdong People's Publishing House 廣州:廣東人民出版社. (pp. 229).

[97] In March 1978, Bao'an County, which then encompassed the present-day area of Shenzhen, was reorganized into Shenzhen City. During that time, Fucheng Commune included what is now Futian District. In March 1978, Shatou Commune was separated from Fucheng Commune. The following month, Shatou Commune was renamed Futian Commune and placed under the administration of

selection of these drought-resistant tree species was a necessary choice for afforestation after the change in habitat, stemming from historical disruptions to the native vegetation and its habitat. Of course, in the warm and humid climate conditions of Guangdong Province, under conditions where human interference is stopped or reduced, the secondary vegetation undergoes a successional succession—plant communities transition from low complexity to high complexity—quite rapidly.[98]

In 1985, after the Guangdong Provincial Party Committee issued the decision to "Green Guangdong in Ten Years (十年綠化廣東)," and with the efforts of various levels of government departments in Shenzhen, the city achieved its greening targets two years ahead of schedule in 1993. During this period, due to the tight timeline and heavy tasks, afforestation often relied on drought-resistant fast-growing tree species such as *Eucalyptus* and *Acacia mangium Willd.* Additionally, the practice of planting a single species was common, leading to low biodiversity and poor stand quality.[99] Although Bao'an County introduced and domesticated more than 20 excellent tree species after the founding of the PRC, the species selected for afforestation were mostly drought-resistant and tolerant of poor soil conditions. Broadleaf trees included *Eucalyptus*, *Acacia delavayi*, *Acacia mangium Willd.*, and *Albizia falcataria (Linn.) Fosberg*, while coniferous trees were mainly *Pinus massoniana*, *Cunninghamia lanceolata (Lamb.)Hook.*, and *Pinus elliottii Engelmann.*[100]

In summary, on the one hand, the plant communities predominantly composed of drought-resistant tree species on the mountain and hills of Shenzhen and Futian District before the 1980s were not entirely naturally formed. Instead, they were the result of the dynamic interaction between human history and the natural environment, demonstrating a historical context. On the other hand, within the historical context, this is also related to the strategies for plant cultivation and forest structure transformation in subsequent urban greening projects.

### 5.4.2 From Greening to Colorization: Botanical Landscapes in International Cities

In February 1993, the Shenzhen Municipal Party Committee Policy Research Office (深圳市委政策研究室), Shenzhen Social Science Research Center (深圳社會科學研口岸建設究中心), Shenzhen University (深圳大學), Shenzhen Economic

---

Luohu District. The boundaries of Futian Commune at that time roughly correspond to the present-day Futian District. See Shenzhen Futian District Local Chronicles Compilation Committee 深圳市福田區地方誌編纂委員會 (2012). *Futian District Annals, Shenzhen 1979–2003*深圳市福田區志1979–2003. Beijing: Fangzhi Publishing House 北京:方志出版社. (pp. 66–67).

[98] Guangdong Institute of Botany 廣東省植物研究所(1976). *Vegetation of Guangdong*廣東植被. Beijing: Science Press 北京: 科學出版社. (pp. 39–40).

[99] Wang, B. Z.王炳真(2001). Ecological Landscape Forest: The Guarantee of a Garden City 生態風景林:花園城市的保障. *Chinese Forestry Science and Technology*中國林業, 24, (pp. 14–15).

[100] Bao'an County Local Chronicles Compilation Committee 寶安縣地方誌編纂委員會(1997). *Chronicles of Bao'an County* 寶安縣誌. Guangzhou: Guangdong People's Publishing Hous 廣州: 廣東人民出版社. (pp. 226).

Daily (深圳商報), and the editorial office of the journal *Special Zone Theory and Practice* (特區理論與實踐) among other institutions, jointly held the "International City·Shenzhen Strategic Seminar[101] (國際性城市·深圳戰略研討會)." Over time, many government departments reached a consensus to transform Shenzhen into an international city. This dedication was evident in other domains, including copyright commerce, higher education, and port development. The government even dispatched personnel to Hong Kong and overseas for relevant inspections in different fields to assist in decision-making.[102] The primary objective of constructing an international city is to bring Shenzhen in line with global benchmarks in all areas, with a specific emphasis on the tourism sector:

*Implement good planning and reasonable layout… to create a comprehensive system of cultural tourism, encompassing various aspects such as historical replication, religious exploration, local customs, entertainment, urban sightseeing, natural scenery tourism, shopping tourism, modern technology entertainment tourism, coastal and sports tourism, as well as business and conference tourism.*[103]

Undoubtedly, amid the process of building an international city, urban sightseeing and natural scenery tourism of the city are poised to become a calling card extended to the world. Shenzhen was granted the distinction of "National Garden City (國家園林城市)" in 1994, and in 1997, it was bestowed with the inaugural title of "National Environmental Model City (國家環保模範城市)." However, these national-level honors did not fully meet Shenzhen's ambitious goals in terms of urban greening for constructing an international city. In January 1998, the Shenzhen Municipal Party Committee has put forward the goal of transforming Shenzhen into a "garden-style international city (園林式、花園式國際性城市)." This statement demonstrates Shenzhen's comprehension of constructing an international city in the aspect of urban greening improvement.

In October 1999, Hangzhou hosted the Asia-Pacific regional conference of the International Federation of Parks and Recreation Administration (國際公園與遊憩管理聯合會), providing the urban greening authorities in Shenzhen with an opportunity to benchmark against international standards. During that period, the association had recently launched the *Nations in Blooms* (國際花園城市) competition, and no Chinese city had previous involvement in participating. In May 2000, Shenzhen submitted the necessary documentation to participate in the competition for E-class cities (with a population of over one million) for that year. Shenzhen's competitors

[101] Lin, Z. J. 林祖基(Eds.). (1993). *Shenzhen - Collection of Essays on an International City* 深圳-國際性城市論文集. Beijing: China Economic Publishing House 北京:中國經濟出版社.

[102] Shenzhen Foreign Affairs Training Leadership Group 深圳市涉外培訓領導小組. (1997). *Research Series on Successful Management Experience of International Cities: Hong Kong Urban Management Research* 國際性城市成功管理經驗研究叢書:香港城市管理研究. Preserved at the Local Documents Room of Shenzhen Library 深圳圖書館地方文獻室.

[103] Wu, D., & Qin, Q. L. 吳鏑,秦群力 (1993). Vigorously Developing Trade and Tourism, Accelerating the Pace of Shenzhen's Construction as an International City 大力發展貿易、旅遊業，加快深圳建設國際性城市步伐. In Lin, Z. J.林祖基 (Eds.), *Shenzhen - Collection of Essays on an International City* 深圳-國際性城市論文集. Beijing: China Economic Publishing House 北京:中國經濟出版社.(pp. 120–126).

included cities such as Chicago (芝加哥) and Phoenix (鳳凰城) in the United States, Novgorod in Russia, and Pretoria in South Africa. These cities all boasted excellent urban landscape and natural scenery. The results were announced in December 2000, and among the E-class cities, Shenzhen took the lead, surpassing the second-place Phoenix[104] in the United States. In the application documentation for the competition, there was a special emphasis on annually increasing the planting of flowering trees in urban greening, implementing scenic forest planting plans, and constructing specialized breeding bases for distinctive flowering trees.[105]

Shenzhen, located in the coastal area of South China, has a subtropical monsoon climate characterized by warmth, abundant rainfall, and ample sunlight, making it suitable for cultivating various flowers and flowering trees. Not far away, Guangzhou is renowned domestically and internationally as the "City of Flowers (花城)." Employing flowers, colorful bushes, and blossoming trees to create urban spectacle has become a crucial strategy for Shenzhen in molding urban landscapes and building an international city. The urban strategy gained momentum after the 26th Summer Universiade (世界大學生夏季運動會)[106] in 2011. It achieved its highest point with the implementation of the *World-Famous Flower City Three-year Action Plan (2017–2019)*(世界著名花城三年行動計畫) and *World-Famous Flower City Five-year Action Plan (2021–2025)*(世界著名花城五年行動計畫) initiated by Shenzhen. During the period from 2017 to 2019, Shenzhen implemented actively a three-year plan intended for establishing itself as a globally renowned city for its floral beauty. This endeavor involved the planting of 138,000 flowering trees and bushes, as well as the construction of 60 roads adorned with flowers. However, in the official view, the pursuit of building a "World-Famous Flower City" comparable to world-class advanced cities, Shenzhen still has some shortcomings. Therefore, a new "Five-year Action Plan"[107] was formulated.

In light of this, let's reconsider the original intent behind the formulation of the *Interim Measures for the Administration of Urban Landscaping and Greening in Shenzhen* (深圳市城市園林綠化管理暫行辦法) in 1982:

---

[104] *Recollection: Unforgettable Experience of Shenzhen Competing for the International "Garden City" Title* 回憶:深圳角逐國際"花園城市"的難忘經歷. Shenzhen Special Zone Daily 深圳特區報, 2010-12-06.

[105] *Application Text for Nations in Blooms*國際花園城市申報文本, preserved at the Local Documents Room of Shenzhen Library.

[106] Shenzhen City Appearance and Environmental Improvement Command Office, Shenzhen Urban Village Comprehensive Renovation Leadership Group Office深圳市市容環境提升行動指揮部辦公室,深圳市城中村綜合整治工作領導小組辦公室, Shenzhen Urban Management Bureau深圳市城市管理局. (January 2010). *Compilation of the Action Plan for Shenzhen City Appearance and Environmental Improvement*深圳市市容環境提升行動工作方案彙編. Archived at the Local Documents Room of Shenzhen Library.

[107] Shenzhen Municipal Urban Management and Comprehensive Law Enforcement Bureau深圳市城市管理與綜合執法局. (2021). *Compilation Explanation of the Five-Year Action Plan (2021–2025) for Building Shenzhen as a "World-Famous Flower City."* 深圳市打造"世界著名花城"五年行動計畫. Retrieved from http://cgj.sz.gov.cn/attachment/0/766/766058/8645699.pdf

*This regulation has been specifically established to improve urban greening, and to develop Shenzhen into a modern city with a meticulously designed layout, aesthetically pleasing surroundings, and clean air...the greening department shall conform to the standards of socialist urban greening, while also considering the unique characteristics of the city and adjusting accordingly...this entails developing superior tree species and flowers that are highly compatible with the local growth conditions, thereby accelerating the greening process....*[108]

It can be observed that in the early stages of the establishment of the SSEZ, the primary emphasis of greening endeavors was directed toward strengthening the city's local environment rather than prioritizing the enhancement of its international image, the emphasis was on "greening" rather than "colorization." During the "Three-Processing and One Compensation" period, which followed the cooperative model of front shops and rear factories with Hong Kong, except for the commercial center in the Luohu area, the industrial areas in Shenzhen primarily emphasized product quality and cost reduction rather than giving excessive importance to spatial environmental features. Nevertheless, Shenzhen sought to surpass these goals by aspiring to establish itself as an international city and cultivate a headquarters economy that would entice multinational corporate headquarters and their talented workforce to relocate to Shenzhen. To achieve this, it is crucial to create a distinctive and competitive spatial environment. Capitalizing on Shenzhen's unique environment, the transition from greening to colorization took place gradually, aiming to gradually build the "World-Famous Flower City" to benchmark other cities in terms of political and economic significance.

The shift in the concept from greening to colorization first occurred in the late 1990s when the government, aiming to construct an international city, dispatched personnel to Hong Kong and developed cities abroad for professional investigations in various fields. Based on these investigations, personnel in the field of urban greening proposed that Shenzhen's urban greening was "green but lacking beauty" and "lacking color and landscape features."[109] It was suggested that a new style emphasizing "texture, volume, linearity, dynamics, color, and other comprehensive factors"[110] needed to be developed. The government's construction of an interna-

---

[108] Shenzhen Municipal People's Government Office深圳市人民政府辦公廳. (1982). *Compilation of Documents on Urban Construction and Economic Management by Shenzhen Municipal People's Government*深圳市人民政府城市建設和經濟管理檔彙編. (pp. 130–138). Archived at the Local Documents Room of Shenzhen Library.

[109] Wang, X. M.王曉明(1997). Hong Kong Urban Garden Greening Funds and Utilization香港城市園林綠化經費及使用. In Shenzhen Municipal Foreign Training Leadership Group深圳市涉外培訓領導小組 (Eds.), *Research Series on International Successful Urban Management Experiences: Hong Kong Urban Management Studies*國際性城市成功管理經驗研究叢書:香港城市管理研究. (pp. 21–28). Archived at the Local Documents Room of Shenzhen Library.

[110] Li, R. K. 黎瑞坤 (1997). Discussion on Modern Urban Greening Construction and Management 現代化城市園林建設與管理淺談. In Shenzhen Municipal Foreign Training Leadership Group深圳市涉外培訓領導小組 (Eds.), *Research Series on International Successful Urban Management Experiences: Hong Kong Urban Management Studies*國際性城市成功管理經驗研究叢書:香港城市管理研究. (pp. 62–74). Preserved at the Local Documents Room of Shenzhen Library

tional city is reflected in its official ideology as well as the use of professional and aesthetic discourse, permeating aesthetic concepts of urban greening into the practice domain. Following the turn of the millennium, the Shenzhen Municipal Government's increased financial capacity has led to a shift in the selection of plant species in urban greening projects toward emphasizing colorization.

Furthermore, the inclusion of vast forested regions on the outside of urban area were also playing a significant role in the city's endeavors to establish itself as an international city. Under the goal of constructing a "garden-style international city," the government's forestry department has initiated the Ecological Scenic Forests Construction Project. The objective is to enhance both the "ecological and investment environment" while concurrently "elevating Shenzhen's status as a modern international city."[111] In accordance with this overarching goal, the government intended to allocate 10 years toward the establishment of an expansive Ecological Scenic Forests spanning a planned total of 54,072.5 ha., surpassing one-fourth of the city's territorial expanse. Within this proposal, the goals for the years 2002–2005 were focused on "beautifying the visible environment," specifically completing the forest structure transformation of urban background mountains in secondary city center areas, and the first-range mountains on both sides of green corridors. From 2006 to 2011, the objective shifted to improving the ecological environment, with comprehensive requirements to complete the transformation of suburban forests and distant mountain forests, forming an overall harmonious and locally distinctive characteristic forest landscape. In the long-term plan (2012–2030), the aim is to catch up with the level of international urbanization.[112]

Prior to the implementation of the Ecological Scenic Forests Construction Project (生態風景林建設工程), the primary forest vegetation types in Shenzhen included needle-leaved artificial forests dominated by *Pinus massoniana* and *Cunninghamia lanceolata (Lamb.)Hook.*, exotic tree species artificial forests dominated by *Eucalyptus* and *Acacia confusa*, economic forests mainly consisting of lychee and longan trees, secondary broad-leaved forests composed of indigenous tree species such as *Castanopsis fissa, Castanopsis hystrix, Schima superba, Schefflera heptaphylla*, and *Sapium discolor*, as well as mangrove forests, low forests, and evergreen broad-leaved shrub forests. According to the research and judge from *General Plan for Ecological Scenic Forests Construction in Shenzhen* (深圳市生態風景林建設總體規劃) (2002–2011), Shenzhen at that time had a higher proportion of man-made forests compared to natural forests, a greater number of

[111] Zhou, G. M.周光明 (1999). Preface.前言 In Wei X. Q. 魏小琴 (Eds.). *The Covenant of the Century: Collected Works on the Construction of Shenzhen Ecological Landscape Forest*世紀之約:深圳市生態風景林建設文集Beijing: China Forestry Publishing House北京:中國林業出版社. (pp. 1–3).

[112] Project Summary.規劃提要(2002, December). In Research Institute of Tropical Forestry Chinese Academy of Forestry中國林業科學研究院熱帶林業研究所. Shenzhen Agricultural, Forestry and Fisheries Bureau深圳市農林漁業局. Shenzhen Greening Committee Office深圳市綠化委員會辦公室. *Overall Planning of Shenzhen Ecological Scenic Forest Construction Project*深圳市生態風景林建設工程總體規劃: 2002–2011 (Revised), preserved at Shenzhen Yantian District Library Local Documents Room.

single-layer forests as opposed to multi-layer forests, a larger quantity of pure forests rather than mixed forests, a higher concentration of young forests in comparison to middle-aged forests, and a prevalence of monotonously colored forests over seasonally changing forests. Faced with such a forestry situation and the gap between the construction goals mentioned above, the issues to be addressed were multi-level and multifaceted. However, the issue of tree species colorization has been frequently raised by technical experts and government officials involved in this project. They have consistently linked its significance to the construction goals of an international city.[113,114,115] According to the overall layout outlined in the *General Plan for Ecological Scenic Forests Construction in Shenzhen* (2002–2011), the tree species are divided into two categories: Landscape Tree Species (景觀樹種) and Ecological Tree Species (生態樹種). And the tow words "Landscape Tree Species" and "Ecological Tree Species" are given similar importance in the plan, highlighting the considerable focus on visually appealing mountain vistas in this forestry transformation initiative. To enhance comprehension of Ecological Scenic Forests construction by the executing departments, a formula has been introduced: Ecological Scenic Forests[116] = Landscape Tree Species + Original Vegetation. This formula indicates that the construction of Ecological Scenic Forests is essentially about adding color foliage and flowering tree species on the foundation of existing vegetation. The ecological importance of conserving water sources, regulating climate, and retaining dust is indeed important in this extensive forestry transformation. Nevertheless, this does not provide a satisfactory justification for the widespread utilization of color foliage and flowering tree species. This rationale is reflected in both the *General Plan for Ecological Scenic Forests Construction in Shenzhen: 2002–2011* and the *Outline of the Ecological Scenic Forests Construction Project in Shenzhen* (深圳市生態風景林建設工程規劃大綱): Firstly, the government views "captivating colors and picturesque landscapes (迷人的色彩和綺麗景觀)" as

[113] Chen, T. 陳濤 (2000). Reflections on Building Shenzhen as an "International Garden City." 建設深圳"國際花園城市"的思考.*Ecological Science*生態科學. (pp. 91–93).

[114] Jiang, Y. X.蔣有緒 (1999). Urban Forestry in the New Century: On the Construction of Ecological Landscape Forests and the Demonstrative Significance of Shenzhen City新世紀的城市林業方向:試論生態風景林建設兼論深圳市的示範意義. In: Wei X. Q. 魏小琴 (Eds.), *A Covenant of the Century: Collected Works on the Construction of Ecological Landscape Forests in Shenzhen*世紀之約:深圳市生態風景林建設文集. Beijing: China Forestry Publishing House. (pp. 8–15).

[115] Zhang, J. 張軍(1999). Grand Projects Bringing Benefits to Descendants: Reflections on the ShenzhenEcologicalLandscapeForestConstructionProject功在當代福蔭子孫的巨集偉工程——深圳市生態風景林建設工程隨想. In: Wei X.Q. 魏小琴 (Eds.), *A Covenant of the Century: Collected Works on the Construction of Ecological Landscape Forests in Shenzhen* 世紀之約:深圳市生態風景林建設文集. Beijing: China Forestry Publishing House 北京: 中國林業出版社. (pp. 16–21).

[116] Wang, D. Y.王定躍(Eds.) (2011). *Investigation and Planning of the Scenic Forest in Wutong Mountain Scenic Area*梧桐山風景區風景林調查與規劃. Shenzhen: Shenzhen Press Group Publishing House深圳:深圳報業集團出版社. (pp. 193).

having the potential to drive industries such as foreign trade, finance, and real estate[117]; Secondly, at that time, government officials observed that developed countries such as Germany, the Netherlands, and Japan were all engaged in urban forestry. Consequently, Shenzhen deemed it necessary to "keep the same pace to create Ecological Scenic Forests."[118]

This Ecological Scenic Forests project promoted by the government was completed in 2011, but this did not signify the termination of the colorization of tree species on Shenzhen's mountains and hills. It continued until the *World-Famous Flower City Three-year Action Plan* initiated in 2017 which continued the basic idea of constructing Ecological Scenic Forests. For example, in terms of goals, it still emphasizes "building an international ecological livable city." In terms of philosophy, there is a shift from "simple greening" to "coexistence of greening and beautification." The focus on background mountains such as Wutong Mountain, Sanzhoutian Mountain, Julongshan Mountain, and others remains. In terms of methods, there is an emphasis on "increasing the planting ratio of color-foliage and flowering tree species" and so on.[119]

## *5.4.3 Colorization of Mountain Park*

### 5.4.3.1 Meilin Park

In March 2002, the city's Ecological Scenic Forests construction officially began to be implemented after more than four years of discussion. At this time, the competition for the planning and design scheme of Meilin Park in northern Futian District was also underway. The plan that finally won the competition had two important features, namely, highlighting "colorful" and "fragrance."[120] The impact of Ecological Scenic Forests construction on the planning and design of Meilin Park thus can be seen.

---

[117] Project Summary. 規劃提要(2002, December). In Research Institute of Tropical Forestry Chinese Academy of Forestry中國林業科學研究院熱帶林業研究所. Shenzhen Agricultural, Forestry and Fisheries Bureau深圳市農林漁業局. Shenzhen Greening Committee Office深圳市綠化委員會辦公室. *Overall Planning of Shenzhen Ecological Scenic Forest Construction Project*深圳市生態風景林建設工程總體規劃: 2002–2011 (Revised), preserved at Shenzhen Yantian District Library Local Documents Room.

[118] Zhou, G. M.周光明 (1999). Preface.前言 In Wei X. Q. 魏小琴 (Eds.). *A Covenant of the Century: Collected Works on the Construction of Ecological Landscape Forests in Shenzhen*世紀之約:深圳市生態風景林建設文集Beijing: China Forestry Publishing House北京:中國林業出版社. (pp. 1–3).

[119] Yi, X. H. 伊宵鴻(2017, April 15). *Shenzhen Creates a New Business Card as the "World Famous Flower City."*深圳打造"世界著名花城"新名片 Shenzhen Evening News深圳晚報.(B08).

[120] Gong, J. N. 龔江南(2002, March 1). *Recommended Plan for Meilin Park Born Last Night*梅林公園推薦方案昨晚誕生. Shenzhen Evening News深圳晚報.(A10).

The westernmost and largest area of the four parts was designated as "Colorful Meilin (彩色梅林)" or Caimei Zone, while the three eastern zones were collectively known as "Green Melin (綠色梅林)," which includes the Guli Zone, Tingtao Zone, and Fangxiang Zone. As seen in Fig. 5.8, the Caimei Zone is strategically positioned on the western slope of Meishan Middle School, as well as in some northern and southern parts and the higher elevation western area. From the vision analysis, these locations correspond to the viewpoints of high-rise buildings in Meilin first Village and Meishan Yuan Community. They are, in fact, the logical results of the transformation and acceptance of financial, commercial, administrative, and other functions by Futian District in the 1990s. In other words, through simple spatial analysis, we can observe that the overall context to construct an international city governs and integrates urban spaces in multiple ways. The process commences with the introduction of governmental administrative functions from Luohu District to Futian District, followed by the development of residential areas, such as Meilin first Village, designated for government officers in the northern area of Futian District. Finally, the professional practices of urban park planning and design have specifically provided color-leaf forest landscapes for these residential areas.

The *Eucalyptus* and *Acacia* species found in different zones of Meilin Park prior to its creation were predominantly the consequence of the "Green Guangdong in Ten Years" that took place from 1985 to 1995. During that period, in order to rapidly accomplish afforestation goals, fast-growing species such as *Eucalyptus* and *Acacia* were primarily utilized for monoculture afforestation.[121] At the start of the new millennium, considering the historical and social context of building an international city, the four foothills of Meilin Mountain that reach into the city were officially

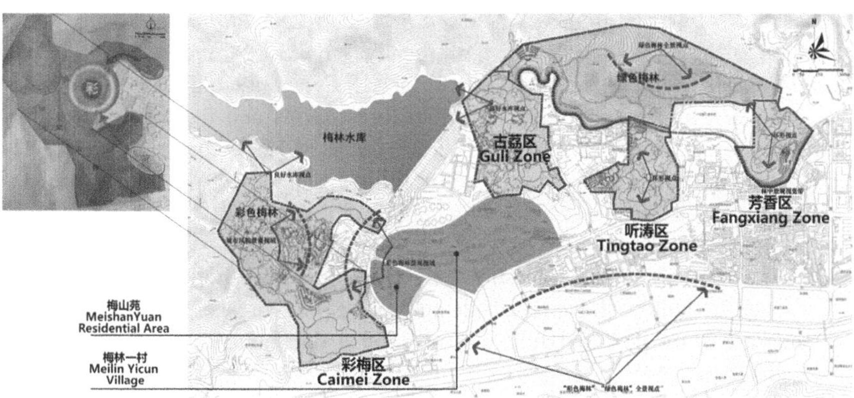


**Fig. 5.8** Visual analysis to the Caimei Zone [Left: Diagram of the color-foliage landscape structure in the Caimei Zone, provided by designer Fu Keqin; Right: Visual analysis diagram from the overall planning of Meilin Park, courtesy of the Shenzhen Park Management Center.]

[121] Wang, B. Z. 王炳真 (2001). Ecological Landscape Forest: The Guarantee of a Garden City 生態風景林:花園城市的保障. *Chinese Forestry Science and Technology*中國林業, 24, 4–15.

declared as urban parks. As a result, the plant communities on the four foothills experienced corresponding alterations.

The initially opened GuLi Zone, prior to construction, was predominantly covered with *Eucalyptus* trees which were retained during the construction process, but they were considered aesthetically inadequate. Therefore, the park construction department added color foliage and flowering plants such as *Bauhinia blakeana Dunn*, *Bauhinia variegata L.*, *Ilex rotunda*, and *Castanopsis fissa,* and the situation in the Camei zone was quite similar.[122] Before the construction of the Tingtao Zone, the vegetation was mainly composed of artificial forests and secondary shrub grasslands, with the dominant tree species being *Acacia auriculiformis A. Cunn. ex Benth.* and *Eucalyptus citriodora Hook.* The vegetation was lush and evergreen throughout the year. However, the designers considered them as having indistinct seasonal changes in vegetation landscape and the overall vegetation hierarchy is monotonous. Similarly, in Tingtao Zone, ornamental and colorful foliage plants, such as *Handroanthus impetiginosus (Mart. ex DC.) Mattos,* and *Ceiba speciosa (A.St.-Hil.) Ravenna,* were extensively used.[123]

#### 5.4.3.2 Bijiashan Park

The area where Bijiashan Park is located belonged to the state-owned Shenzhen Forest Farm for a long time. The state-owned forest farms established after the founding of the PRC aimed to eliminate barren mountains and establish timber-producing forests. In the era of socialist industrialization in the 1950s and 1960s, the primary objective of the Shenzhen Forest Farm was the management of timber-producing forests. As a result, the vegetation on Bijiashan Hill has predominantly consisted of *Pinus massoniana*. *Shenzhen Bijiashan Park Planning Manual* (深圳市筆架山公園規劃說明書)(1992) specifically emphasizes the conservation of these *Pinus massoniana* forests. While in fact, designers were aware of the differences between timber-producing forests in forest farms and the requirements for plant communities in urban parks. Even in the early 1990s, they had sensitively recognized the necessity of using color foliage and flowering plants, such as *Rhododendron simsii Planch.*and *Bougainvillea spectabilis Willd.*, to create "*a vibrant landscape for the hill* (山花爛漫的熱烈景觀)." However, at this time, the space for using color foliage and flowering plants mainly emphasized the "under-forest space," "shallow slopes, roadside and sparse pine forests" to enrich the internal landscape of the park and did not involve the visual connection with the broader urban landscape.[124]

---

[122] According to the interview with Mr. Wang Xiaoming, 2023-06-06.

[123] Shenzhen Garden Corporation, Shenzhen Garden Planning and Design Institute (Eds.) 深圳園林股份有限公司 深圳園林規劃設計院. *Design Proposal for Tingtao Zone of Meilin Park*梅林公園聽濤區方案設計. Preserved at the Meilin Park Archive Room, Shenzhen.

[124] Zeng, X., & Mao, X. 曾憲均,毛小琪(1992). *Shenzhen Bijiashan Park Planning Manual*深圳市筆架山公園規劃說明書. Landscape Department, Architectural Design Institute, Shenzhen

In December 1992, the Bijiashan Park Construction Office and the state-owned Shenzhen Forest Farm signed the *Land Acquisition Agreement for Bijiashan Park in Shenzhen* (深圳市筆架山公園征地協議書). The area acquired in this land acquisition was 80.62 ha. The plants recorded in the agreement included 116,662 *Pinus massoniana* trees, 19,862 *Cunninghamia lanceolata (Lamb.) Hook.* trees, 4114 *Pinus elliottii Engelm* trees, 1995 *Eucalyptus* trees, 765 bamboo clumps, and 13,496 seedlings in the nursery.[125] It is evident that, until 1992, the mountainous section of Bijiashan Park was predominantly populated with *Pinus massoniana* and *Cunninghamia lanceolata (Lamb.) Hook.* However, the unexpected outbreak of the nematode disease in 1994 abruptly accelerated and altered the process of vegetation colorization.

This outbreak of pine wilt disease, seemingly a natural disaster, is intricately linked to the characteristics of economic special zones and its opening policies at that time. Pine wilt disease was initially discovered in Japan in 1905, and it subsequently emerged in the United States in 1979. In the late 1980s, the SSEZ had a substantial volume of inbound and outbound goods, facilitating the introduction of biological vectors such as *Monochamus alternatus Hope* carrying the pine wood nematode from overseas. Following its initial discovery in Shenzhen in July 1988, the disease rapidly spread.[126] To prevent the invasion of this disease into the inland areas, a broad-leaved tree isolation belt, stretching 86 kilometers long and 4 kilometers wide, was constructed along the second line border in 1991. Additionally, 12 timber inspection checkpoints were established.[127] Within the isolation belt, coniferous tree species such as *Pinus massoniana* were extensively cleared.[128] In 1994, the *Pinus massoniana* forests on Bijiashan Hill, located within the second line border, were entirely infected. Following comprehensive cleaning and eradication efforts by the park construction department, a total of 168,000 broad-leaved tree seedlings,[129] including *Acacia confusa Merr.*, *Ficus altissima Blume*, *Ficus concinna Miq.*, *Syzygium hainanense Chang et Miau*, *Sapium sebiferum*, *Cinnamomum Camphora (L.) Presl.*, and *Cinnamomum burmanni (Nees et T.Nees) Blume.*, were replanted on Bijiashan Hill. Had it not been for the pine wilt disease, in the Ecological Scenic Forests construction several years later, Bijiashan Hill might have faced challenges

University深圳大學建築設計院園林部. (pp. 018–019). Preserved at the the Archive Office of Shenzhen Park Management Center, with Document Number: 全宗號GY9;分類號0916;案卷號003;件號001.

[125] *Land Acquisition Agreement of Bijiashan Park, Shenzhen* 深圳市筆架山公園征地協議書, preserved at the Archives office of Shenzhen Park Management Center with Document Number: 全宗號GY20;年度1992;件號16.

[126] Huang, H. H.黃煥華 (1990). Pine tree devastating epidemic in Shenzhen—Pine Wilt Nematode Disease深圳市松樹毀滅性疫病—松材線蟲萎蔫病. *Guangdong Forestry Science and Technology(Forestry and Environmental Science)*廣東林業科技, (06), 34–36.

[127] Wang, S. Y.王淑英 (1993). Achievements in the research on controlling pine wilt disease in Shenzhen深圳控制松材線蟲病研究取得階段性成果. *Forest Pest and Disease Communication (Forest Pest and Disease)*森林病蟲通訊, (02), 31.

[128] According to the interview with Mr. Wu Qingxiang, 2023-08-01.

[129] According to the interview with Mr. Huang Chongyuan, 2023-06-15.

in planting extensive areas with color foliage and flowering plants due to its dense forest coverage.

In 1997, the *Outline of the Ecological Scenic Forests Construction Project in Shenzhen* was issued. The same year, the Urban Management Bureau's Park Department organized experts to assess and deemed it necessary to plant colorful floral belts. As a result, tens of thousands of *Bauhinia variegata L.* were planted on the slopes near Beihuan Road, creating a floral belt landscape. At that time, there was a requirement to plant floral belts on every visible part of the Bijiashan Hill from the Beihuan Road. Around 2000, Shenzhen officially initiated the Ecological Scenic Forests project. In the Huangmugang area and places along the Futian River and the artificial lake, more than 1000 of *Bauhinia variegata L.*(宫粉紫荊) and *Magnolia soulangeana Soul.-Bod* (二喬玉蘭. were planted to create a floral landscape.[130] At this time, the concept of planting color foliage and flowering plants had evolved beyond the 1992 approach that focused solely on enriching the internal landscape of the park, while the new concept explicitly emphasized interaction with the broader urban landscape.

#### 5.4.3.3 Lianhuashan Park

In 1998, prior to the formal initiation of the citywide Ecological Scenic Forests project, Lianhuashan Park, situated in the central region of Futian, was chosen as a pilot[131] for demonstrating afforestation due to its moderate elevation, height, slope, and open slope space. In the period before the mid-1990s, Lianhuashan Park was mostly barren in terms of vegetation, with only some terraced lychee forests and *Eucalyptus* on the southern slope of the main peak. During the park construction process, these lychee forests were preserved, while the construction department has planted around 10,000 trees of 10–20 different species, including *Delonix regia (Boj.) Raf.*, *Albizia falcataria (Linn.) Fosberg*, *Jacaranda mimosifolia D. Don*, *Castanopsis fissa*, *Bauhinia blakeana Dunn,* and *Bauhinia variegata L.*[132]

By downplaying the productive function of lychee cultivated by local farmers and introducing ornamental flowering trees, this round of greening exhibited a clear emphasis on colorization in terms of plant configuration. In this pilot project for Ecological Scenic Forests, the government greening department deemed that Lianhuashan Park primarily featured lychee, *Eucalyptus,* and *Acacia*, resulting in a less satisfactory forest structure and visual effect. Therefore, to align with the overall planning and construction of Futian Central Business District, it is necessary to implement forest structure transformation in stages and phases. In consideration of plant coloration, a variety of color foliage and flowering trees such as *Koelreuteria*

---

[130] According to the interview with Mr. Huang Chongyuan, 2023-06-15.

[131] Chen, T., Li, G., & Liang, Z.陳濤,李剛,梁正陽 (1999). Tree species selection and demonstration forest construction in Shenzhen's ecological landscape forests深圳生態風景林樹種選擇與示範林營造. *Chinese Landscape Architecture*中國園林. (04), 44–46.

[132] According to the interview with Mr. Wang Xiaoming, 2023-06-06.

*bipinnata Franch., Bombax malabarica DC., Ceiba speciosa St.Hill., Delonix regia (Boj.) Raf., Lagerstroemia speciosa (L.)Pers., Erythrina variegata L., Bauhinia blakeana Dunn, Albizia falcataria (Linn.) Fosberg, Albizia chinensis (Osbeck) Merr, Spathodea campanulata Beauv., Jacaranda mimosifolia D. Don,* and many others have been incorporated into Lianhuashan Park.[133] While the government emphasizes ecological aims such as climate regulation, dust absorption, and soil conservation in the development of Ecological Scenic Forests, it is clear that this initiative has additional interests beyond ecological considerations. Otherwise, the steady replacement of the current *Eucalyptus* forest year by year would not be necessary.

Furthermore, the selection of Lianhuashan Hill with a slope facing the Futian Central Business District as a pilot for demonstrating afforestation had profound underlying reasons. The Futian Central Business District symbolizes the "second venture (二次創業)" of Shenzhen in the 1990s and the urban goal of constructing an international city. Therefore, providing a favorable visual landscape for this new city center became a significant motivation for the forest structure transformation in Lianhuashan Park.[134] During the forest structure transformation, on the one hand, the productive function of the lychee trees originally planted by villagers was weakened, on the other hand, it gradually replaced the existing *Eucalyptus* forest with a large number of color foliage and flowering plants. Through such forest structure transformation, the government aims to achieve characteristics of "flowers in all seasons, seasonal changes, and colorful diversity" in Lianhuashan Park, thereby contributing to the construction of Shenzhen as a "modern and first-class garden-style international city."[135]

The symbolic and aesthetic roles of color are not innate or static. Michel Pastoureau, a historian, has conducted a series of research on color, which serves as a reminder that colors are not abstract object. As symbols, all colors carry meanings that have been assigned within the context of society and history, contributing to the understanding we have today.[136] The increasing quantity and proportion of color foliage and flowering tree species chosen for urban greening in Shenzhen is not merely a matter of afforestation; rather, it reflects an aesthetic question on urban greening shaped by globalization. When Saskia Sassen recognized Shenzhen as a new center in the economic geography of globalization, the city's greening efforts had been advancing toward this goal for many years. This new center is precisely what the official authorities in Shenzhen began exploring and pursuing as an "international city" in the early 1990s. In Sassen's view, it is not solely globalization

[133] Chen, T., Li, G., & Liang, Z.陳濤,李剛,梁正陽 (1999). Tree species selection and demonstration forest construction in Shenzhen's ecological landscape forests深圳生態風景林樹種選擇與示範林營造. *Chinese Landscape Architecture*中國園林. (04),44–46.

[134] According to the interview with Mr. Yang Yibiao, 2023-06-18.

[135] Chen, T., Li, G., & Liang, Z.陳濤,李剛,梁正陽 (1999). Tree species selection and demonstration forest construction in Shenzhen's ecological landscape forests 深圳生態風景林樹種選擇與示範林營造. *Chinese Landscape Architecture*中國園林. (04), 44–46.

[136] Pastoureau, M. (2014). Green: The history of a color. Princeton University Press. (pp.1–6).

shaping local urban construction but, to a large extent, local construction that molds the city into a platform for globalization.[137] From this perspective, Shenzhen's greening and urban park construction have also shaped the city's globalization. The color of plants on hilly terrain is indeed a subject within the realm of modern botany. However, upon historical examination, it becomes evident that this matter is intricately linked to the ambitious endeavor of building an international city. Subsequently, the colors of plants in the city became a visual image disseminated and legitimized in the public discourse of urban ecological construction, and the aim is to construct Shenzhen's urban image within the global context.[138] According to Olmsted, urban parks and urbanization are in opposition; the plant landscapes in parks need to embody a picturesque elegance to resist the barbarism and coarseness brought by urbanization. The homogeneous and heterogeneous relationship between the bourgeois ideology behind picturesque aesthetics and the urban problems caused by industrial capitalism, however, is not acknowledged or realized by Olmsted. Olmsted's perspective contrasts noticeably with the contemporary experience in Shenzhen. Today, the flowering and colorful leafy plants growing on the hills of Shenzhen, represented by Meilin Mountain, Bijiashan Hill, and Lianhuashan Hill, are both a consequence and a reason of Shenzhen's urbanization process. They are not in opposition to this process but are interconnected, a conclusion drawn by retracing the historicity of these plants.

## 5.5 Summary

From before the era of reform and opening up to the "Three-Processing and One Compensation" period and then to the initial phase of establishing an international city, there have been significant landscape changes in the Meilin-Gangxia area. Firstly, the farmland and hilly topography in front of the mountains in the Meilin-Gangxia area have transitioned from rice and fruit cultivation to industrial zones and residential areas. Secondly, driven by Shenzhen's development of finance and high-tech industries, the area has gradually transformed, with industrial zones shifting from the "Three-Processing and One Compensation" model to high-tech industries, while residential areas need to be supplemented with supporting facilities to form mid-to-high-end living areas. Finally, the government has scattered administrative functions from Shangbu-Luohu to this area, improving the appearance of the city and clearing illegal construction. In this context, the urban landscape of the Meilin-Gangxia area has gradually emerged from its former agricultural landscape. Then, three mountain parks, including Meilin Park, Bijiashan Park, and Lianhuashan Park, have been successively constructed and opened. The mountains and hills on

[137] Sassen, S. (2018). Cities in a world economy. Sage Publications. (pp. 1–13; 101–103).

[138] Liu, X.Y., & Wang, X.J.劉新宇,王曉俊(2021). A Study of F. L. Olmsted and His Designing Ideas from the Perspective of American History in 19th Century 19世紀美國史視角下的奧姆斯特德及其設計思想. *Chinese Landscape Architecture*中國園林. 37(02), 106–111.

which these three parks are built mostly come from collective forests near surrounding rural villages such as Shangmeilin, Xiameilin, and Gangxia, except for the majority of Bijiashan Park, coming from the state-owned Shenzhen Forest Farm. From the mid-1990s to the early 2000s, after being requisitioned by the government, these mountains and hills, once forested and cultivated with fruits in specific social and historical contexts, were transformed into urban parks through specific planning and design. The original plants on the mountains and hills have also been transformed into plant communities more in line with urban park aesthetic principles through various forest structure transformation projects.

In the 1990s, Futian District, where these three mountain parks are located, had the highest population density among all administrative districts within the economic special zone. To address the issue of resettling temporary residents, the government constructed temporary settlement sites in Bijiashan Park and Lianhuashan Park, as well as in their vicinity. Additionally, numerous temporary residents spontaneously settled in the marginal mountainous areas of the economic special zone, including Lianhuashan Hill, Bijiashan Hill, and Meilin Mountain. Whether through vegetable and fruit cultivation or poultry and livestock farming, many temporary residents earn their livelihoods from the natural environment of these marginal mountainous areas. Simultaneously, such natural environments provided them with shelter from government inspections. Under the economic and political goals of Shenzhen's development in finance and high-tech industries and the construction of an "international city," both governmental settlement sites and self-formed habitations were dismantled by the government, restored to greenery, and reintegrated into urban parks. The people who once lived in these mountainous areas had to leave. The rough nature of the marginal mountainous areas within the economic special zone was redesigned to become lush and exquisite. From collective ownership by rural villages to government ownership, from temporary occupation by temporary residents to being restored as urban parks, these mountainous areas underwent continuous reproduction, and the cultural form of nature underwent continuous transitions. Behind this transformation lie not only the local government's imagination and efforts toward deeper globalization but also the brutality and complexity of the global economy.

The destruction of the original vegetation in the Lingnan region has corresponding social and historical reasons. The degradation of plant habitats forced Guangdong Province to use drought-resistant and barren-tolerant tree species for its afforestation projects after the founding of the People's Republic of China. In 1985, to balance the economic disparities between southern and northern Guangdong and improve the overall greenery of the province, the Guangdong Provincial Committee decided to implement the "Green Guangdong in Ten Years" afforestation project across the province. To meet this tight timeline and heavy task plan, fast-growing drought-resistant and barren-tolerant tree species such as *Acacia confusa* and *Eucalyptus* were extensively planted on the hillsides and hills throughout the province. In the mid-to-late 1990s, as the concept of "garden-style international city" became Shenzhen's understanding of building an international city in terms of urban greening, the plant communities on these hills were considered to be monotonous and lacking in aesthetics. On the one hand, urban greening departments continuously enhanced the use of color foliage and flowering plants in urban greening through activities such as the "Nations in Blooms" competition, the 26th Summer Universiade in 2011, and the construction of the "World Famous Flower City." On the other hand, the forestry department implemented colorization of plant communities in the hills within the Shenzhen city scope through the "Ecological Scenic Forest" construction project. Both aspects of creating natural spectacles aim to construct urban natural scenery and spatial environments that are more competitive, with the goal of improving the investment environment and enhancing Shenzhen's status as a modern international city. Meilin Park, Bijiashan Park, and Lianhuashan Park—these three mountain parks—have undergone corresponding forest structure transformations in this context. The purpose is not only to enrich the internal landscape of the parks but also to provide the city with more competitive visual landscapes, which represent urban greening aesthetics in the era of globalization.

In the discourse of urban environmental history, nature often appears as a victim. However, for many mountainous areas in Shenzhen, nature is not necessarily the victim of urban development. On the contrary, these mountainous areas have become more lush, colorful, and "natural" due to urban development. It can be seen that the relationship between urban development and nature depends on specific practices and historical contexts, and it is not appropriate to adhere rigidly to the view of urban–nature opposition, nor should one generalize.

The Meilin-Gangxia region is not primarily characterized by mountainous and hilly terrain but rather plains. However, the Meilin-Gangxia Park Group is based on mountainous and hilly terrain, while the plains are mainly utilized for the development of various urban industries, a situation closely related to the spatial resource constraints within the Special Economic Zone. From the 1990s to the early 2000s, as the Meilin-Gangxia Park Group developed, these mountains and hills blended into the time's social and historical context in new ways, presenting a human–land relationship different from that of the Shangbu-Luohu Plains area. Moreover, for these mountains and hills, nature is not necessarily the "victim" of urban development; on the contrary, these areas have become richer in vegetation, more vibrant, and more "natural" due to urban development. This demonstrates that the relationship between urban development and nature depends on specific practices and historical contexts, and it is inappropriate to adhere rigidly to the dichotomous view of the city versus nature or to generalize.

# Chapter 6
# Shenzhen Bay Park Group: From Insulative Nature to Developmental Nature

## Contents




**Abstract** This chapter examines the urban parks developed and opened on reclaimed land in the Shenzhen Bay area. Approaching from the perspective of Cold War history, it first discusses the transformation of this area from an "insulative nature" to a "developmental nature." It then reveals how these coastal parks, as a form of upscale and refined natural space, are deeply integrated with the surrounding high-end residential complexes, luxury commercial hubs, and corporate headquarters. This integration, in fact, closely correlates with the regional integration of the Pearl River Delta within the context of globalization.



**Keywords** Shenzhen Bay · Cold War · Gentrification · Regional integration





X. Liu, *Fabricating Nature*, The Urban Book Series,
https://doi.org/10.1007/978-981-95-8208-2_6

## 6.1 Landscape History of the Shenzhen Bay Area

### *6.1.1 Before the Opening-Up Policy*

Shenzhen Bay was once a partially enclosed shallow water bay, characterized by a broader entrance and a narrower inner, with a typical depth of less than 5 m. The bay is supplied with water by three rivers: Dashahe River (大沙河) from the north, Shenzhen River from the east, and Yuen Long River from the south. Moving north from the northern shoreline, the terrain gradually rises, successively forming mud-flat coastal plains, terraces, hills, and low mountains. The terrace area is substantial, yet the slope is generally gradual, with elevations varying from 10 to 30 m formed by sedimentation from the downstream Shenzhen River, with elevations near the estuary ranging from approximately 1.3 to 1.8 m. The southern shore consists of the Yuen Long River delta's sedimentary plain and the granite hills of the New Territories. From the early twentieth century to the 1970s, Shenzhen Bay underwent minimal changes in both its seabed topography and coastal morphology (Fig. 6.1).[1]

Historically, Shenzhen Bay was initially referred to as Houhai (後海), in opposition to Qianhai (前海), named for its location behind the county town at that time, which was the Ancient Town of Nantou (南頭古城). As reported in the *Xin'an County Chronicle* (新安縣誌): *Houhai, five miles from the county town, connects to the sea, extending from west to east, and is adjacent to Wutong Mountain, encircling and protecting the county town*. Although the Ancient Town of Nantou, not far north of Shenzhen Bay, had established as county town as early as the year of 331 AD, villages along the western and northern shorelines of Shenzhen Bay began to appear mainly after the late Song Dynasty, especially from the Ming Dynasty onward. For generations, these villages along Shenzhen Bay have traditionally depended on agriculture and fishery as their primary economic sectors, taking advantage of the flat terrain and coastal features near the coastline (Table 6.1).

During the agricultural era, besides staple food crops like rice, lychee, sweet peaches, and oysters in the Nanshan area were collectively known as the "Three Treasures of Nanshan (南山三寶)." The lychees grown in the Nantou Peninsula benefit from the southwest monsoon originating from Lingdingyang Bay, as well as the presence of trace elements in the soil. These factors contribute to the lychees having a dense meat, plentiful juice, and a delightful, sweet flavor. Given the consistent soil properties, the combination of extended solar exposure and high cumulative temperatures created optimal circumstances for developing sweet peaches. In addition, during the mid to late Ming Dynasty, the freshwater supply in the Pearl River Basin increased significantly. The characteristics of Shenzhen Bay, which connects to the estuary of the Pearl River, where freshwater and saltwater meet, and the abundance of plankton-rich waters make it conducive for oysters to thrive. This

[1] Liu, X., Zhao, H., Zheng, D., et al. 劉杏玲,趙煥庭,鄭德延等 (1988). Evolution and Development of Shenzhen Bay 深圳灣的演變與開發利用. *Journal of Tropical Oceanography*熱帶海洋: 48–56.

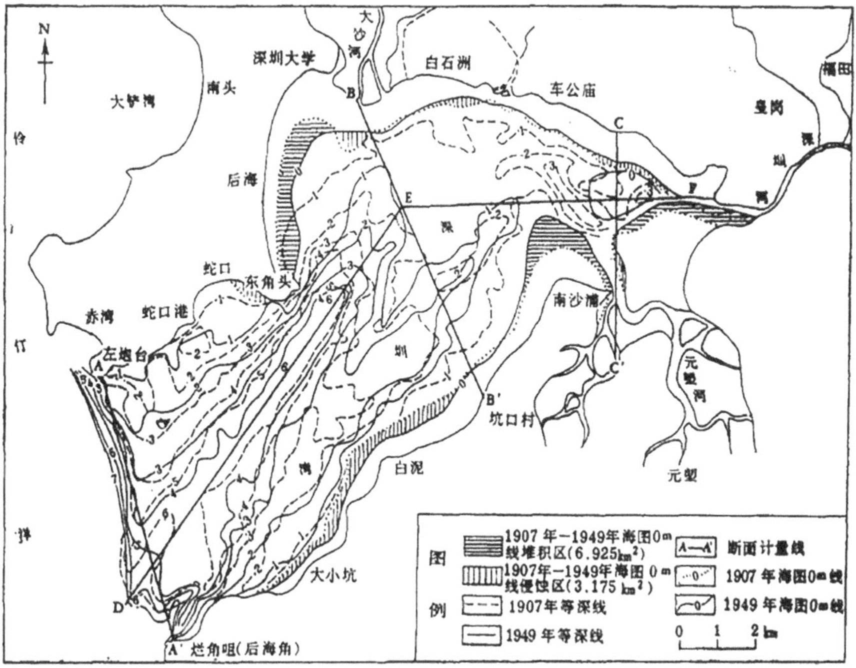


**Fig. 6.1** The underwater topography of Shenzhen Bay [Liu, X., Zhao, H., Zheng, D., et al. 劉杏玲,趙煥庭,鄭德延等 (1988). Evolution and Development of Shenzhen Bay 深圳灣的演變與開發利用. *Journal of Tropical Oceanography* 熱帶海洋:48–56]

also led to the flourishing development of fisheries in the area. During the agricultural age, when land was the main source of production, the productive function of nature unquestionably held a prominent position.

Under specific natural conditions such as topography, monsoons, and soil, the Three Treasures of Nanshan alongside rice cultivation and fisheries became the ancestral heritage of farmers in this region, serving as their significant source of livelihood for centuries. Throughout this process, farmers also altered the landscape of their farmland and beaches to suit their specific agricultural and lifestyle requirements, such as establishing orchards near mountains and reclaiming oyster fields near the sea. During the period of reform and opening up, with the exception of few village such as Beitou Village (北頭村) that exclusively engaged in oyster fishing, most villages possessed a designated area of farmland.,[2,3] The fundamental statistics pertaining to forestry, agriculture, oyster farming, fishing, and fruit cultivation in four villages among them in 1979 are shown in Table 6.2.

[2] Jiang, R., & Nan, Z. 蔣榮耀,南兆旭(2019). *Time, Mountains, and Rivers: A Century of Shenzhen Historical Images* 歲月山河:百年深圳歷史影像紀. Shenzhen: Haitian Publishing House 深圳:海天出版社. (pp246–247).

[3] Nanshan District Annals Editorial Committee Office深圳市南山區區志編委會辦公室(2012). *Annals of Nanshan District, Shenzhen*深圳市南山區志. Beijing: Fangzhi Publishing House北京:方志出版社. (pp. 410–411; 416–417).

**Table 6.1** Villages inhabited for generations along Shenzhen Bay[a]

| Subdistrict (present) | Village | Built year | Major surname | | Traditional economic industries |
|---|---|---|---|---|---|
| | | | Start-up clan | Local clan | |
| Shahe Subdistrict | Baishizhou 白石洲 | Mid Qing Dynasty | Wu | Wu/Gao/Chen/Guo/Wen | Crop farming/Fishery |
| | Tangtou 塘頭 | 1959 | – | Chi/Chen | Crop farming |
| | Shangbaishi 上白石 | Qing Dynasty | Zheng | Chen/Zhong/Zeng | Crop farming/Oyster farming/Herb gathering |
| | Xiabaishi 下白石 | Late Ming Dynasty | Su | Zeng/Chen/Zhang | Crop farming/Oyster farming/Herb gathering |
| | Xintang 新塘 | Early Qing Dynasty | Zheng | Zheng/Zhuo | Crop farming/Oyster farming/Herb gathering |
| Yuehai Subdistrict | Dachong 大沖 | Late Song Dynasty | – | Zheng/Zhong/Lin/Long/Li | Crop farming/Oyster farming |
| | Yuehaimen 粵海門 | Qing Dynasty | Zhang/Jiang | Zhang/Jiang | Crop farming/Oyster farming/Fishery |
| | Guimiao 桂廟新村 | Qing Dynasty | Ye | Ye/Liang | Crop farming/Oyster farming/Fishery |
| | Houhai 後海村 | Late Ming Dynasty | – | Zhao/Zeng/Zheng/Wu | Crop farming/Fishery |
| | Ruanwu 阮屋村 | Ming Dynasty | Ruan | Ruan | Crop farming/Oyster farming |
| | Wuwu 吳屋村 | Qing Dynasty | Wu | Wu/Huang | Crop farming/Oyster farming |
| Shekou Subdistrict | Wanxia 灣廈村 | Late Ming Dynasty | Fan | Fan/Liu/Mo/Chen/Qu | Crop farming/Oyster farming/Fishery |
| | Haiwan 海灣村 | Late Ming Dynasty | – | Fan/Liu/Mo/Qu | Fishery/Oyster farming |
| | Yuyi 漁一村 | 1950s | – | Zhang/Zhou/Wang/Li | Fishery |
| | Yuer 漁二村 | 1950s | – | Ao/Huang/Wu | Fishery |
| | Nanshui 南水村 | Late Qing Dynasty | Wu/Zhang | Wu/Zhang/Yang/Chen/Zheng | Crop farming/oyster farming/Fishery |
| | Dachan 大鏟村 | Late Qing Dynasty | – | Wang/Zhang/Liang | Transport/Crop farming/Fishery |

(continued)

**Table 6.1** (continued)

| Subdistrict (present) | Village | Built year | Major surname: Start-up clan | Major surname: Local clan | Traditional economic industries |
|---|---|---|---|---|---|
| Nanshan Subdistrict | Nanshan 南山村 | Early Ming Dynasty | Chen | Chen/Liang | Crop farming//Fruit farming/Oyster farming |
| | Nanyuan 南園村 | Mid Nansong Dynasty | Wu | Wu | Crop farming//Fruit farming |
| | Beitou 北頭村 | Yuan Dynasty | Huang | Huang/Chen | Crop farming//Fruit farming/Oyster farming |
| | Xiangnan 向南村 | Ming Dynasty | Zheng | Zheng/Ye | Crop farming/Fruit farming/Oyster farming |
| | Nanguang 南光村 | Qing Dynasty | Wu | Wu/Zhang/ Chen | Crop farming/Oyster farming |

[a]Shenzhen Municipal Chronicles Office深圳市史志辦公室(2020). *Overview of Shenzhen Villages, Volume II: Futian Nanshan Volume*深圳村落概覽·第二輯: 福田南山卷. Guangzhou: South China University of Technology Press廣州:華南理工大學出版社. (pp. 223–240; 244–270; 283–304).

**Table 6.2** Basic data on forestry, agriculture, oyster farming, fishing, and fruit cultivation for the four villages in 1979[a]

| Village | | Dachong | Beitou | Nanshan | Nantou |
|---|---|---|---|---|---|
| Forest area (mu) | | 2165 | – | 1000 | 605 |
| Forest product (dan) | | 30 | – | – | – |
| Farmland (mu) | Year-beginning area | 2037 | – | 3320 | 4970 |
| | Year-end area | 1990 | – | 3272 | 4727 |
| Fish and oyster production(dan) | Sea Fish | 50 | 23 | 40 | – |
| | Shrimp and crab | 20 | 5 | – | – |
| | Oyster | 30 | 132 | 80 | 220 |
| | Pond fish | 65 | 5 | 40 | 474 |
| Aquaculture area (mu) | Oyster farm | 810 | 1594 | 908 | 300 |
| | Fishpond | 90 | 4 | 28 | 148 |
| Fruit cultivation area (mu) | | 715 | 670 | 1914 | 1114 |
| Fruit production (dan) | | 155 | 500 | 1749 | 239 |

[a]Nantou Commune 南山公社 (1979). *The annual report of Nantou Commune, 1979* 1979年南頭公社年報統計數字, preserved at the Nanshan District Archives, Shenzhen, with Archive Number 檔案號: C42; 2014;Y 2507.

### *6.1.2 Three-Processing and One Compensation Period*

This pattern of production and livelihood continued until a fundamental transition commenced with the initiation of reform and opening up. Following the reform and opening-up policy, substantial urban development with large-scale topographic transformations occured in three specific locations along the coastline of Shenzhen Bay: Shekou Industrial Zone (蛇口工業區), Nanyou Development Zone (南油開發區), and the Overseas Chinese Farm (華僑農場), as shown in Fig. 6.2.

During the "Three-Processing and One Compensation" period, Shenzhen heavily relied on the "front-shop, back-factory" model formed with Hong Kong to integrate into the external market. Therefore, establishing a harbor in Shenzhen was deemed imperative. Initially, the site selectors held the belief that Shekou, due to its shallow water and vast mud, was more suited for oyster farming rather than harbor construction, unlike the more appropriate locations of Dapeng Bay (大鵬灣) and Yantian. Nevertheless, in comparison to the other two potential sites at that moment, Shekou has the fundamental prerequisites for development, including access to water and power, adequate living arrangements, and a well-established system of roads.[4] At the time of weighing the pros and cons, engineers familiar with the topography insisted that Shekou could indeed be developed into a harbor. The issue of shallow water could be addressed by reclaiming land for the construction of waterfront piers, dredging deep harbor pools, and berths. This was because the Earth's rotation historically caused the Pearl River Estuary to experience eastward erosion and westward siltation. Consequently, following the process of dredging, Shekou, situated on the eastern side, was able to effectively sustain an adequate water depth, thereby reducing the risk of serious sediment accumulation.[5]

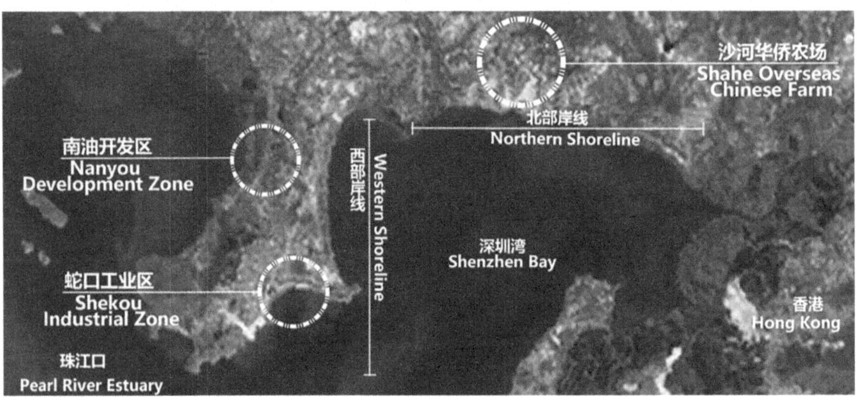


**Fig. 6.2** Shenzhen Bay Area in 1979 [Sourced from Google's historical imagery, and the author has redrawn it.]

[4] Zhang, Z. X.張振聲(1999). Unforgettable One Year and Three Months 難忘的一年零三個月. In Witness Shekou Compilation Committee見證蛇口編委會 (Eds.) *Witnessing Shekou* 見證蛇口. Guangzhou: Huacheng Publishing House廣州: 花城出版社.(pp. 14–20); Zhu, S. X. 朱士秀 (1999). The Arrival of the Golden Phoenix_ A Documentary on the Site Selection of Shekou Industrial Zone 金鳳飛來兮——蛇口工業區選址紀實. In Witness Shekou Compilation Committee (Eds.) 見證蛇口編委會 編, *Witnessing Shekou*見證蛇口. Guangzhou: Huacheng Publishing House 廣州: 花城出版社. (pp. 21–25).

[5] Yan, F.顏鳳迎 (2018). Special Zone Window: The Three Lives and Three Worlds of Shekou Port特區視窗: 蛇口港的三生三世. In Xu, Y.J., & Liu, W. (Eds.) 許永軍 劉偉 編. *Shekou: The Place Where Dreams Began - Paying Tribute to 40 Years of Reform and opening-up*蛇口夢開始的地方——致敬改革開放40年Beijing: People's Daily Press北京:人民日報出版社.

Aside from the influence of topography on the choice of location, it is possible to see how the selection of a site shapes the topography. Previously, these areas had extensive oyster fields. The ground in fifth Bay (五灣) and sixth Bay(六灣) was leveled by the Shekou Industrial Zone starting in July 1979. By May 1981, the "Five Connections and One Leveling (五通一平)" project had reached its final stages, with the excavation and filling of earth totaling over two million cubic meters, covering an area of over 100 ha. In addition, starting in December of the same year, advanced techniques such as water-gel explosive blasting and electric blasting were employed to accelerate the process of transforming the topography. In 1984, a comprehensive hill-cutting and land reclamation project was launched to modify the coastal area on a greater scale. In 1984, a comprehensive hill-cutting and land reclamation project was launched to modify the coastal area on a greater scale. At first, 20 ha. of land were formed in the first Bay and second Bay. Subsequently, a protruding embankment was constructed, effectively elongating the coastline by 7 km, and enlarging the operational area of the port to 2 km$^2$. In August 1985, the first Bay and second Bay projects was completed, creating an artificial plain between Shekou and Chiwan (赤灣). This added 33 ha. of industrial land and over 1000 m of coastline for the Shekou Industrial Zone, making it the largest land reclamation project in the SSEZ at that time.[6] All these intensively developed land foundations were previously oyster fields and farmland (Figs. 6.3 and 6.4).[7]

In 1983, the Guangdong Provincial Government and the Shenzhen Municipal Government collaborated to extensively develop the Nantou Peninsula, known as the Nanyou Development Zone, with an original planned area of 38 km$^2$. By 1987, the construction institute had developed an area of 3.3 km$^2$ and constructed 323,000 km$^2$ of houses. Between 1988 and 1991, the Nanyou Development Zone prioritized the construction of the Moon Bay (月亮灣) and Mawan Bay Port Areas (媽灣港區), erecting 1.05 million square meters of buildings. By March 1992, a cumulative total of 1.61 million square meters of industrial buildings had been completed.[8] In March 1992, the land area of the Nanyou Development Zone was reconfirmed to be 23.01 km$^2$. Out of this, 9.17 km$^2$ of land, accounting for 40%, with 4.72 km$^2$ specifically designated as developable land, while the rest comprising mountains and lychee orchards; the combined expanse of the sea area and mudflats

(pp. 225–236); Sun, S. X.孫紹先 (1984). Port Construction in Shekou Industrial Zone蛇口工業區的港口建設. *Port & Waterway Engineering*水運工程,1984. 1984(Z1).(pp.78–81).

[6] Nanshan District Annals Editorial Committee Office 深圳市南山區區志編委會辦公室(2012). *Nanshan District Annals, Shenzhen*深圳市南山區志. Beijing: Fangzhi Publishing House 北京: 方志出版社. (pp. 345).

[7] Gu, X.J. 辜曉進(Eds.) (2018). *In the Name of News III: Selected Works of News Graduation Design from Shenzhen University - Frontier Observations* 新聞的名義III:深圳大學新聞畢業設計作品選·前沿觀察. Beijing: Renmin University of China Press 北京:中國人民大學出版社. (pp. 53).

[8] Nanshan District Annals Editorial Committee Office 深圳市南山區區志編委會辦公室(2012). *Nanshan District Annals, Shenzhen*深圳市南山區志. Beijing: Fangzhi Publishing House 北京: 方志出版社. (pp. 363).

**Fig. 6.3** Oyster beds in Shekou (current Sea World) [Nanshan District Annals Editorial Committee Office 深圳市南山區區志編委會辦公室 (2012). *Nanshan District Annals, Shenzhen*深圳市南山區志. Beijing: Fangzhi Publishing House 北京: 方志出版社. (pp. 417).]

was 13.84 $km^2$, or 60% of the total area.[9] The government has designated a large area for the Nanyou Development Zone, but it is primarily agricultural land, and at that time, the actual ownership of numerous lands did not belong to Nanyou Development Zone. To undertake development and construction, the only viable approach is to initiate land reclamation from the sea.[10]

Simultaneously, numerous indigenous villages in the northern region of the peninsula, including Dachong Village (大沖村), Houhai Village (後海村), Yuehaimen Village (粵海門村), Guimiao New Village (桂廟新村), Ruanwu Village (阮屋村), etc., constructed diverse "Three Pocessing and One Compensation" industrial

[9] Lin, L., & Yin, K. 林淩,殷克勝 (1993). *Economic Development Strategy and Outline Planning of Nanyou Development Zone (Excerpt)* 南油開發區經濟發展戰略與規劃綱要(摘編). China Opening Journal開放導報. 1993(04):41–44.

[10] Gu, X. (Eds.) 辜曉進 編. (2018). *In the Name of News III: Selected Works of News Graduation Design from Shenzhen University - Frontier Observations* 新聞的名義III:深圳大學新聞畢業設計作品選·前沿觀察. Beijing: Renmin University of China Press 北京:中國人民大學出版社. (pp. 50–52).

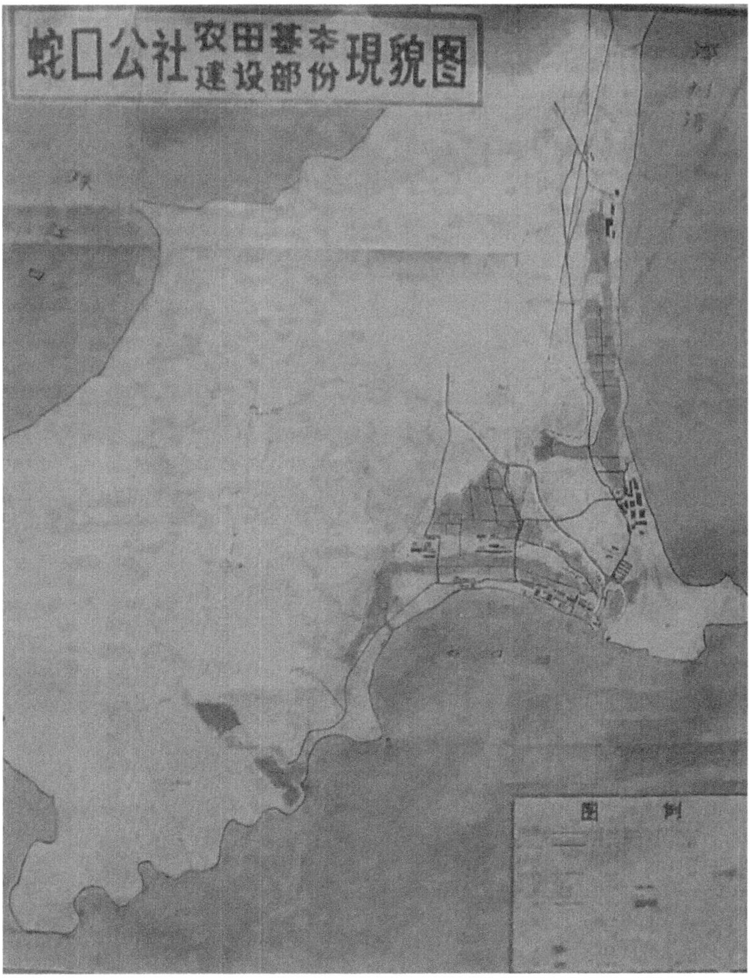


**Fig. 6.4** Site plan of the basic construction of farmland in Shekou Commune (1979) [Shenzhen Nanshan District Shekou Sub-district Office Compilation深圳市南山區蛇口街道辦 (2022). *Strolling in Shekou*蛇口漫步. Shenzhen: Shenzhen Press Group Publishing House 深圳: 深圳報業集團出版社. (pp. iii).]

factories on the preexisting collective agricultural land to foster the collective economy.[11]

[11] Shenzhen Municipal Chronicles Office 深圳市史志辦公室 編. (2020). *Overview of Shenzhen Villages, Volume II: Futian and Nanshan* 深圳村落概覽·第二輯:福田南山卷. Guangzhou: South China University of Technology Press 廣州: 華南理工大學出版社. (pp. 285–303).

The Shahe Overseas Chinese Farm of Guangdong Province is located on the northern shoreline of Shenzhen Bay.[12] In 1979, the area saw the establishment of Shahe Industrial Zone, and the Shahe Overseas Chinese Farm shifted its focus from agriculture to industry, setting up processing factories and engaged in activities such as sand business.[13] During that period, the topographic features of the Shahe Overseas Chinese Farm in the coastal region south of Shennan Avenue were as follows: the estuary of the Dashahe River in the west had a height of roughly 1.8–2.7 m, while Baishizhou (白石洲) had an elevation of around 3.2–3.4 m. The space separating them consisted primarily of rice fields. The region extending from Baishizhou to the central coastal area near Xialong (下蘢) has varying heights between 3.6 and 16.6 m. This area is characterized by the presence of oyster beds, fishponds, and *Eucalyptus* forests, among other features. The mangrove elevation in the area stretching from Xialong to the estuary of the Shenzhen River ranged from 2.3 to 4.0 m, with sandy beaches and sand fields, excluding the 10.2-meter-high hill located on the northern island of Beizhou, as shown in Fig. 6.5.[14]

The topography of this area underwent a rapid transformation during the development of Shenzhen Bay throughout the subsequent decades, even resulting in the

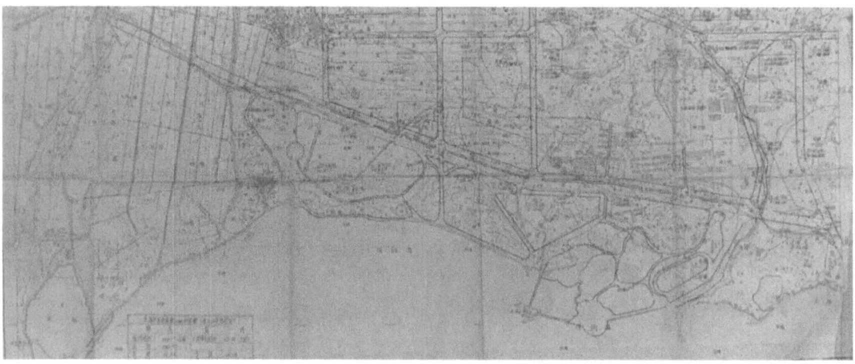

**Fig. 6.5** Topography of Shenzhen Shahe Overseas Chinese Enterprise Company (Partial, 1983) [Design and Research Institute of the Ministry of Mechanical Industry, Shenzhen Branch 機械工業部設計研究總院深圳分院. (1983). *Overall Planning Map of Shenzhen Shahe Overseas Chinese Enterprise Company (1:5000)* 深圳沙河華僑企業公司總體規劃圖1:5000. Archive No.: 0919830350101. Preserved at the Shenzhen Urban Planning and Natural Resources Bureau Archives.]

---

[12] It was initially referred to as the Sand River Branch of the Guangming Overseas Chinese Livestock Farm. However, in 1981, its name was changed to its current one.

[13] Nanshan District Annals Editorial Committee Office 深圳市南山區區志編委會辦公室 編. (2012). *Nanshan District Annals, Shenzhen*深圳市南山區志. Beijing: Fangzhi Publishing House 北京: 方志出版社. (pp. 157–159).

[14] Design and Research Institute of the Ministry of Mechanical Industry, Shenzhen Branch 機械工業部設計研究總院深圳分院. (1983). *Overall Planning Map of Shenzhen Shahe Overseas Chinese Enterprise Company (1:5000)* 深圳沙河華僑企業公司總體規劃圖1:5000. Archive No.: 0919830350101. Preserved at the Shenzhen Urban Planning and Natural Resources Bureau Archives.

coastline moving around 1–2 km southward. In the early 1980s, Overseas Chinese Town decided to abandon the growth plan for the industrial zone and, instead, innovatively embraced a new strategy. By capitalizing on the topographic characteristics of the central hilly area and the southern seaside area, they adjusted their strategy to focus on the establishment of theme parks and contemporary residential communities. The strategic shift was represented by the master plan formulated in 1986.[15] In the late 1980s, areas located to the south of Shennan Avenue were specifically allocated for the purposes of tourism, the creation of natural areas, and residential development.[16] In 1996, the updated master plan specifically allocated the land on the eastern bank of the Dashahe River for tourism purposes.[17] While both version of the Overseas Chinese Town master plan incorporated industrial property, none of them designated any in the coastal area south of Shennan Avenue.

During the 1980s, the Shekou Industrial Zone, Nantou Peninsula, and Overseas Chinese Farm along the Shenzhen Bay experienced a transition from mostly engaging in agricultural to becoming centers of industrial activity and related sectors. The functions and physical forms of the land are intricately woven into such transitions. Following 1991, the government of Nanshan District ceased reporting grain production,[18] signifying the end of agricultural progress along the Shenzhen Bay coastline and the commencement of a new era.

### *6.1.3 Initial Phase of Establishing an International City*

In the early 1990s, Shenzhen actively observed changes in the global economic environment, and the development of high-tech industries was essential for the city's transformation into an international city, sometimes hailed as the "primary driving force."[19] As early as 1987, prior to the first Lychee Festival, Shenzhen had

[15] China Urban Planning and Design Research Institute Shenzhen Consulting Center, Shenzhen Engineering Design Consulting and Advisory Company, Overseas Chinese Town Construction Headquarters Design Office中國城市規劃設計研究院深圳諮詢中心, 深圳市工程設計諮詢顧問公司, 華僑城建設指揮部設計室 (1986). *Explanation of the Overall Planning of SSEZ Overseas Chinese Town.* 深圳特區華僑城總體規劃編制說明, Archive Number: Y00–09-19,860,007. Preserved at the Shenzhen Urban Planning and Natural Resources Bureau Archives.

[16] The main developments include the completion of Splendid China in 1989 and the construction of Oriental Garden Residential Area in 1988.

[17] Referring to the already completed Window of the World area in 1994.

[18] Nanshan District Annals Editorial Committee Office 深圳市南山區區志編委會辦公室(2012). *Nanshan District Annals, Shenzhen*深圳市南山區志. Beijing: Fangzhi Publishing House 北京: 方志出版社. (pp. 407).

[19] Lin, Z. J. 林祖基(1993). Actively and steadily building Shenzhen into an international city based on reality 從實際出發積極穩步地把深圳建設成為國際性城市. In Lin, Z. J. 林祖基 (Eds.). *Essays on Shenzhen - An International City* 深圳-國際性城市論文集. Beijing: China Economic Publishing House 北京:中國經濟出版社.(pp. 4–11); Mi, P. M. 米鵬民 (2004, January 19). *High-*

already implemented the *Interim Regulations on Encouraging Scientific and Technical Personnel to Establish Private Scientific and Technological Enterprises* (關於鼓勵科技人員興辦民間科技企業的暫行規定). In February 1994, the Shenzhen Municipal Government released a notice titled *Notice on Ceasing the Approval of the "Three-Processing and One Compensation" and Other Projects in the Special Econominc Zone* (關於經濟特區停止審批"三來壹補"等項目的通知), which left a significant amount of urban space for the development of high-tech enterprises.[20] In May 1996, the municipal government decided to establish the Shenzhen High-tech Industrial Park (High-tech Park, 高新園). In June 2000, recognizing that the limited space in the High-tech Park was insufficient to accommodate the demand for high-tech industries across the entire city, the government issued the *Outline of the Plan for the Shenzhen High-tech Industry Belt* (深圳市高新技術產業帶規劃綱要). This plan designated 11 areas for the development of high-tech industries throughout the city, including Qianhai, High-tech Park, Liuxiandong Area (留仙洞片區), University Town (大學城), Kuiyong-Dapeng Area (葵涌-大鵬片區), and several more.[21]

The High-tech Park along the Shenzhen Bay is a product of the integration of many existing institutions, including the Shenzhen Science and Technology Industrial Park (深圳科技工業園), Chinese Academy of Sciences Institute of Advanced Technology (中國科技開發院), Shenzhen University (深圳大學), and the Fifth Industrial Zone (第五工業區). The High-tech Park spans 11.5 $km^2$ and is segmented into three distinct zones, namely, the north, central, and south zones, which are demarcated by Beihuan Avenue and Shennan Avenue. Prior to its construction, the land composition of the High-tech Park was as shown in Table 6.3:

In the late 1990s, the area along the Shenzhen Bay, including the High-tech Park, began a process of "transitioning from the secondary sector to the tertiary sector (退二進三)," and even "transitioning from the primary sector to the tertiary sector (退一進三)," as illustrated in Fig. 6.6.

Initially, following the establishment of the High-tech Park in 1996, measures were taken to regulate construction projects that deviated from the development objectives of the High-tech Park. In the same year, all the fishing and farming land located in Dachong Village, which is situated to the south of Shennan Avenue was requisitioned. After a period of two years, the land used by the Zhengdakangdi Company's pig farm was also reclaimed. Through such measures, the High-tech Park has implemented a series of projects including the Innovation Square, greening

*tech Industry: The Primary Driving Force for Constructing an International City* 高新產業是建設國際化城市第一動力. Shenzhen Business Daily深圳商報. (T00).

[20] Shenzhen Municipal People's Government深圳市人民政府(1994, March 30). *Notice of the Shenzhen Municipal People's Government on Suspending the Approval of "Three Processing and One Compensation" Projects and Others within the Special Economic Zone*深圳市人民政府關於在特區內停止審批三來一補等專案的通知. Retrieved from http://www.sz.gov.cn/zfgb/1994/gb51/content/post_10090800.html

[21] Zhao, C. 趙川 (2002). *Ten Major Areas of High-tech Industries in Shenzhen* 深圳高新技術產業的十大片區. Shenzhen: Haitian Publishing House 深圳:海天出版社. (pp. 01–31).

**Table 6.3** Land composition before the establishment of the High-Tech Park[a]

| Zoning | Original institute | Area | Land use |
|---|---|---|---|
| North zone | The fifth industrial park | 1.32 km$^2$ | As of 1997, 18 companies have been established, including Olympus and Haiking Pharmaceutical. In 1996, the total industrial output value reached 2.78 billion |
| | Songpingshan Residential Area | 1.25 km$^2$ | Residential area and the supporting living area of the fifth industrial zone |
| Middle zone | Science industrial park | 1.02 km$^2$ | There were already 150 companies, including Great Wall, Huawei, Haiman, Jinko, etc. In 1996, the total industrial output value reached 6.3 billion yuan |
| | Maqueling industrial area | 0.20 km$^2$ | Established in the early 1980s, mainly engaged in dyeing and printing circuit board enterprises. In 1996, the total industrial output value was 110 million yuan |
| | Dachong Village | 0.90 km$^2$ | Residential area for villagers with some individual enterprises, but the industrial foundation is weak. |
| | Continental Grain Company | 1.60 km$^2$ | Primarily engaged in pig farming, with a total industrial output value of 790 million yuan in 1996 |
| South zone | Newly-built area | 1.24 km$^2$ | Planned to be dominated by the electronic information industry, with the establishment of institutions such as the Shenzhen Tsinghua University Research Institute and Jingshan Civilian Technology |
| | Shenzhen University | 1.48 km$^2$ | There were nearly 10 university-run enterprises with an annual industrial output value of nearly 50 million. There is also reserved land of 4 ha. for high-tech industries |
| | Sea reclaimation area | 2.26 km$^2$ | Located between Houhai Avenue and Binhai Avenue, the land reclamation project started in 1997 |

[a]Shenzhen Yearbook Compilation Committee深圳年鑒編纂委員會.深圳年鑒 (1997). *Shenzhen Yearbook 1997*深圳: 深圳年鑒社1997. Shenzhen: Shenzhen Yearbook Press. (pp. 521–522).

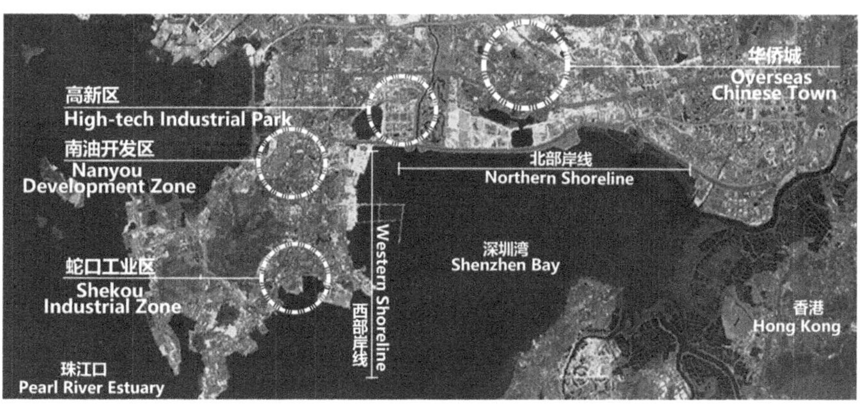


**Fig. 6.6** Shenzhen Bay Area in 2000 [Sourced from Google's historical imagery, and the author has redrawn it.]

projects, residential buildings, drainage systems, firefighting infrastructures, and electricity facilities.[22]

In addition, in 1997, the Shekou Industrial Zone implemented the *Interim Regulations on Investment Promotion and Attraction in the China Merchants Shekou Industrial Zone* (招商局蛇口工業區招商引資工作暫行規定), which specifically prohibited the establishment of processing industries and environmentally non-compliant industries that cause significant pollution. The aim was to accelerate the introduction of high-tech firms, business centers, exhibition centers, design centers, and other similar facilities, with the purpose of facilitating industrial upgrading and transformation.[23]

Furthermore, following the year 1997, the Nanyou Development Zone had a modification in its growth strategy, shifting toward high-tech industries and the establishment of corporate headquarters. The zone was relocated and converted polluting businesses, such as textiles and printing and dyeing, with a focus on building an industrial zone predominantly involved in light manufacturing.[24] By 2003, the entire development zone had utilized and developed a total area of 5.94 $km^2$, including 3.5 $km^2$ of reclaimed land. The industrial arrangement of the area was primarily characterized by real estate, port, and warehouses, supported by industries such as light manufacturing, printing and dyeing, textiles, and electronics with a population exceeding ten thousand people.[25]

Subsequently, Overseas Chinese Town diverged from the development trajectory of the Shekou and Nanyou Development Zone in Nanshan area, following the implementation of a master plan in 1986. In the new *Master Plan for Overseas Chinese Town* (華僑城總體規劃)(1996–2005) in 1996, the amount of land designated for industry was decreased, while there was a greater focus on allocating land for residential and tourism purposes.[26]

In the mid-1990s, the urban landscape of the Shenzhen Bay area underwent considerable transformations as part of an initiative to create an international city. These transformations involved a strategic approach known as "transitioning from the

---

[22] Nanshan District Annals Editorial Committee Office 深圳市南山區區志編委會辦公室(2012). *Nanshan District Annals, Shenzhen* 深圳市南山區志. Beijing: Fangzhi Publishing House 北京: 方志出版社. (pp. 377–378).

[23] Nanshan District Annals Editorial Committee Office 深圳市南山區區志編委會辦公室(2012). *Nanshan District Annals, Shenzhen*深圳市南山區志. Beijing: Fangzhi Publishing House 北京: 方志出版社. (pp. 351).

[24] Nanshan District Annals Editorial Committee Office 深圳市南山區區志編委會辦公室(2012). *Nanshan District Annals, Shenzhen*深圳市南山區志, Shenzhen. Beijing: Fangzhi Publishing House北京: 方志出版社. (pp. 363).

[25] Nanshan District Annals Editorial Committee Office 深圳市南山區區志編委會辦公室(2012). *Nanshan District Annals, Shenzhen*深圳市南山區志. Shenzhen. Beijing: Fangzhi Publishing House 北京: 方志出版社. (pp.364–365).

[26] Dong, G., & Zhang, Y董觀志 張穎(2008). *Tourism + Real Estate: Business Model of Overseas Chinese Town* 旅遊 + 地產:華僑城的商業模式. Sun Yat-sen University Press中山大學出版社. (pp. 158).

secondary sector to the tertiary sector" across four important locations. In addition, the reclamation project of the Binhai Avenue (濱海大道) commenced on the northern shoreline of Shenzhen Bay one year prior to the establishment of the High-tech Park. Five years later, the main lanes of the Binhai Avenue were opened to traffic. The Binhai Avenue spans a total length of 9.6 km, with 7.6 km constructed on the tidal flats located on the northern side of Shenzhen Bay. Initially, Shenzhen Bay consisted of a muddy shoreline, with the mud extending to a maximum depth of 12 m. The construction of the Binhai Avenue necessitated significant earthwork, resulting in the leveling of numerous hills in the Xili (西麗) area. According to the government, the construction of Binhai Avenue served two main purposes: it created a ring expressway that reduced traffic congestion from Luohu to Nanshan, and it also created additional land to the north of Binhai Avenue and south of Oversea Chinese Town,[27] laying the groundwork for the next phase of development in the Shenzhen Bay area.

### 6.1.4 The PRD Regional Integration Period

In terms of planning and positioning, the urban space north of the Binhai Avenue, once known as the "South Overseas Chinese Town Reclamation Area (華僑城南填海區)," has undergone a transition from a singular "Coastal City Residential Zone (2001)" to a diversified "Coastal Cultural and Business Center (2004)" and then to a "Super Headquarters Base (2007)."[28] In March 2014, the municipal government approved the *Control Detailed Planning of the Shenzhen Bay Super Headquarters Base* (深圳灣超級總部基地控制性詳細規劃) submitted by the planning department. The planning objectives encompass (Fig. 6.7):

*The Super Headquarters Base exemplifies the city's prominent role in the worldwide economic industry chain, serving as a central hub for Shenzhen's future development into an international city. The planning capitalizes on its superlative location advantage to attract top-tier economic activities, creating an image of an outstanding city. This is intended to make it an urban area that provides an experience of the new coastal lifestyle in Shenzhen, showcasing the core area for the city's social appearance and construction standards. Simultaneously, it will become a platform to address Shenzhen's cultural innovation and the demands of the new*

[27] Li, Z. 李子彬(2020). *My Time as the Mayor of Shenzhen* 我在深圳當市長. CITIC Publishing Group 中信出版社. (pp. 251–256).

[28] Shenzhen Bay Super Headquarters Base Development and Construction Command Office深圳灣超級總部基地開發建設指揮部辦公室(2019, June). *International Consultation for Overall Planning, Development and Operation of Shenzhen Bay Super Headquarters Base (Consultation Requirements)* 深圳灣超級總部基地片區統籌開發建設與運營國際諮詢(諮詢要求). Retrieved from https://www.renrendoc.com/paper/217209080.html

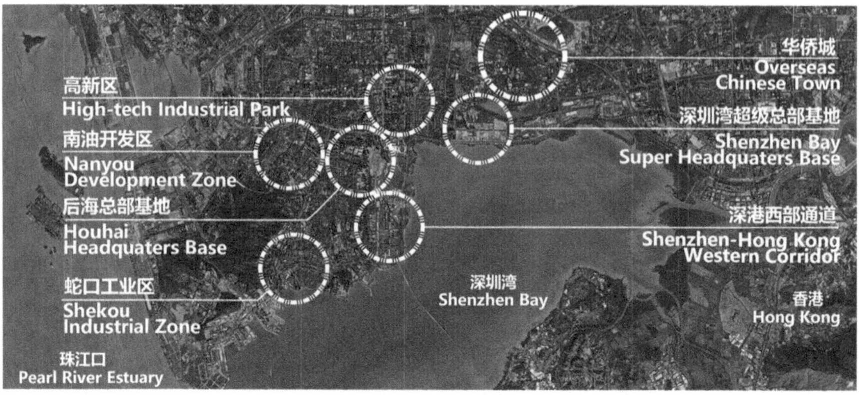


**Fig. 6.7** Shenzhen Bay Area in 2020 [Sourced from Google's historical imagery, and the author has redrawn it.]

*economy, serving as an effective medium for Shenzhen to engage in competition and cooperation among regions and cities.*[29]

For Shenzhen during that period, the issue of competition and cooperation among different regions and cities was not a novel one. In 2005, Guangdong Province implemented the *Plan for the Coordinated Development of the PRD metropolitan area* (珠江三角洲城鎮群協調發展規劃) (2004–2020) to address the issues of excessive competitiveness and unnecessary construction in the PRD metropolitan area. Subsequently, a series of specific plans were implemented, outlining planning requirements for the infrastructure, industrial layout, environmental protection, urban planning, and public service of the Guangzhou-Foshan-Zhaoqing, Shenzhen-Dongguan-Huizhou, and Zhuhai-Zhongshan-Jiangmen urban clusters, along with the entire PRD metropolitan area.[30]

From a global perspective, the rapid rise of the Asia-Pacific region has led to a shift in the center of world economic growth. China's rapid development has positioned it as a pivotal economic entity in the Asia-Pacific region. Its urban system has significantly elevated its status within the Asia-Pacific and even the global city system. Over the course of several decades, the PRD metropolitan area has progressively formed its unique industrial characteristics, with a particular emphasis on major cities and specialized towns as growth poles, and the industrial spaces are interconnected, forming belts or clusters, which in turn serve as a driving force for regional development. Following China's entry into the WTO, the economic paradigm that heavily depended on labor-intensive sectors has progressively exhibited

[29] Shenzhen Urban Planning and Land Resources Commission深圳市規劃和國土資源委員會. (2013, August). *Controlling Detailed Plan for the Shenzhen Bay Super Headquarters Base*深圳灣超級總部基地控制性詳細規劃. Retrieved from https://www.360docs.net/doc/4a18825764.html

[30] China Development Research Foundation中國發展研究基金會(2022). *Research on the Integration of city clusters in China*中國城市群一體化研究. China Development Press中國發展出版社. (pp. 141–159).

its negative effects. A considerable proportion of labor-intensive products exported globally have been subjected to increasing scrutiny because of anti-dumping investigations. Exporting products can be challenging due to technical constraints, environmental restrictions, and the need for intellectual property protection. Meanwhile, the PRD metropolitan area, despite its strong local economic ties and high degree of autonomy, is comparatively vulnerable to external factors. This has exposed the regional economy to certain risks. In order to achieve successful transformation in the PRD, it is imperative to integrate the regional economy for the purpose of reconstructing the industrial structure. Furthermore, it is crucial to focus efforts on creating Guangzhou and Shenzhen as hubs for independent innovation, harnessing the driving force of independent innovation in the development of the PRD metropolitan area.[31]

Metropolitan areas represent a new spatial form emerging in the global information era.[32] To concentrate resources for independent innovation, companies utilize information technology to adjust organizational structures. They strategically position headquarters in well-endowed large cities to carry out high-value-added activities such as strategy, operations, and research and development. Meanwhile, low-value-added manufacturing plants can be distributed in medium and small cities, and the coordination between the two is facilitated through information technology. In the past decade, the focus of new manufacturing development in the PRD metropolitan area has gradually shifted from the eastern bank of the Pearl River Estuary to the western bank. This shift is facilitated through a collaborative relationship characterized by the "headquarters + base" model with the two major central cities of Shenzhen and Guangzhou. As manufacturing industries relocate, it frees up space in the central cities for modern service industries and high-tech sectors. Such headquarters economy has fueled the spatial concentration and diffusion of urban functions, supporting the interconnected development of the PRD metropolitan area. It has also contributed to Shenzhen emerging as the engine driving the development of advanced manufacturing in the PRD metropolitan area.[33]

In January 2008, the Shenzhen Municipal Government released the *Several Opinions on Accelerating the Development of Headquarters Econom*y (關於加快總部經濟發展的若干意見), which outlined clear objectives for the growth of the headquarters economy in Shenzhen. Therefore, the Houhai area, located on the eastern bank of the Pearl River Estuary along the Shenzhen Bay, was specifically

[31] Shenzhen Municipal People's Government 深圳市人民政府. (2009). *General Urban Plan of Shenzhen City (2009–2020) Explanatory Booklet*.深圳市城市總體規劃(2009–2020)說明書. December 2009. (pp. 42–44).

[32] Castells, M. (2011). *The Rise of the Network Society: The Information Age: Economy, Society, and Culture Volume I (second ed., Vol. 12)*. West Sussex_ Wiley-Blackwell. (pp.407–423).

[33] Development Planning Committee of Guangdong Province 廣東省發展計畫委員會(2003). *Compilation of Plans for the Preliminary Achievement of Modernization in the PRD*珠江三角洲率先基本實現現代化規劃彙編. (pp. 117); Zhao, H. 趙弘 (2014). *A New Study on Headquarters Economy: The New Driving Force for Urban Transformation and Upgrading* 總部經濟新論:城市轉型升級的新動力. Nanjing: Southeast University Press 南京:東南大學出版社. (pp. 1–6).

designated to fulfill the role of the headquarters economy.[34] The former South Overseas Chinese City Reclamation Area has been converted into the Shenzhen Bay Super Headquarters Base. Enterprises situated in this area must meet the requirements outlined in the *Interim Measures for Encouraging the Development of Headquarters Enterprises in Shenzhen*[35] (深圳市鼓勵總部企業發展暫行辦法). Furthermore, the moniker "Cloud City (雲城市)" aptly symbolizes the pivotal significance of information technology in the Shenzhen Bay Super Headquarters Base.[36]

Furthermore, due to the favorable synergy between Shenzhen and Hong Kong in terms of urban functions and industrial structures, the continuous enhancement of Shenzhen's economic strength has made it realistically feasible to jointly establish a global city with Hong Kong. The cooperation between Shenzhen and Hong Kong has been elevated to a critical position in regional economic development due to the integrated development in the PRD metropolitan area.[37] Therefore, the construction of the Shenzhen-Hong Kong Western Corridor (深港西部通道) commenced in August 2003. It comprises two primary components: a 5545 m-long bridge that spans the sea, connecting Dongjiaotou (东角头), Shekou in Shenzhen to Aokanshi (鳌勘石), Yuen Long in Hong Kong, and a port area situated east of Dongjiaotou, covering an area of 117.9 ha.[38] In order to comply with planning requirements, the construction of the port area necessitates a process of land reclamation, which involves creating new land over an area of around 150 ha. that was then covered by shallow waters and oyster fields. The water depth here generally ranges from 2.67 to 5.61 m, with an underwater terrain showing a gentle slope from north to south

[34] Shenzhen Urban Planning and Natural Resources Bureau 深圳市規劃與自然資源局. (2014, April 14). *Zoning ordinance for Shenzhen Nanshan 01–01, 02–03 & 04 Areas [Houhai Bay - Dongjiaotou Area]* 深圳市南山01–01、02–03&04號片區[後海灣-東角頭地區]法定圖則. Retrieved from http://pnr.sz.gov.cn/ywzy/fdtz/cggbcx/nsq/content/post_5838399.html

[35] Peng, Y. 彭琰(2021, June 24). *Encouraging the Development of Headquarters Enterprises: Shenzhen Introduces Supportive Policies*鼓勵總部企業發展 深圳出臺政策支援. Shenzhen Business Daily深圳商報.

[36] Urban Planning, Land and Resources Commission of Shenzhen Municipality, Shenzhen Center for Design 深圳市規劃和國土資源委員會 深圳市城市設計促進中心(2014). *Collection of Shenzhen Bay Cloud City International Design Competition Entries* 深圳灣雲城市國際競賽作品集. Guangzhou: Lingnan Fine Arts Publishing House 廣州: 嶺南美術出版社.

[37] Wang, Y., Zhang, J., Wang, Z., et al. 王雨,張京祥,王梓懿等 (2022). *Rescaling and Cross-border Governance Transformation of the Northern Metropolis of Hong Kong Under the Background of "City-Regions"* 城市—區域"背景下香港北部都會區的尺度重構和跨界治理轉型. Urban Planning International國際城市規劃, doi:10.19830/j.upi.2022.054; Zhang, Y. G. 張玉閣 (2017). *Deepening Cooperation between Shenzhen and Hong Kong: The Key to the Construction of the Guangdong-Hong Kong-Macao Greater Bay Area* 深港合作:粵港澳大灣區建設的關鍵. China Opening Journal 開放導報, 2017(04): 3; Zhang, Y., Wen, Y., Guo, W., et al. 張玉閣,文雅靖,郭萬達 等. (2022). *Construction of the Greater Bay Area and Forty Years of Shenzhen-Hong Kong Cooperation* 大灣區建設與深港合作四十年. Social Sciences Academic Press 社會科學文獻出版社. (pp. 137–157).

[38] *Shenzhen-Hong Kong Western Corridor Project Construction Office* 深圳市深港西部通道工程建設辦公室.深港西部通道. (2005, December). Shenzhen-Hong Kong Western Corridor. Preserved at Shenzhen Library Local Archive Room.

and from west to east, characterized by a relatively gradual incline. Furthermore, the subaquatic sediment possesses a range of thicknesses, spanning from 5 to 24 m, with an average thickness of 10 m.[39] Beneath the layer of mud, there exists alluvial sand and gravel, and cohesive soil, or residual granite. The building of the Western Corridor has required the utilization of 17 million cubic meters of soil and stone for the purpose of land reclamation on the specific physical topography.[40]

The top-down urban policies have resulted in the submergence of these layers of land beneath reclaimed areas, with the once-flooded soil now buried beneath the newly created land. Beyond the physical changes, what has also been submerged is the former way of life and the production activities that once characterized this area (Fig. 6.8).

Prior to 2007, the Houhai Village, Haiwan Village (海灣村), Tangxia Village (塘廈村), Yu'er Village (漁二村), and Dachan Village (大鏟村) in the Shenzhen Bay area were home to oyster farmers, with each village having different quantities of them (Fig. 6.8). The *Qianhai Oyster Farming Facility Clearance Plan*[41] (前海蠔業

**Fig. 6.8** Oyster fields and oyster farmers in Shenzhen Bay [Figure on the left from: Gu, X.J. 辜曉進(Eds.). (2018). *In the Name of News III: Selected Works of News Graduation Design from Shenzhen University - Frontier Observations* 新聞的名義III:深圳大學新聞畢業設計作品選·前沿觀察. Beijing: Renmin University of China Press 北京:中國人民大學出版社. (pp. 54–55); Figure on the right from: Kong, L. J. 孔令軍 編(Eds.). (2019). *Shenzhen Private Visual History: Zheng Dongsheng Volume* 深圳私人視覺史: 鄭東升卷. Shenzhen: Shenzhen Press Group Publishing House 深圳: 深圳報業集團出版社. (pp. 149).]

[39] Shenzhen Land Investment and Development Center 深圳市土地投資開發中心. (2004, November). *Technical Analysis Report on Soft Foundation Treatment for Reclamation and Foundation Treatment Project of Shenzhen-Hong Kong Western Corridor* 深港西部通道填海及地基處理工程軟基處理技術分析報告. Preserved at Shenzhen Library Local Archive Room.

[40] Shenzhen-Hong Kong Western Corridor Project Construction Office 深圳市深港西部通道工程建設辦公室. (2005, December). *Shenzhen-Hong Kong Western Corridor* 深港西部通道. Preserved at Shenzhen Library Local Archive Room.

[41] Chen, Y. 陳玉(2004, December 22). Demolition of the Oyster Fields in Qianhai 清拆前海蠔田. Retrieved from https://news.sina.com.cn/c/2004-12-22/16234599655s.shtml

養殖設施清理工作方案) was implemented by the government of Nanshan District in November 2004. Despite the plan's objective of clearing the area within a two-month timeframe, the achieved outcomes were not totally satisfying for the government, as some offshore aquaculture facilities that had vanished during the operation reappeared.[42] In April 2007, the Shenzhen Municipal Government introduced the *Comprehensive Improvement Plan for the Marine Environment in the Shekou Area of Shenzhen Bay* (深圳灣蛇口片區海洋環境綜合整治工作方案). Its issuance was closely related to the construction of the Western Corridor and the establishment of the Shenzhen Bay Coastal Leisure Belt (深圳灣濱海休閒帶), also known later as Shenzhen Bay Park. Moreover, it also concerns the fishermen and oyster farmers who have traditionally lived along the Shenzhen Bay coastline:

*The comprehensive cleanup and rectification of the marine environment in the Shekou area of Shenzhen Bay are closely related to the overall effectiveness of major projects such as the Shenzhen-Hong Kong Western Corridor and the Shenzhen Bay Coastal Leisure and Sightseeing Belt. Ensuring the navigational safety of the waters in the Shekou area of Shenzhen Bay and thoroughly addressing the marine environmental issues are pressing concerns… The work objectives entail devising a plan to achieve the thorough enhancement of the maritime environment in Shenzhen Bay by May 30, 2007. Consolidate the results of the cleanup, ensuring that no new aquaculture facilities or illegal structures appear in the area…sign Dismantling and Compensation Agreements(清拆補償協議書) with fishermen and oyster farmers before April 15th. All fishermen and oyster farmers must completely evacuate the sea areas subject to this cleanup operation. The Shenzhen Municipal Agriculture, Forestry, and Fisheries Bureau (深圳市農林漁業局) will enter the field to dismantle aquaculture facilities…After the dismantling work, the subsequent management responsibilities will be transferred to the Shenzhen Municipal Fishery Administration and Harbor Supervision Office (深圳市漁政漁港監督管理處)…Enhance patrols in the sea area, resolutely preventing the resurgence of illegal aquaculture facilities.*[43]

Back then, the Shenzhen Municipal Finance Bureau (深圳市財政局) allocated a budget of 5.20 million yuan and compensation of 15.88 million yuan to undertake the comprehensive rectification of local oyster farmers and oyster field.[44] Nevertheless, providing compensation for the aquaculture facilities owned by

[42] Tang, Y. 唐毅(2007, May 30). Shenzhen Bay: The Disappearing Indigenous Oyster Farmers 深圳灣漸遠的原住蠔民. [Online]. Retrieved from https://news.sina.com.cn/o/2007-05-30/095211923480s.shtml

[43] Shenzhen Municipal People's Government深圳市人民政府(2007,May 9).*Comprehensive Remediation Plan for the Marine Environment in the Shenzhen Bay Shekou Area*深圳灣蛇口片區海洋環境綜合整治工作方案.Retrieved from http://www.sz.gov.cn/zfgb/2007/gb546/content/post_4988477.html

[44] Shenzhen Municipal Finance Bureau 深圳市財政局. (2007, July 13). *Opinions on Relevant Issues Regarding the Comprehensive Regulation of Sea Areas in the Shekou Area of Shenzhen Bay* 關於深圳灣蛇口片區海域綜合整治工作有關問題的意見. In Shenzhen Ocean Bureau Office (Eds.), Compilation of Documents on the Comprehensive Regulation of Sea Areas in the Shekou Area of Shenzhen Bay (Volume One) 深圳市海洋局辦公室. 深圳灣蛇口片區海域綜合整治工作檔彙編(一). Preserved at Shenzhen Library Local Archive Room.

ecdemic oyster farmers proved to be more challenging. A letter addressed to the Shenzhen Municipal Agriculture, Forestry, and Fisheries Bureau revealed that they were facing a more difficult situation than the local oyster farmers:

*The local oyster farmers have received corresponding compensation, but should we (from Yangjiang, Xinyi) who were engaged in oyster farming in Shekou also receive fair compensation?... Our aquaculture facilities have already been dismantled by the government, rendering us unemployed after more than twenty years of hard work...no return on investment... our entire investment is tied up in the aquaculture facilities... and our lands at hometown has been lost over the long years...there is no more land to cultivate, leaving us with no way forward and no retreat...In this vast world, there is no place for us oyster farmers to stand. Beneath the vast expanse of the sky, we don't want to become beggars....*[45]

In the government's view, compensating ecdemic oyster farmers posed a broad challenge and had the potential to trigger additional compensation demands from local oyster farmers. The final resolution was that those aqua culturists who had not registered in Shenzhen official departments were not eligible for compensation. Registered individuals were compensated according to the criteria set for local oyster farmers, with the funds being raised by the government of Nanshan District.[46] In May 2014, a 23-square-kilometer area in the Shenzhen Bay waters, under the jurisdiction of Shenzhen, was officially declared as a no-fishing zone for a period of five years, later extended to a duration of ten years.[47]

The Nanshan Oysters, which are part of the "Three Treasures of Nanshan," have vanished indefinitely, regardless of the situation. Oyster farmers confronted a predicament: either shift to other occupation or persist in oyster farming in other locations. Subsequently, the pertinent divisions of the Shenzhen Municipal Government conducted a reassessment of the living conditions of oyster farmers, aiming to provide insights for government decision-making. They also organized representatives of oyster farmers to visit places like Yangjiang (陽江), Zhuhai (珠海), Danshui (淡水), and Zhanjiang (湛江), guiding them toward developing oyster farming in

[45] Oyster Farmers from Yangjiang& Xinyi, Guangdong Province 陽江、信宜養蠔戶. (2007, May 22). Request Regarding Compensation for Foreign Aquaculture Facilities 關於外來養殖設施補償問題的請示. In *Compilation of Documents on the Comprehensive Regulation of Sea Areas in the Shekou Area of Shenzhen Bay (Volume One)* 深圳灣蛇口片區海域綜合整治工作檔彙編(一). Preserved at Shenzhen Library Local Archive Room.

[46] Office of the People's Government of Nanshan District, Shenzhen 深圳市南山區人民政府辦公室. (2007, July 9). *Reply Regarding Urgent Inquiry on Relevant Issues of Comprehensive Regulation of Sea Areas in the Shekou Area of Shenzhen Bay*關於深圳灣蛇口片區海域綜合整治工作有關問題緊急請示的復函. In Compilation of Documents on the Comprehensive Regulation of Sea Areas in the Shekou Area of Shenzhen Bay (Volume One) 深圳灣蛇口片區海域綜合整治工作檔彙編(一). Preserved at Shenzhen Library Local Archive Room.

[47] Administration of Ocean and Fisheries of Guangdong Province 廣東省海洋漁業局(2014, May 1). *Announcement on the Establishment of Fishing Prohibited Areas in Parts of the Sea in Shenzhen Bay* 廣東省海洋與漁業局關於在深圳灣部分海域設置禁漁區的通告. Retrieved from http://ocean.nbt.edu.cn/show/showdetail.php?id=241586

different locations.[48] To summarize, in the process of global industry–space restructuring, the robust integration of the PRD metropolitan area has made the native land and ancestral occupation no longer compatible, and the Shenzhen Bay has gradually distanced itself from productive natural activities.

## 6.2 Alternations in Ownership and Morphology

### *6.2.1 The Current Status and Historical Administration of the Sea Areas in Guangdong Province*

The maritime management in the Shenzhen area can be traced back to the establishment of a military garrison on the southeast coast of Ancient Town of Nantou in the year 736 AD. In the year 1394 AD, defense and garrison posts were established in Dongguan and Dapeng, further strengthening maritime defense and local security in Guangdong Province, especially in Shenzhen and along the Pearl River coast. Subsequently, with Westerners embarking on the Great Navigation and opening up new routes to the East, the Ming and Qing dynasties continued to enhance maritime defense forces to resist external threats.[49] These were part of the pre-modern maritime management methods, primarily aimed at repelling external enemies rather than systematically utilizing marine resources. Beginning in the Republic of China era, the Guangdong Provincial Government placed the Xisha Islands (西沙群島) under the jurisdiction of Yaxian County (崖縣) on Hainan Island. It was only then that administrative management related to island construction, maritime development, and approvals for the South China Sea islands gradually began. Simultaneously, during the era of the Republic of China, there were numerous incidents of illegal invasions by foreign powers.[50]

After the establishment of the People's Republic of China, in June 1950, the *Land Reform Law of the People's Republic of China* (中華人民共和國土地改革法) was promulgated. However, it did not address the issue of maritime ownership. In September 1958, the Chinese government made a formal declaration to the international community stating that the width of China's territorial waters is 12 nautical miles. In May 1993, the Ministry of Finance (财政部) and the State Oceanic Administration (国家海洋局) promulgated and enforced the *Interim Regulations*

[48] Tang, Y. 唐毅(2007, May 30). *Shenzhen Bay: The Disappearing Indigenous Oyster Farmers* 深圳灣漸遠的原住蠔民. Retrieved from https://news.sina.com.cn/o/2007-05-30/095211923480s.shtml

[49] Chen, H. X. 陳海先(2013, April 24). "*Establishment of Coastal Defense in Shenzhen Throughout the Ages.*" 歷代深圳海防的設立 Retrieved from https://www.szzx.gov.cn/content/2013-04/24/content_8987085.htm

[50] Guangdong Provincial Local Chronicles Compilation Committee 廣東省地方市志編纂委員會 (2000). *Chronicles of Guangdong Province: Maritime and Island Chronicles* 廣東省志·海洋與海島志. Guangzhou: Guangdong People's Publishing House 廣州: 廣東人民出版社. (pp. 23).

*on the Administration of the Use of National Sea Areas* (國家海域使用管理暫行規定), with Article 4 stipulating for the first time that the sea areas are the property of the state.[51] In October 2001, the People's Republic of China promulgated the *Law of the People's Republic of China on the Administration of the Use of Sea Areas* (中華人民共和國海域使用管理法), with Article 3 explicitly states: "Sea areas belong to the state, and the State Council exercises ownership of sea areas on behalf of the state." According to Article 18, all initiatives that involve reclaiming an area of 50 ha. or larger in the sea must be submitted to the *State Council for approval. The Property Law of the People's Republic of China* (中華人民共和國物權法) was enacted by the People's Republic of China in March 2007. Simultaneously, its Article 48 specifies that "natural resources such as forests, mountains, grasslands, wasteland, and tidal flats belong to the state, except as otherwise provided by law for collective ownership." The alterations in the ownership system of coastal tidal flats are relatively complex. Following the enactment of the *Land Reform Law of the People's Republic of China* in 1950, tidal flats have experienced a multifaceted transformation in ownership structures. Initially, there was coexistence of state and private ownership after land reform, followed by coexistence of state and collective ownership after the cooperative movement. In 1962 and 1988, the central government made two attempts to fully nationalize tidal flats. However, the *Law of the People's Republic of China on the Administration of the Use of Sea Areas* in 2001 stated that "sea areas belong to the state," without addressing the property rights of tidal flats. This has established the foundation for numerous controversies and continues to be a topic of discussion in legal academics.[52] The fundamental cause is determined by the inherent attributes of intertidal areas situated in the intermediary region between the ocean and the coast, as well as their intricate interplay with matters such as national security, environmental protection, and other factors.

At the provincial level in Guangdong, the South China Sea Branch of the National Oceanic Administration (國家海洋局南海分局) was established in 1965. In 1984, it was separated from the navy's organization and became an institution responsible for the construction and management of maritime affairs.[53] In July 1991, following the deployment of the State Planning Commission (國家計劃委員會), maritime development planning was conducted in Guangdong province under the leadership

[51] Zhang, Q., & Feng, L. 張清勇,豐雷(2020). Who are the Owners of China's Coastal Wetland? The Institutional Change and Controversy of the Ownership of Coastal Wetland. 誰是中國沿海灘塗的所有者?——灘塗所有權的制度變遷與爭議. *China Land Science*中國土地科學. 34(09):24–31.

[52] Beijing Zai Ming Law Firm 北京在明律師事務所(2023, March 3). How to Define Collective Ownership of Coastal Tidal Flats? Experts Say Relevant Laws Are Missing 沿海農民灘塗集體產權如何界定?專家:相關法律缺失亟待完善. Retrieved from https://baijiahao.baidu.com/s?id=1761876830377141777&wfr=spider&for=pc

[53] Guangdong Provincial Local Chronicles Compilation Committee 廣東省地方市志編纂委員會. (2000). *Chronicles of Guangdong Province: Resources and Environment Volume* 廣東省志資源·環境卷. Guangzhou: Guangdong People's Publishing House 廣州: 廣東人民出版社. (pp. 269).

of the Guangdong Provincial Land Planning Leading Group (廣東省國土規劃領導小組). The calculated maritime area within this plan was 41.93 $km^2$.[54]

In May 1996, the *Regulations on the Administration of Sea Area Utilization in Guangdong Province* (廣東省海域使用管理規定) were promulgated and implemented, making it the first governmental regulation on sea area utilization in the nation. Article 4 of the regulations stipulates: "Sea areas belong to the state, and other provisions shall apply as specified by laws and regulations." In January 2007, Guangdong Province promulgated the *Rules on the Administration of Sea Area Utilization in Guangdong Province* (廣東省海域使用管理條例). Article 3 reaffirms: "Sea areas belong to the state," while Article 15 stipulates that any initiatives reclaiming less than 50 ha. of the sea must obtain approval from the provincial administration.

Regarding the ownership of tidal flats, this study finds it challenging to provide an in-depth discussion. Certainly, but what can be confirmed is that the parks built on reclaimed land in the Shenzhen Bay area were constructed after a significant advancement of urban construction toward the sea, as illustrated in Fig. 6.9. These parks were not built on original tidal flats but rather on the reclaimed sea area. As the sea areas entirely belong to the state and do not involve any so-called "ancestral collectively owned sea,"[55] the ownership of the Shenzhen Bay sea area has no direct connection with the surrounding villages. Thus, while urban development, like as the construction of parks, has indeed altered the production mode and lifestyle of the villages along the Shenzhen Bay coast, the construction of these four parks on reclaimed land did not involve acquiring land from rural collectives.

### 6.2.2 *Government-led Land Reclamation*

The process of land reclamation in the Shenzhen Bay area commenced during the initial phases of the establishment of the SSEZ. Up until the early 1990s, the focus of land reclamation was primarily on the construction of the port and the Shekou Industrial Zone at the southern tip of the peninsula. Commencing in the mid-1990s, there was a significant upsurge in land reclamation activities in the Shenzhen Bay area. The areas of the north of the Binhai Avenue, west of the Western Corridor, and in the Houhai Central District were all filled during this period. After the promulgation of the *Law of the People's Republic of China on the Administration of the Use of Sea Areas* in 2001, the extensive land reclamation trend was somewhat restrained but not entirely halted. The previously initiated land reclamation projects in the

[54] Guangdong Provincial Local Chronicles Compilation Committee 廣東省地方市志編纂委員會. (2000). *Chronicles of Guangdong Province: Resources and Environment Volume* 廣東省志資源·環境卷. Guangzhou: Guangdong People's Publishing House 廣州: 廣東人民出版社. (pp. 207).

[55] Xu, H. 許暉(2020, May 6). The Sea Belongs to the State; There Is No So-Called "Ancestral Sea Owned Collectively."海域屬於國家所有不存在所謂"集體所有的祖宗海." Zhuhai Special Zone News 珠海特區報. (02).

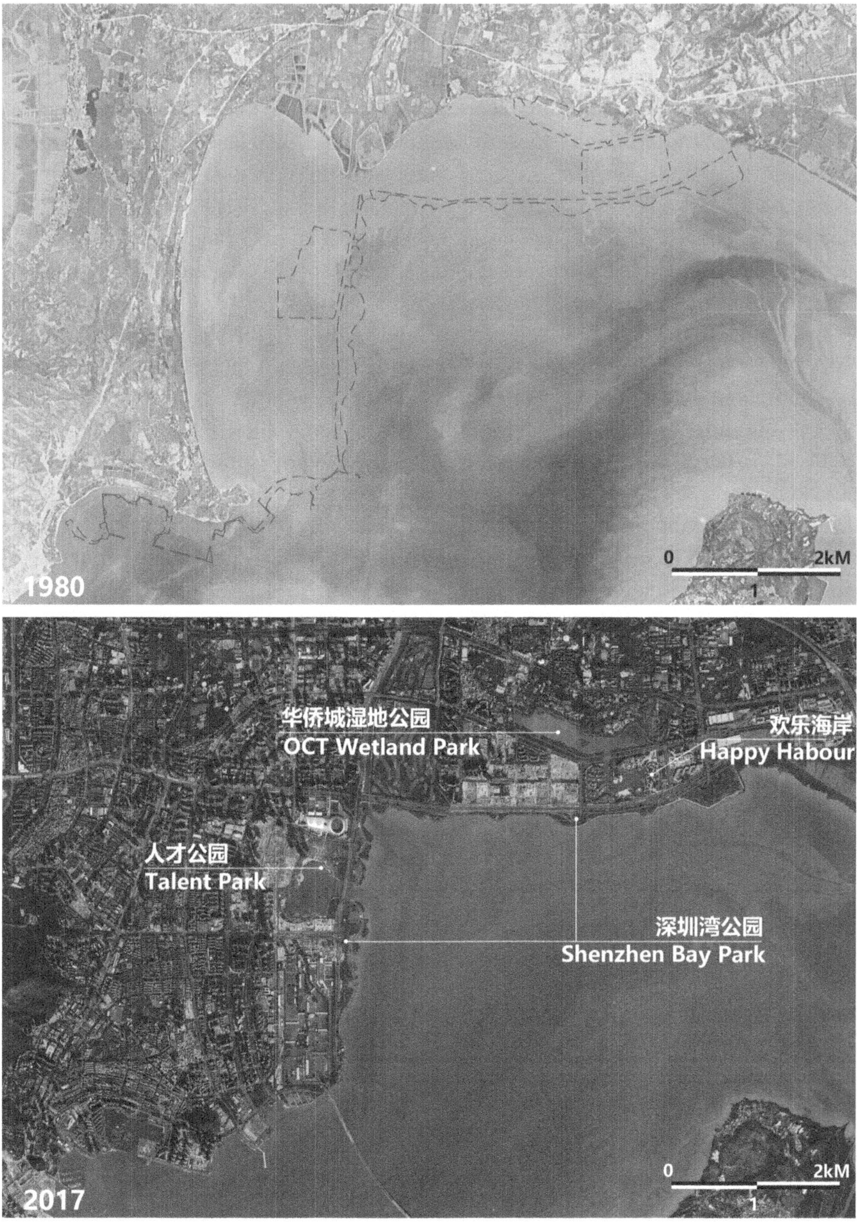


**Fig. 6.9** Location of four parks in 1980 and 2020 [Source from: USGS and Google Historical Image, redrawn by the author.]

Houhai Central District and around the Central River were completed between 2005 and 2012. Following that, relatively regulated land reclamation projects for the Shenzhen Bay Coastal Leisure Belt began. During this period, there was a notable emergence of reclaimed lands that were still vacant. These lands covered substantial regions, stretching from the eastern section of the Nanshan Central District to Shenzhen Bay, and from the southern section of the Nanshan Central District to Dongjiaotou. From 2012 to 2020, the land reclamation in Shenzhen Bay entered its concluding phase, with a smaller increase in reclaimed land. The two aforementioned reclaimed but then vacant areas were, respectively, built for commercial and service use and residential purposes.[56]

In contrast to the "Unified Acquisition" and "Unified Transformation" processes both within and outside the SSEZ in 1992 and 2004, land reclamation, as a means to acquire land, had a relatively smaller impact on the villagers, providing more land but causing more noticeable effects on the coastal ecosystem. The land reclamation in the Shenzhen Bay area was an action taken by the government due to a shortage of land resources. The entirety of the obtained land has been transformed into urban building and development land, which was under the direct control of the government. A portion of this area was utilized for the establishment of the four urban parks, as indicated in Table 6.4.

## 6.3 Insulative Nature: From Cold War Frontier to Urban Parks

### *6.3.1 Formation of the Cold War Frontier*

In March 1946, following the end of World War II, the former British Prime Minister Winston Churchill delivered the "Iron Curtain Speech" at Westminster College in the United States. This event marked the commencement of the Cold War, a

**Table 6.4** Basic information of the four reclaimed urban parks

| Park Name | Construction Time | Open Time | The total of the park (Ha.) |
|---|---|---|---|
| Talent Park | 2012 | 2017–11 | 77.0 |
| Shenzhen Bay Park | 2006 | 2011–08(East part)<br>2017–06(West part) | 128.7 |
| Happy Coast | 2008–05[a] | 2011–08 | 125.0 |
| OCT Wetland Park | 2007–04[b] | 2012–05 | 68.5 |

[a]The construction of Happy Coast commenced in May 2008.
[b]Overseas Chinese Town took over the wetland from the government in April 2007.

[56]Zheng, H. X. 鄭浩欽. (2020). *Characteristics and Mechanisms of Spatiotemporal Evolution of Shenzhen Reclamation from 1975 to 2020* 1975-2020年深圳填海空間時空演變的特徵及其機制研究.Shenzhen University 深圳大學. (pp. 30–84).

protracted geopolitical confrontation lasting over four decades between the Capitalist Bloc led by the United States and the Socialist Bloc led by the Soviet Union. At that time, the atmosphere of ideological confrontation not only pervaded the two leading military powers but also penetrated into all the countries belonging to the two blocs.

In October 1949, the People's Republic of China, a member of the Socialist Bloc, was officially established. On its southern border, the two major towns of Ancient Nantou Town and Shenzhen Town in Bao'an County were successively taken over by the Communist Party's army in October of that year. In April 1950, the entire Bao'an County came under the control of the Communist Party.[57] At this time, the New Territories of Hong Kong to the south of the Shenzhen River had been under British colonial rule for 52 years (since 1989). The southern part of the Kowloon Peninsula has remained under British governance for a period of 90 years (since 1860). In addition, Hong Kong Island has been under British governance for 108 years (since 1842). The political upheaval on the northern bank of the Shenzhen River was something the British Hong Kong government could not ignore:

*But my concern as Governor of Hong Kong was how the communist take-over of China would affect the Colony...The communist forces reached our border in October 1949, but for months previously we had been in consultation with London...Whitehall wanted us to wire the frontier and to resist by force of arms any who tried to cross into British territory...in our argument with London as to whether or not the border should be closed...more and more crossed the frontier so we had to close it, which was done without fuss or bother in May 1950....*[58]

*...in pre-communist days the relationship between the Chinese and Hong Kong authorities was, generally speaking, that of normal friendly states, that no barriers existed at the frontier, and that any Chinese was free to come and go as he pleased...It was only later, May 1950, that we imposed restrictions on people coming into the Colony. This we did, not for political reasons to keep out communists, but because we were being flooded out with refugees....*[59]

Indeed, it was more than a mere barbed wire fence. The Korean War commenced in June 1950, China and the United Kingdom were belligerent countries. Following the imposition of sanctions by the United Nations, the Hong Kong border became the only frontier for China to access international commodities. Consequently, in November 1950, the British Hong Kong government decided to establish restricted areas by completely sealing off the border, with the objective of thwarting the illegal transportation of commodities. The restricted area encompassed the regions of Sha Tau Kok (沙頭角), Mai Po (米埔), Ao Tou (凹頭), and Shenzhen Bay, as shown in

---

[57] Bao'an County Local Chronicles Compilation Committee 寶安縣地方誌編纂委員會(1997). *Chronicles of Bao'an County*寶安縣誌. Guangzhou: Guangdong People's Publishing House廣州: 廣東人民出版社. (pp. 24–[25]

[58] Grantham, A. (2011). *Via Ports: From Hong Kong to Hong Kong (Vol. 1)*. Hong Kong University Press. (pp. 138; 140–147).

[59] Grantham, A. (2011). *Via Ports: From Hong Kong to Hong Kong (Vol. 1)*. Hong Kong University Press. (pp. 149–152).

Fig. 6.10.[60] Similarly, for the newly established government on the northern bank of the Shenzhen River, the international impact brought about by a large-scale exodus of the population and espionage activities were deemed unacceptable. In July and August of 1950, Guangdong Province established a Border Defense Bureau (邊防局) and enforced stringent control measures by entirely sealing off the Shenzhen River. As a consequence, individuals on either bank of the river were subjected to inspections when entering or exiting the region.[61] In February 1951, Guangdong Province further strengthened border management by establishing border defense lines and tightening border control. Travelers were required to present specific

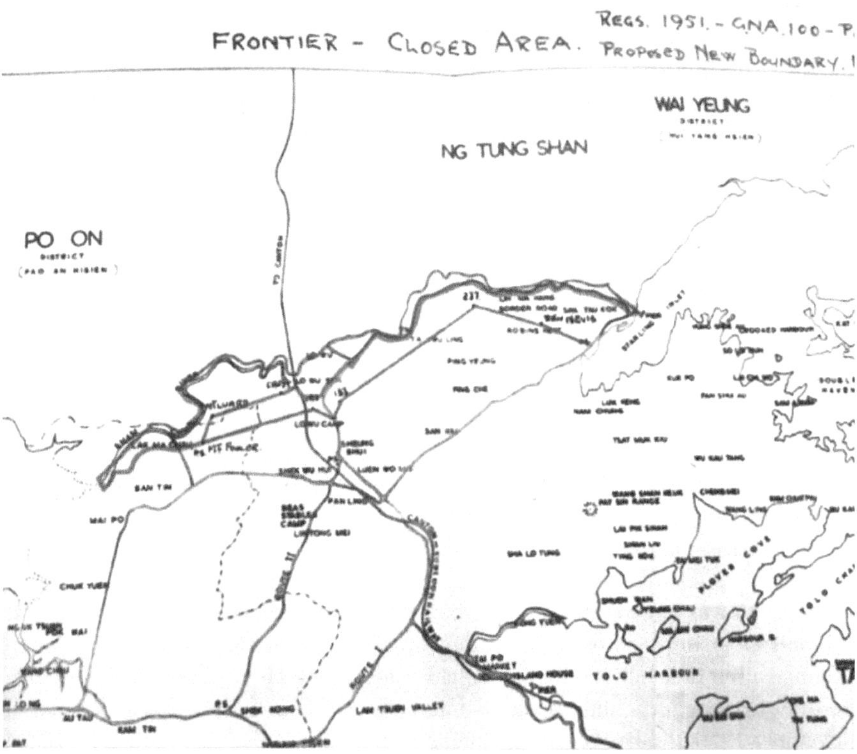


**Fig. 6.10** Restricted areas designated by the British Hong Kong Government in 1951 (in Red) and 1962 (in Blue) [Ruan, Z. 阮志 (2014). *Entering Forbidden Zones: A History of Hong Kong's Border Restricted Areas*入境問禁 香港邊境禁區史. Hong Kong: Joint Publishing (Hong Kong) Co., Ltd香港:三聯書店(香港)有限公司. (Illustration page插圖頁).]

[60] Ruan, Z. 阮志(2014). *Entering Forbidden Zones: A History of Hong Kong's Border Restricted Areas* 入境問禁 香港邊境禁區史. Hong Kong: Joint Publishing (Hong Kong) Co., Ltd. 香港:三聯書店(香港)有限公司. (pp. 39–42).

[61] Bao'an County Local Chronicles Compilation Committee 寶安縣地方誌編纂委員會(1997). *Chronicles of Bao'an County* 寶安縣誌. Guangzhou: Guangdong People's Publishing House 廣州:廣東人民出版社. (pp. 24–25).

documents at designated ports for entry and exit, while all other coastal and border areas were strictly prohibited from entry and exit. In February 1956, the 49 townships and two towns within Bao'an County were officially designated as border management areas, with a restricted zone established along the Shenzhen River, extending 3–5 km inland.[62]

At this point, the Guangdong–Hong Kong border was almost completely closed. Prior to the signing of the "Nanjing Treaty," Hong Kong and Bao'an County were both part of Xin'an County.[63] Whether it be the "Nanjing Treaty (南京條約)" of 1842, the "Beijing Treaty (北京條約)" of 1860, or *The Convention Between Great Britain and China Respecting an Extension of Hong Kong Territory* (拓展香港界址專條) of 1898, the three rounds of relative changes in the scope between Xin'an County and Hong Kong have almost always used natural features such as maritime areas, islands, rivers, and mountain ranges as boundaries. Shenzhen Bay experienced its first marine transgression around 30,000 to 36,000 years ago and underwent a second one approximately 8000 years ago. After these events, it essentially took its current form.[64] However, as a concept in political geography, borders are not predetermined by nature but rather continuously constructed, reconstructed, and dissolved throughout historical processes. In the early 1950s, with the onset of the Cold War, Shenzhen Bay-Shenzhen River-Dapeng Bay were constructed as a political frontier. In comparison to a mere border, it possessed a distinct magnitude and functioned as a form of isulative nature, separating the lives and beliefs of the populations in the two major camps of the Cold War and between Guangdong and Hong Kong.

### 6.3.2 Shenzhen Bay and the Wave of "Escape to Hong Kong"

In 1978, the average yearly income for farmers in the entire Guangdong province was only 77.4 Chinese yuan. In nearby Shenzhen, just across the river from Hong Kong, farmers had an average income of only 134 Chinese yuan. Meanwhile, across the river in the Hong Kong New Territories, farmers earned an income of 13,000 Hong Kong dollars during the same period. The significant disparity in living standards made "Escape to Hong Kong" temporarily the "ultimate dream" for farmers,

[62] Shenzhen Museum 深圳博物館(2019). *Research on the Historical Changes of the Management Line (Second Line) in SSEZ* 深圳經濟特區管理線(二線)歷史變遷研究. Shenzhen: Haitian Publishing House 深圳: 海天出版社. (pp. 24–29).

[63] In 1914, to avoid duplication with Xin'an County in Henan, Xin'an County was renamed Bao'an County.

[64] Liu, X., Zhao, H., Zheng, D., et al. 劉杏玲,趙煥庭,鄭德延等 (1988). Evolution and Development of Shenzhen Bay 深圳灣的演變與開發利用. *Journal of Tropical Oceanography* 熱帶海洋. 1988(01), (pp.48–56).

workers, citizens, and even civil servants residing in the Bao'an County area.[65] In reality, the wave of emigration to Hong Kong was already approaching its end by 1978, this ebb and flow of cross-border migration, marked by periods of tension and relaxation, had persisted for nearly three decades during the Cold War.

Following the closure of the Guangdong–Hong Kong border, occasional incidents of unauthorized and illegal border crossings occurred from time to time. The situation was particularly tense in the years 1957, 1962, and 1978. The *Guangdong Provincial Annals Public Security Annals* (廣東省志·公安志) records that in July 1957, thousands of people gathered at the Shenzhen border daily, preparing for illegal border crossings. There were even incidents where some fleeing individuals forcibly attacked the Shenzhen checkpoints. During the months of April and May in 1962, a substantial crowd of approximately four to five thousand individuals gathered on a daily basis in the vicinity of Bao'an and Shenzhen Town, eagerly anticipating an opportunity to gain entry into Hong Kong. On the busiest day, the number surpassed eight thousand individuals. In 1978, the number of illegal border crossings sharply increased once again, with a total of over 48,000 instances of outbound illegal migration from both within and outside the province. Throughout the year, more than 60,000 individuals attempting illegal border crossings were intercepted.[66]

Furthermore, in May 1979, there were rumors circulating in the regions of Huiyang (惠陽), Dongguan (東莞), and Bao'an that Queen Elizabeth II would assume the throne (or, in alternative versions, commemorate her birthday) on the same day. Shenzhen was rumored to have a "grand opening of the river checkpoint (大放河口)" on May 6, while Hong Kong would also implement a general amnesty, allowing individuals stranded in Hong Kong to be permanent residents within a three-day timeframe. On May 7, 1979, the second day, hundreds of corpses were discovered adrift in the waters close to Hong Kong.[67]

The massive influx of emigrants to Hong Kong resulted in numerous villages in Bao'an County seeing a significant number of vacant residences. This situation greatly shook the central authorities, prompting the commencement of nationwide reform and opening up.[68] Conversely, it also offered inexpensive workforce for the flourishing industrial economy in Hong Kong during that period.[69]

---

[65] Wang, D. J. 王地久(2018). Analysis of the Agricultural Situation in Shenzhen in 1979 1979年深圳農民狀況分析. In Shenzhen Municipal Chronicles Office (Eds.) 深圳市史志辦公室. *Chronicles of Shenzhen's Reform and opening-up, Volume II* 深圳改革開放實錄. 第二輯. Shenzhen: Shenzhen Press Group Publishing House 深圳: 深圳報業出版社. (pp. 1–10).

[66] Guangdong Provincial Local Chronicles Compilation Committee 廣東省地方市志編纂委員會 (2000). *Chronicles of Guangdong Province: Public Security* 廣東省志·公安志. Guangzhou: Guangdong People's Publishing House 廣州: 廣東人民出版社. (pp. 312–313).

[67] Chen, H. 陳宏(2006). *1979–2000: Significant Decisions and Civil Observations of Shenzhen* 1979-2000 深圳重大決策和時間民間觀察. Wuhan: Changjiang Literature and Art Publishing House 武漢: 長江文藝出版社(pp. 003–009).

[68] Chen, B. A. 陳秉安(2010). *The Great Escape to Hong Kong* 大逃港. Guangzhou: Guangdong People's Publishing House 廣州: 廣東人民出版社. (pp. 327–344).

[69] Lu, S. C., & Lu, D. Q. 盧受采 盧冬青 著(2002). *A History of Hong Kong Economy* 香港經濟史. Hong Kong: Joint Publishing (Hong Kong) Co., Ltd. 香港:三聯書店(香港)有限公司.

As shown in Fig. 6.11, the routes for escaping to Hong Kong were divided into three lines: the central line involved taking a train or walking along the Guangzhou-Kowloon Railway to enter Buji Town, then finding opportunities to cross the border into areas like Wutongshan, Luofang Village, etc.; the eastern line involved climbing Wutongshan from the eastern part of Bao'an County, then crossing the Shenzhen River or Sha Tau Kok Sea near Luofang Village or Sha Tau Kok, or swimming across Dapeng Bay to reach the Sai Kung (西貢) area in Hong Kong; the western line involved crossing the sea from Shekou or near Baishizhou under the mangroves, swimming across to areas like Yuen Long (元朗) in Hong Kong. Among them, the western line was the most preferred choice.[70] However, swimming 4–5 km across Shenzhen Bay is no easy task, and the area has become a deadly zone for many attempting to escape to Hong Kong. Numerous escapees who entered the sea from Shekou faced turbulent waters, and many lost their lives at sea, with their bodies washed back to the Shekou beach by the tides. The maritime police station in Shekou commissioned some older individuals to bury the bodies, and locals referred

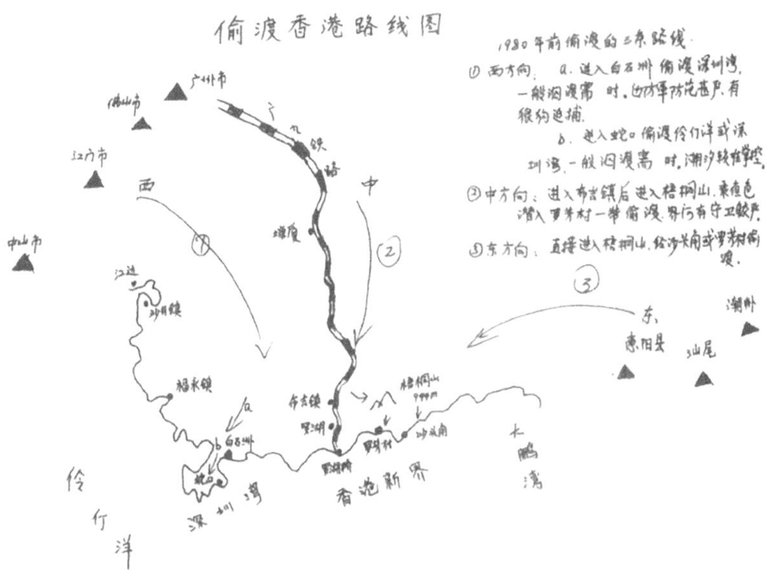


**Fig. 6.11** Three routes for illegal immigration to Hong Kong during the Cold War [Chen, B. A. 陳秉安 (2010). *The Great Escape to Hong Kong* 大逃港. Guangzhou: Guangdong People's Publishing House 廣州: 廣東人民出版社. (pp. cover page).]

[70] Lin, T. H. 林天宏 (2010, December 8). *People Will Vote with Their Feet* 人民會用腳投票. China Youth Daily 中國青年報.

to them as "body carriers (拉屍佬)." For each body buried, they could receive a labor fee of 15 yuan, which could increase to 20 yuan if the body was decomposed and emitted a foul odor.[71]

The significant reason for the preference for the western line was the natural conditions it provided. On the one hand, escapees were considered "traitors" or "defectors," and before the 1960s, border soldiers had the authority to open fire at any time. The dense mangrove forests along the shores of Shenzhen Bay at that time offered visual cover. On the other hand, Shenzhen Bay is adjacent to the Pearl River Estuary, and the inflow of a large amount of freshwater reduced the salinity of the seawater.[72] Consequently, compared to Dapeng Bay, there were fewer occurrences of sharks in Shenzhen Bay.[73] In addition, the success of escapees was indeed related to their understanding of the natural conditions of the maritime area:

*At that time, Mr. Wang was a member of the West Sea People's Commune Fisheries Second Brigade. Local fishermen in this brigade typically owned their own fishing boats. During the process of illegally crossing to Hong Kong … their success rate was often high due to their familiarity with the sea area. Local residents who were familiar with the situation often chose to escape to Hong Kong by entering the water near current areas such as Shekou Old Street Blue Bay Garden (蛇口老街藍漪花園), Building D of Yu'er Village (漁二村D座), and the vicinity of Dongjiaotou Oil Depot (東角頭油庫). At that time, these locations were still sandy beaches and relatively secluded, making them popular starting points for local stowaways.*[74]

*Relatively speaking, those who from inland cities came here and intended to stowaway to Hong Kong were not so fortunate. In Mr. Wang's recollection, these individuals, upon arriving in Shekou for the first time, were unfamiliar with the locals and not well-acquainted with the sea area…despite their successfully entry into the water, their unfamiliarity with variables like water currents significantly reduced their chances of success. Many individuals, driven by impatience or lacking swimming skills, would hastily enter the water without proper consideration…often resulting in fatalities. To successfully swim to Hong Kong, aside from choosing the optimal entry points… stowaways often had to assess the positions of sentinels and the direction of water currents. Selecting the right entry point was crucial for maximizing the chances of a successful stowaway attempt.*[75]

---

71 Tu, Q. 塗俏(2016). *Biography of Yuan Geng: Scenes of Reform 1978–1984* 袁庚傳: 改革現場 1978–1984. Shenzhen: Hai Tian Press深圳: 海天出版社. (pp. 70).

72 Huang, D. H. 黄東漢(2013). *Everlasting Memories of the Mangrove Forest*永志難忘紅樹林. Remembrance記憶. 2013, 93(02). (pp.35–38).

73 Li, J. Z. 李建中(2013). *From "Illegal Crossing" to "Setting Sail."* 從"督卒"到"起錨" .Remembrance 記憶.93(02): 3–10.

74 He, Y. H. 何有華 (2012, August 29). *Former Temptations: Crossing the Water to the Right* 從前的誘惑，涉水向右. Retrieved from http://www.szdag.gov.cn/dawh/szzg/content/post_99164.html

75 He, Y. H. 何有華 (2012, August 29). *Former Temptations: Crossing the Water to the Right* 從前的誘惑，涉水向右. Retrieved from http://www.szdag.gov.cn/dawh/szzg/content/post_99164.html

*The locals have coined a proverb: "On the third and the eighteenth day, the water flows smoothly." During this period, entering the water was considered going with the flow, and swimming to the opposite shore of Hong Kong requires much less effort. During periods of ebbing tide, when the tide was retreating, the sea area significantly narrows, only waterway in the middle, and it's possible to swim across in just over an hour. Nevertheless, there are pros and cons. Once the tide recedes, a wide stretch of muddy and deep mud flat emerges near the coastline, posing challenges for stowaways. Walking makes the target larger, and some stowaways were forced to crawl through the mud flat. If a stowaway encounters oysters, their thighs and bodies could be cut and injured all over....*[76]

Nature does not inherently possess the presumed goodwill and morality as commonly understood in contemporary contexts. During the Cold War era, Shenzhen Bay, as a specific form of nature, stirred up political tensions between China and Britain as well as between Guangdong and Hong Kong. It even directly resulted in a large-scale loss of life. Of course, it was not always as harsh as Cold War politics. For those familiar with this sea area, Shenzhen Bay could also serve as a means to escape hunger and reach the desired opposite shore.

### 6.3.3 Parks, Reclamation, and Border Defense Line

In August 1950, the Guangdong Provincial Public Security Department (廣東省公安廳) established the Border Defense Bureau (邊防局) along with related police stations and checkpoints. In December 1951, the military began jointly managing the province's border affairs in collaboration with local governments.[77] After several reorganizations and changes, in April 1989, the Guangdong Provincial Public Security Department Border Defense Bureau established the Sixth Brigade (referred to as Border Defense Sixth Brigade, 邊防六支隊) responsible for border patrols from Yantian Port (鹽田港) in the east to Nantou Daxin Wharf (南頭大新碼頭) in the west.[78] The 70-km border line has undergone changes, and as shown in Fig. 6.12, with the progress of land reclamation projects along the Nantou Peninsula and Shenzhen Bay, as of October 2008, it had extended to 83.14 km.[79] The rapid

[76] Li, J. Z. 李建中 (2013). *From "Illegal Crossing" to "Setting Sail."* 從"督卒"到"起錨." Remembrance記憶. 93(02): 3–10.

[77] Guangdong Provincial Local Chronicles Compilation Committee廣東省地方市志編纂委員會. (2000). *Chronicles of Guangdong Province: Public Security*廣東省志·公安志. Guangzhou: Guangdong People's Publishing House廣州: 廣東人民出版社. (pp. 293).

[78] Shenzhen Municipal Chronicles Compilation Committee 深圳市地方誌編纂委員會. (2009). Chronicles of Shenzhen City: Volume on Political Parties and Political Power深圳市志·政黨政權卷. Beijing: Fangzhi Publishing House 北京: 方志出版社. (pp. 520).

[79] Li, G. L., Hao, X. T., & Feng, L. F. 李國良,郝雪濤,馮立峰 (2008, October 30). *My Experience: The Extended Border Defense Line*我的親歷: 延長的邊防線. Retrieved from https://news.sohu.com/20081030/n260343578.shtml

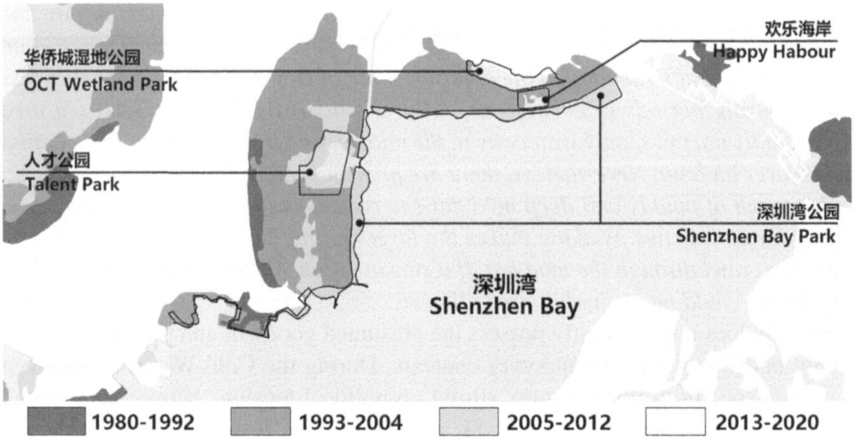


**Fig. 6.12** Diagrams of reclamation projects along Shenzhen Bay at different periods [Drawing References from Google Historical Imagery.]

development of the Shenzhen Bay reclamation project means that the topography is constantly changing. Sometimes, after a few days of rotation, a group of sentinels may find that their assigned position has moved significantly. Nevertheless, the principle of border defense remains steadfast: for every meter the reclamation advances, the sentinels advance one step.[80]

In addition, a sequence of sentry towers along the border is also undergoing changes. In the 1990s, land reclamation initiatives caused the northern coastline of Shenzhen Bay to shift 1.0–1.5 km toward the south, resulting in the transformation of certain coastal areas into an inland lake. The OCT Wetland Park, which was formerly a section of this coastline, still preserves three border defense guard posts. The border defense guard posts, measuring 5–7 m in height and with a wall thickness of around 55 cm, have been transformed from their original military structures into landscaping buildings in the wetland park,[81] as depicted in Fig. 6.13.

Changes made in the structure and function of the three guard posts in the OCT Wetland Park are by no means accidental. Around the turn of the new millennium, guard posts, outposts, and border defense lines along the whole Guangdong–Hong Kong front began to undergo changes.[82] In November 2008, a total investment of

[80] Liu, J. S., & Song, J. T. 劉家順 宋京濤.(2007, August 17). *Border Defense Sentiments along the Shenzhen Bay*深圳灣畔戍邊情.Retrieved from https://news.cctv.com/military/20070817/109034.shtml

[81] Meng, X. W., et al.孟祥偉等 (Eds.) (2021). *In the City Center: Operation and Management of Overseas Chinese Town Wetland and the Model of Nature Education* 在城中央—華僑城濕地運營管理及自然教育模式. Beijing: China Forestry Publishing House 北京: 中國林業出版社. (pp. 43).

[82] Li, G. L., Guo, X., & Feng, L. F. 李國良,國喜,馮立峰 (2008, October 24). *My Experience: Border Guard Posts Have Become Scenic Spots* 我的親歷:邊防哨所成了風景線.Retrieved from https://news.sohu.com/20081024/n260234835.shtml

**Fig. 6.13** The three pavilion posts distributed from west to east in the park and the sentinel regulations, taken by the Author

400 million RMB was allocated for the commencement of the *Outpost Transformation Plan* (哨所改造計畫) in Shenzhen and Hong Kong. As per this proposal, the border between Shenzhen and Hong Kong would involve the removal of barbed wire, the erection of high-tech border facilities and the establishment of 32 additional outposts.

*...will completely abandon the old image of "resembling a military encampment or a watchtower"...farewell to the "bunker-style" architecture, replaced by the stylish and modern "trapezoid-shaped" design...in the border areas adjacent to the urban district, the original barbed wire has been replaced with landscape fencing...the coastline is no longer heavily fortified, and the iron mesh installed in some sections will be completely removed...today, in popular areas like Shenzhen Bay and Dapeng Bay, where stowaways once entered the sea, all barbed wire has been removed and transformed into a leisurely waterfront promenade for recreational activities...it has become a sought-after destination for residents of Shenzhen and Hong Kong to enjoy a "zero-distance sea view."*[83]

The alteration of the outward facade of guard posts, outposts, and border defense lines, from a stern military image to a more amiable and accessible appearance, mirrors the shifts in the fundamental power dynamics that influenced their construction. In March 1979, Deng Xiaoping had already evaluated, "It appears unlikely that a world war will occur in the next decade."[84] After the disintegration of former Soviet Union in December 1991, the international situation became more evident. The previously antagonistic Cold War ideology disintegrated, paving the way for a new wave of globalization that fostered tighter diplomatic ties between nations. Consequently, the theme shifted from "war and revolution" to "peace and development." The warm breeze of dialogue and cooperation between nations and regions

[83] Li, H. Y. 李海元(2008, November 3). Massive Transformation Along the Guangdong-Hong Kong Border: 4 Billion Investment in Building a New Border粵港一線大變革4億建新邊界. Retrieved from https://news.sohu.com/20081103/n260399819.shtml

[84] Central Compilation and Translation Bureau of the Communist Party of China中共中央文獻研究室 編 (2004). Chronicles of Deng Xiaoping 1975–1997 (Volume I) 鄧小平年譜 一九七五-一九九七(上). Beijing: Central Compilation and Translation Press北京: 中央文獻出版社. (pp. 494).

gradually dispelled the Cold War clouds that once hung over Shenzhen Bay. This brought about corresponding changes, transforming the border defense line from a sentinel's "no mercy shotting" stance in the 1960s to an urban park where citizens can enjoy a "zero-distance sea view."[85]

The most prominent physical features of Shenzhen Bay Park are twofold: (1) Linear space. Starting from the intersection of Qiaocheng East Road and Binhai Avenue, it extends westward along the coastline, then southward to Shekou Sea World, with an actual length of 16.2 km. The width varies, reaching approximately 380 m at its widest and around 20 m at its narrowest, forming a long and narrow linear structure. (2) Double-deck platform-style space. Within the 16.2 km, approximately 9.5 km from Qiaocheng East Road to the central river adopts a double-deck platform-style section.

The linear spatial form can be traced back to the patrol area of the Border Defense Sixth Brigade, established in 1989: spanning from Yantian Port in the east to the Nantou Daxin Wharf in the west. The current 16.2-km coastal line from Qiaocheng East Road to Shekou Sea World is located within this range, albeit continuously extending seaward over the past two decades due to ongoing reclamation projects since the 1990s. Despite the continued existence of its border defense function under the principle of "one country, two systems," it has undergone continuous adjustments and changes throughout Shenzhen's decades-long urbanization process. When the Shenzhen Municipal Government decided to halt large-scale reclamation and development in Shenzhen Bay around the year 2000, a new vision for the role of this border defense line emerged, namely, the construction of a 15-km coastal leisure belt.

However, transforming the straight border patrol road into a coastal leisure belt still poses significant challenges. As a public space, the long and narrow linear spatial form lacks depth for citizen activities. If additional activity space is to be provided, it would inevitably require further reclamation. The scale of land reclamation was a major hurdle throughout the planning and construction of Shenzhen Bay Park. The originally projected reclamation area underwent adjustments, reducing it from the initial 150 ha. to 57 ha., and ultimately settling at 49.74 ha. The provincial government has jurisdiction to authorize reclamation areas up to a maximum of 50 ha.[86] Any reclamation area exceeding this limit necessitates approval from the National Oceanic Administration. Apart from the issue of area, the reasons for reclamation also needed approval. At that time, the construction of an urban park alone was not considered a sufficient justification for reclamation based on the approval principles. Consequently, urban transportation development became the submitted justification for official review. "Wherever there are transportation facilities and overpasses, we reclaim land from the sea… The additional space is designated for recreational activities, aiming to change the original single linear form. The original

[85]According to the interview with Mr. Zhu Rongyuan, 2023-09-14.

[86]According to the interview with Mr. Xie Chaoping, 2023-09-10.

linear form was too straight…,"[87] and the previously straight patrol road was filled in with new earthworks.[88]

One can imagine that, without the construction of the South Overseas Chinese City Reclamation Area and the Binhai Avenue, the border patrol road would most likely go along the south of the Window of the World (世界之窗) and Splendid China (錦繡中華) to the estuary of Dashahe River. Nevertheless, the construction of the costal leisure belt will still be closely aligned with the patrol road and run parallel to it, as both share a logical commonality: they must closely follow to the maritime and land borders. As is well known, whether it's the transportation efficiency pursued by the Binhai Avenue, the engineering efficiency required for reclamation, or the military action efficiency demanded by the border defense, a straight line is undoubtedly the optimal form to demonstrate these efficiencies. However, when this once-optimal form needs to align with the functions of the costal leisure belt, a contradiction arises: it will result in the costal leisure belt lacking enough space for activities and having a monotonous visual landscape, and so on. The reclamation of 49.74 ha. resulted in the transformation of the straight border patrol road into a curved costal leisure belt.

Another structural physical characteristic of Shenzhen Bay Park—its double-deck platform-style section design (Fig. 6.14)—is closely related to its former role as a Cold War patrol road:

*…in principle, patrol roads should not conflict with visitors… the design on two elevation levels is primarily to allocate a separate patrol route for them… this is a prerequisite that must be met in the design… the patrol road must first exist and maintain its original patrol methods, and only then can permission be granted for visitors and citizens to access it…The current situation is due to a shift in border defense procedures….*[89]

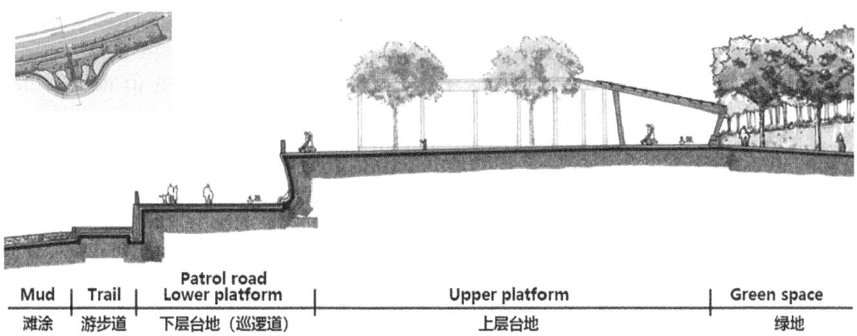


**Fig. 6.14** Typical section of Shenzhen Bay Park [Source from: Management Office of Shenzhen Bay Park.]

[87] According to the interview with Mr. Zhu Rongyuan, 2023-09-14.

[88] According to the interview with Mr. Xie Chaoping, 2023-09-10.

[89] According to the interview with Mr. Zhu Rongyuan, 2023-09-14.

Since the completion and opening of Shenzhen Bay Park, the lower level road has consistently served a dual function as both a patrol road and a scenic pathway. The military nature of the patrol path and the border defense authority's insistence on military operational efficiency make it difficult for it to be completely open to the public like other types of public spaces. There are few steps or ramps connecting the two terraces, often leading park visitors to climb over the railings by hand. This has been a frequent cause of complaints to the park management, but due to the importance of border patrol road, communication between the park management and the border defense authority has not yielded any substantive results, and the addition of stairs and ramps is still not allowed.[90]

The linear configuration and the Double-deck platform-style spatial layout of Shenzhen Bay Park is evidently influenced by its historical significance as a former Cold War frontier. The construction of Shenzhen Bay Park exemplifies the shift of power structure from "war and revolution" to "peace and development." It also represents the evolution of the relationship between different nations and the regions. According to Jussi Laine, borders are not just limits of national sovereignty but have also expanded into the realm of social practices within the context of globalization.[91] As part of the China–British and Guangdong–Hong Kong border during the Cold War era, Shenzhen Bay is more than just a boundary of territory and political systems; it has also evolved into a social practice among local residents filled with joys and sorrows, separations and reunions, emotions, and memories. During the later phases of the Cold War, the inflexible border started to be gradually broken down. The cross-border flow of capital, people, and information has become the norm, whether it was the migration of Hong Kong capital to the north in the 1980s or the current integration in the Guangdong–Hong Kong–Macao Greater Bay Area. These cross-border mobilities represent the phenomenon of globalization, on the one hand, they actively advance the reclamation projects along the coastlines of Shenzhen Bay, persistently modifying the physical structure of the boundary, on the other hand, the border no longer projects a militaristic image and is instead becoming more integrated into the everyday lives of citizens, giving rise to new social practices. In this process, Shenzhen Bay, as a border, transforms from an isulative nature into another cultural form of nature—developmental nature. That is, Shenzhen Bay serves as a symbolic environment for the construction of internationalized communities along its shores, promotes the development of headquarters economy, and draws in multinational professionals.

[90] Shenzhen News Website深圳新聞網 (2021, February 1). *Why Are There So Few Stairs on the Seaside Boardwalk of Shenzhen Bay Park? Latest Response!* 深圳灣公園濱海不到上下階梯為什麼這麼少?最新回應!Retrieved from https://wxd.sznews.com/BaiDuBaiJia/20210201/content_460695.html; Wu, Z.吳臻 (2023, February 28). *What a Predicament! In a Famous Park in Shenzhen, Tourists Have to "Climb" In to Take Photos!* 好狼狽!深圳一網紅公園，遊客要"翻"進去打卡!Retrieved from https://new.qq.com/rain/a/20230228A01L4Q00

[91] Laine, J. P. (2017). Borders and borderscapes under contemporary globalization. *Progress in Geography*, 36(12), 1531–1539.

## 6.4 Gentrified Nature: Talent, Park, and Gentrification

### *6.4.1 Talents, Industry, and Urban Space*

During the initial phase of the establishment of the Special Economic Zone, Shenzhen faced a severe shortage of talent in various industries. Taking the construction industry as an example, in 1980, the entire city possessed a solitary design office with little more than ten experts, in handling the architectural planning of structures with a maximum of six stories. There were only two technical professionals with the title of engineer in the entire city.[92] Evidently, there was a notable disparity between the talent situation in this border town and the desired standard for a modernized city, necessitating a major influx of talent resources. Nevertheless, the bulk of professionally educated individuals in China have been graduates from either the period before the Cultural Revolution (pre-1966) or the period after the Cultural Revolution (post-1982) since the establishment of the country. This led to a widespread talent gap during a period of around 10–15 years around the transition to the new millennium.[93]

In Shenzhen, a city characterized by fragile basis and ambitious goals, the problem of Talent discontinuity will result in more severe consequences. As early as 1991, Shenzhen had already recognized the potential consequences of the talent discontinuity and organized relevant studies led by the Municipal Human Resources Bureau (市人事局).[94] In September 2001, China's entry into the World Trade Organization directly spurred Shenzhen's demand for talents with an international vision. In December 2003, the National Talent Work Conference was held, explicitly proposing the strategy of building a talent-strong nation. Shortly thereafter, in February 2004, the Shenzhen Municipal Party Committee Organizational Department established the "Shenzhen Talent Development Strategic Research Group (深圳人才發展戰略調研組)," conducting further research on talent work in the city. This research puts forward the goals and basic principles for future talent work in Shenzhen, which include cultivating a large number of international talents adapted to the characteristics of an internationalized city based on the current status and trends of talent competition at both international and domestic levels. The aim is to enhance the core competitiveness of talents and their international.

Moreover, the study also identified several challenges faced by Shenzhen in attracting talent: (1) Competition for talent has intensified with the Yangtze River Delta Economic Zone, Bohai Economic Zone, other cities in the PRD, as well as regions and countries such as Hong Kong, Macao, and Singapore, leading to a

---

[92] Cao, R. X., & Yan, S. X. 曹榮新,嚴書翔(Eds.). (1993). *Shenzhen Talent Market* 深圳人才市場. Shenzhen: Hai Tian Press深圳:海天出版社. (pp. 3–4).

[93] Qiu, D. 邱大昭 (1990). On the Issue of Talent Fault in Professional and Technical Fields 試論專業技術人才斷層問題. *Chinese Talents*中國人才. (pp.6: 12–14).

[94] Nie, Z.G. 聶政光(1994). Research on Talent Fault in Shenzhen 深圳人才斷層研究. Hai Tian Press 海天出版社. (pp. 3–4).

weakening of Shenzhen's talent attraction efforts; (2) At that time, the proportion of international talents in Shenzhen was only 0.1% of the total population, much lower than Shanghai's 0.4% and Hong Kong's 7.1%. In addition, the number of high-tech talents did not keep up with the development of the industry, and there was a shortage of highly qualified talents familiar with international trade rules. Furthermore, the study also noted the negative impact of low-quality research, education, health, cultural facilities, and the urban environment on talent attraction, believing that these factors "directly affect the learning, communication, and life of talents."[95] However, at that time, promoting Shenzhen's urban environment was identified as one of the top ten strategies for talent attraction. Therefore, the *Urban Master Plan of Shenzhen Municipality (2010–2020)* explicitly stated, in the section on regional cooperation policies, the need to enhance the city's international level and build internationalized communities.

*Create a favorable business service environment, urban natural environment, and cultural environment to attract foreign nationals with international cultural backgrounds and returnees with overseas study experience to settle in Shenzhen and build internationalized communities. Emphasis is placed on the introduction and cultivation of professionals in trade, management, and technology who can grasp global industrial development trends and are familiar with international business rules, with the aim of enhancing international competitiveness.*[96]

In addition to its broader urban planning efforts, Shenzhen released a local standard known as the *Evaluation guidelines for international talent community* (國際人才街區評價指南), which was announced in November 2021. This is the country's first urban block planning guideline specifically formulated for urban talents.[97] The document includes a definition of "international talent": *Individuals with an international vision, possessing high levels of knowledge and professional expertise, including foreign nationals, residents of Hong Kong, Macao, and Taiwan, as well as domestic talents with overseas professional experience or a global perspective who have returned after studying abroad*. Additionally, it outlines the urban spatial types that should be equipped in an international talent district, such as urban parks, libraries, museums, recreational facilities, fitness centers, as well as requirements for cultural salons and venues for gatherings.[98]

---

[95] Shenzhen Talent Work Coordination Group, Organization Department of the CPC Shenzhen Municipal Committee深圳市人才工作協調小組, 中共深圳市委組織部(2005). *Compilation of Research Materials on Talent Work in Shenzhen*深圳市人才工作調研材料彙.(pp.139–145). Preserved at Shenzhen Library Local Archive Room.

[96] Shenzhen Municipal People's Government 深圳市人民政府 (2009). *General Urban Plan of Shenzhen City (2009–2020) Explanatory Booklet*深圳市城市總體規劃(2009–2020)說明書. December 2009, (pp. 71–72).

[97] Shenzhen Talent Work Website深圳人才工作網. (2021, November 15). *The first local standard " Evaluation Guidelines for International Talent Community " released in Shenzhen*全國首部《國際人才街區評價指南》地方標準在深發佈. Retrieved from http://www.sztalent.org/content/2021-11/15/content_24739626.htm

[98] Shenzhen Institute of Standards and Technology深圳市標準技術研究院. (2021, November 16). *"Yi Shenzhen" reports on the Municipal Standards Institute: Shenzhen releases the first national*

Shenzhen officials have been diligently focused on the progress of talent development in the city. After China's accession to the World Trade Organization, there was a clear shift in the demand for talent toward international professionals in high-end service industries and advanced manufacturing, which were urgently needed for the urban industrial transformation. In other words, the Shenzhen Municipal Government, on the one hand, pays attention to the relationship between industry and talent in the process of regional integration in the PRD and, on the other hand, defines the goals of talent work, namely, attracting and cultivating international talents. At the same time, such talent strategies are also reflected in the government's considerations for urban space and its quality.

### 6.4.2 Talent Park and Internationalized Elite Community

Shenzhen marked its first-ever Talent Day on November 1, 2017, which coincided with the enforcement of the *Shenzhen Special Economic Zone Talent Work Regulations* (深圳經濟特區人才工作條例). The recently constructed Shenzhen Talent Park was officially inaugurated on the same day. The government scheduling these three events on the same day holds special significance, which is reflected in the relationship between urban park construction, industrial development, and talent attraction. The letter from the Nanshan District Urban Management Bureau (南山區城管局) to the Shenzhen Urban Park Management Center regarding the construction application for the Talent Park encapsulates this unique significance.

*Nanshan District, as a hub for technology and talent, allocates 5.8% of its GDP to research and development, equivalent to the level of international innovative countries. 75% of individuals with a college degree or higher in the city work or reside in Nanshan. Innovation carriers such as national key laboratories constitute 58% of the city. The "Thousand Talents Plan (千人計畫)," "Peacock Plan(孔雀計畫)," and provincial innovation teams collectively make up over 70% of the city's total. National high-tech enterprises account for 29.7% of the entire city, and the patent ownership per ten thousand people ranks first among national districts and counties. Apart from the Shenzhen Bay Sports Center, our district is home to 15 super headquarters, including China Resources Group, Baidu, Aerospace International Center, CNOOC, Alibaba, and others…In view of the favorable conditions for talent and cultural learning exchange in Houhai Park, our bureau hereby applies to your center to designate Houhai Park as the Shenzhen Talent Park.*[99]

---

*Evaluation Guidelines for International Talent Community* 《壹深圳》報導市標準院:深圳發佈全國首部《國際人才街區評價指南》. Retrieved from https://www.sist.org.cn/xwzx/mtbd/202112/t20211206_2364350.html

[99] Nanshan District Urban Management Bureau南山區城市管理局. *Regarding the request to develop Houhai Park into Shenzhen Talent Park*關於申請將後海公園建設成為深圳市人才公園的請示.(March 1, 2016). Archived at Nanshan District Archives with File Number C10-WS-2016-D30–0126.

The primary objective of the establishment of the Talent Park is to create a recreational, exercise, communication, and visually appealing space for the international and high-tech talents who are employed or living in the Shenzhen Bay area, particularly in the Nanshan District. In fact, this is also the main reason why Nanshan District has become a "hotspot" for popular parks.[100] In the 1991 study organized by the Municipal Personnel Bureau on talent discontinuity, the negative impact of a purely profit-driven social atmosphere on the performance of talents in the special economic zone was mentioned.[101]

At that time, the government had already recognized the importance of establishing a talent honor system and fostering an atmosphere that respects talents. This was seen as crucial for attracting talents to the city and promoting industrial development. During that period, talent policies predominantly emphasized aspects such as household registration, finance, professional titles, and children's education. Over the course of two decades, the rivalry for skilled talents escalated among nations, regions, and cities; these preferential policies in these aspects still appeared somewhat inadequate. Using urban space as a talent incentive strategy was pioneered by Shenzhen. In March 2016, the city introduced the *Several Measures to Promote the Priority Development of Talents* (關於促進人才優先發展的若干措施). In this document, Shenzhen, which had gained fame for building the first theme park in China, proposed the planning and construction of the nation's first "talent" theme park. This park was envisioned as a place to showcase, serve, and motivate talents, aiming to establish and improve an honor system for talents.[102] The initial proposal for this theme came from the Shenzhen Municipal Party Committee Organizational Department, which holds the responsibility for the formulation of personnel and talent policies throughout the city. This organization also contributes to the evaluation of the planning and design of the Talent Park. During that period, the planning and design of the Talent Park were considered quite challenging: "Because no one had ever done a talent-themed park, we searched for a lot of information, but there was none anywhere in the world…".[103]

Prior to 2016, the Talent Park was formerly called Houhai Park. The Talent Park was designated when the construction project was halfway completed. At that time, the Shenzhen Municipal Party Committee Organizational Department was selecting a location to build an urban park with a talent theme across the entire city. There were two reasons why Houhai Park was included in the selection process[104]: (1)

---

[100] Shenzhen Nanshan District Government深圳市南山區政府(2021, February 18). Nanshan Becomes a Gathering Place for "Internet-Famous Parks," Living Up to Its Reputation as the "District of a Hundred Parks." 南山成"網紅公園"聚集地 "百園之區"名副其實. Retrieved from http://www.szns.gov.cn/xxgk/qzfxxgkml/bmdt/content/post_8555918.html

[101] Nie, Z. G. 聶政光 (1994). *Research on Talent Faultage in Shenzhen*深圳人才斷層研究. Hai Tian Press深圳:海天出版社. (pp.24).

[102] Du, Y. 杜䵍 (2018, July 12). *Ten Years of Construction: The Story of Shenzhen Talent Park*深圳人才公園十年煉成記. Retrieved from https://www.sohu.com/a/240668591_161794

[103] According to the interview with Mr. Zhu Jie, 2023-09-19.

[104] According to the interview with Mr. Zhu Jie, 2023-09-19.

Creating a completely new Talent Park would take a relatively long time. During that period, Houhai Park was already halfway through construction, requiring only the addition of some elements to express the talent theme; (2) Houhai Park had favorable location conditions to express the talent theme. Indeed, in terms of location, Houhai Park is situated along Shenzhen Bay. During that period, the vicinity of the location already accommodated a diverse range of multinational company headquarters, higher education institutions, research and development centers, talent residences, and other establishments. The selection of such a location where talented individuals gather, work, and reside accurately aligns with the expectations set for the Talent Park (Fig. 6.15).

First, in terms of overall style, the Talent Park is far removed from traditional Chinese gardens and instead bears a resemblance to the abstract gardens admired by the American modernist landscape theorist James Rose,[105] as shown in Fig. 6.16. The design process of the Talent Park was not carried out in isolation; indeed, it involved extensive study and learning from numerous international cases. Particularly, waterfront public spaces in cities like Toronto and New York were examined. Because the designers believed that the Talent Park needed to showcase "an image with a more international vision, a more modern and artistic, and ecological character."[106]

While the terms "international" and "modern" used by the designers are specifically related to the style and form of the design, on an analytical level, they are directly connected through the medium of modernist aesthetics to the political motivation behind the construction of the Talent Park. This motivation involves using symbolic environments to attract talent with an international vision to settle in Shenzhen and to develop an internationalized community for them. Additionally, the English explanations for many attractions contribute linguistically to the internationalization of the park.

Furthermore, compared to the political motivation conveyed through formal style of the urban park, the visual representation of the talent theme is significantly more straightforward and appealing to the general public. The Wall of the Talents (人才功勳牆), Starlight Bridge (星光橋), Chinese Character Corridor (漢字長廊), Formula Corridor (公式長廊), Academician Grove (院士林), and Hundred Elite Hill (百傑山) all exemplify Shenzhen's emphasis to knowledge and talents, as well as its promotion of innovative spirit through visual and thematic elements. In selecting the types of talents showcased:

*…At the beginning, when setting the basic principle, everyone unanimously agreed that the selected talents might not necessarily be those high up in the hierarchy. Shenzhen is an immigrant city, and its success relies on the efforts of people from all walks of life. It includes high-end professionals as well as ordinary citizens. This selection must be broad in scope…*

[105] Rose, J. (1938). Freedom in the Garden-A Contemporary Approach. *Pencil Points*, 19, 639–644.

[106] According to the interview with Mr. Zhu Jie, 2023-09-19.

**Fig. 6.15** Park attractions depicting the theme of talent

This perception exemplifies Shenzhen's capacity for embracing diversity, while also hinting at its socialist inclination. However, it is evident that only a limited number of attractions in the park's northern section, such as the Wall of the Talents, genuinely embody talents "from all walks of life." Collectives such as migrant worker girl, engineering soldiers, who have made significant contributions to the growth of Shenzhen, are merely abstractly named by collectives. The overall representation in the Talent Park still focuses on relatively "high-end" talents, with their individual names being specifically and accurately expressed. On Hundred Elite Hill (百傑山), sculptures of renowned scientists from ancient to modern times, such

**Fig. 6.16** Abstract composition of forms in the Talent Park

as Cai Lun (蔡倫), Zu Chongzhi (祖沖之), Nicolaus Copernicus, and Dmitri Ivanovich Mendeleev, can be found. The Academician Grove consists of trees planted by contemporary well-known academicians from various fields worldwide. The lamp posts on Starlight Bridge (星光橋) bear the names and portraits of 65 contemporary renowned scientists from both domestic and international backgrounds. On the lawn to the northwest, sculptures of two pioneering figures in Shenzhen, Yuan Geng (袁庚) and Niu Hanben (牛憨笨), stand tall. Such a representation of the talent theme, though selective, is inevitable because only in this way can the Talent Park become a part of Shenzhen's establishment of a talent honor system, echoing its structural demand for high-level talents. Within this demand, the vast majority of migrant workers who had contributed significantly to Shenzhen's urban development do not stand at the forefront. Instead, they become what architectural scholar Zhu Jianfei (朱劍飛) calls "the Anonymous."[107]

Furthermore, the facilities and activities organized within the Talent Park align with corresponding motivations. To assist urban elites in balancing their work and fitness needs, the park has meticulously constructed a high-standard, continuous 2.7-km circular running track. In addition, the park has cleverly implemented shower systems that adhere to Japanese norms, recognizing the fact that: "Many people come here for a run in the morning. After finishing, they need to meet friends or go to work. At this point, if they go home to take shower and change clothes, the time cost is too high."[108] Such design arrangements reflect the lifestyle and work culture of the international elite along the Shenzhen Bay. In addition to the shower systems, the inclusion of a library for reading and learning within the Talent Park

[107] Zhu, J. F. 朱劍飛 (2021). *What is "the Everyday "? Reading across Multiple Spheres for an Architectural Agenda*什麼是"日常生活"? 建築學的遠望與策劃. Time + Architecture時代建築, 2021(05), 14–17.

[108] Wu, Y.Y. 武媛媛 (2017, November 16). *How Was Shenzhen's First Talent-themed Park Created? Chief Designer Explains...* 深圳首個人才主題公園是怎麼煉成的?總設計師說.... Retrieved from https://ilonggang.sznews.com/lgnews/content/2017-11/16/content_17764288.htm

was also unique in Shenzhen at that time. The park's southwest features two buildings, the Pavilion of Talent and the Assembly of Heroes. These structures have been used for various events such as returnees exchange conferences, international cooperation forums, academic lectures, popular science talks, and the World Barista Championship. The Assembly of Heroes was even selected as a 2023 Scientist Spirit Education Base (科学家精神教育基地), marking Shenzhen's transition, under official guidance, from a city that once emphasized entrepreneurial spirit to one gradually embracing the spirit of scientists.

It is evident that Shenzhen Talent Park remains a highly politicized park. However, this political nature does not originate from the nation-state power structure as seen in modern parks such as Zhongshan Park (中山公園). Instead, it is rooted in the interplay between globalization, industry, and the economy, along with the consequential demand for talents. The nation-state elements has progressively receded from the symbolic center.

### 6.4.3 The Nature and Aesthetic Aspects of Gentrification

The Talent Park is the most straightforward and evident example among the Shenzhen Bay Park Group that links Shenzhen's structural demand for international talents and urban park construction. However, concentrating exclusively on the Talent Park may be excessively limited in perspective. While other parks may not be as straightforward and populist as the Talent Park, they are equally strong in establishing an environmental atmosphere for the construction of an internationalized elite community along the Shenzhen Bay.

*Shenzhen Bay Park is an adjective that describes the lands surrounding it. It represents a lifestyle, such as walking, running, birdwatching, and is facilitated by sports fields, various businesses and sports facilities, and cultural amenities. This lifestyle, where you can enjoy these resources within a twenty-minute timeframe, is not very common worldwide….*[109]

As an innovative node in the global city network, along the shores of Shenzhen Bay, the growth of high-income individuals has gradually become a significant consumer force. Their tastes and demands have evolved into the cultural center, not only driving the development of consumption but also giving rise to new environmental aesthetics and lifestyles. This aesthetic and lifestyle are also evident in the Happy Coast on the north side of the Binhai Avenue. It was completed and opened in August 2011, and though not a park in the conventional sense, it similarly features the sea as the main landscape and possesses a considerable degree of public space. The key difference lies in the Happy Coast's dedication to creating a one-stop experience for culture, leisure, tourism, business, commerce, and real estate, rather than being a general urban park. It encapsulates the potential for consumption and

[109] According to the interview with Mr. Zhu Rongyuan, 2023-09-14.

reproduction through the environmental identification of the middle-classification of high-income individuals. In this environmental identity, the expansive water views and maritime areas play a central role.,[110][111]

The adjustment of the urban consumption structure along the Shenzhen Bay has resulted in the production of large-scale high-end commercial and luxury residential areas. Since the opening of Happy Coast, a diverse array of beachfront eateries, upscale lodgings, coffee shops, culinary thoroughfares, cinemas, and international grocery stores have progressively established their presence. The high-end residential area, Happy Coast·Blue Bay (歡樂海岸·藍楹灣), in the western part has also taken shape. Adjacent to Happy Coast, extending along Shenzhen Bay, lie several upscale commercial complexes including Kingkey Baina Plaza (京基百納廣場), Shenzhen Bay MixC (深圳灣萬象城), Bao'an Taigucheng (寶能太古城), Coastal City Shopping Center (海岸城購物中心), Shekou Sea World (蛇口海上世界), and numerous opulent residential districts. This gradually constructs an environment and aesthetic atmosphere of middle-class gentrification along the Shenzhen Bay, catering to a high-income demographic. In this aesthetic, the Shenzhen Bay area becomes a core element, reflected in the planning and design of numerous high-end residences in the Shenzhen Bay area, exemplified by CITIC Mangrove Bay (中信紅樹灣, Fig. 6.17) and Shenzhen Bay No. 1(深圳灣1號, Fig. 6.18):

*CITIC Mangrove Bay...is arranged along the southeast and southwest directions to maximize the enjoyment of the landscape... all buildings are laid out in a zigzag form or "L" shape to ensure that every unit faces south, providing scenic views for each room... to guarantee that every residence is 100% south-facing, a design with a 41-degree angle orientation was adopted... ensuring that each unit enjoys the most ideal landscape. Regardless of which unit one stands in within CITIC Mangrove Bay, the south side offers an unobstructed view of the vast sea. The starting point of the sea is a 15-kilometer-long waterfront promenade under construction, embodying the quality of life in Shenzhen....*[112]

*The design principle of the Apartment Tower of Shenzhen Bay No. 1 is to ensure that all units have the maximum possible view of the Shenzhen Bay landscape, overlooking the ocean and the entire city. In the overall layout, the first row of towers adjacent to the bay is lower than the towers in the rear, ensuring unobstructed water views for all towers behind....*[113]

In the discourse on built environment, design agencies and real estate sales organizations maintain a high degree of consistency. The Shenzhen Bay area, including

---

[110] Thompson, D., Zhang, S., Dalton, H., et al. 大衛·湯普森,張樹儀,亨利·道爾頓等 (2013). Shenzhen OCT Bay 深圳華僑城歡樂海岸. *Landscape Architecture*風景園林.2013 (03), 68–76.

[111] Shenzhen Happy Coast, Shenzhen, China 深圳歡樂海岸中國深圳. (2020). *World Architecture Review* 世界建築導報, 35(02): 74–79.

[112] Song, Y. H., & Liu, L. J. 宋穎慧,劉麗君(2005). Zhongxin Hongshuwan: Precision Ensures Victory中信紅樹灣 精准奠定勝局. *Chinese and Foreign Real Estate Times*中外房地產導報. 2005(09),32–37.

[113] Kohn Pedersen Fox Associates KPF建築設計事務所(2021). Shenzhen Bay No. 1深圳灣1號. *Architectural Practice*建築實踐, (03):176–183.

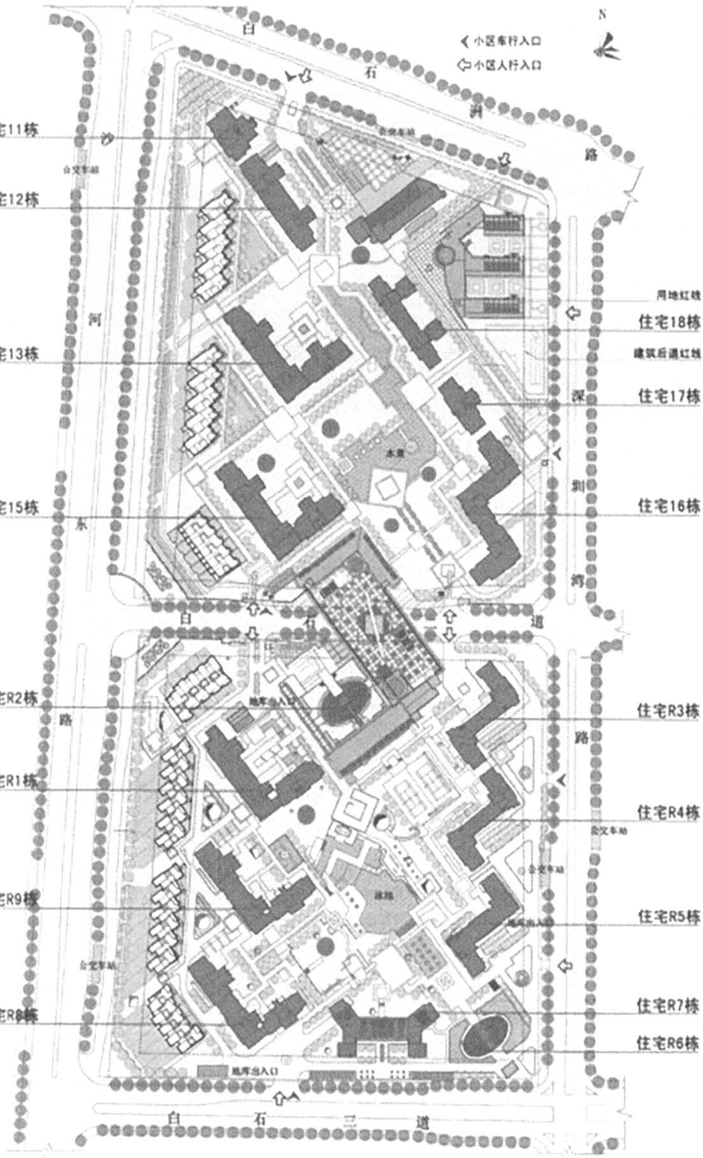


**Fig. 6.17** Site plan of CITIC Mangrove Bay residential area along Shenzhen Bay [Australia PT Design Consultants Limited & Chaoli Sifang Architectural Design Firm, Shenzhen澳大利亞柏濤設計諮詢有限公司, 深圳市朝立四方建築設計事務所(2012). *Architectural Detail Design: Shenzhen Zhongxin Hongshuwan Volume*建築細部設計 深圳中信紅樹灣篇. Dalian: Dalian University of Technology Press大連: 大連理工大學出版社. (pp. 211).]

**Fig. 6.18** Living room of T4-F apartment in Shenzhen Bay No 1, with floor-to-ceiling windows showcasing Talent Park and the waters of Shenzhen Bay [Xu, H. 徐航(Eds.) (2020). *Shenzhen Bay No. 1: Standing Tall with Belief*深圳灣1號 因信念而屹立. Shenzhen: Haitian Publishing House 深圳: 海天出版社. (pp. 150).]

the 15-km coastal leisure belt,[114] serves as both the starting point for designers' creations and a crucial narrative in the marketing strategies of real estate sales organizations.[115] Whether consciously or unconsciously, they become involved and construct physical spaces and marketing narratives based on this. Nevertheless, for them, the Shenzhen Bay is an irreplaceable landscape resource. The sea area of Shenzhen Bay area is imbued with a middle-class aesthetic, and behind this aesthetic lies the inherent land value.

*In the design of Shenzhen Bay Park, one issue we discussed at the time was its relationship with the city. Specifically, how to enhance the value of the surrounding land through the park. The goal was to give these lands higher value compared to further land reclamation… As there is no longer ongoing land reclamation, the land surrounding Shenzhen Bay Park has become the most precious. Consequently, the value of these lands has appreciated. In the conditions of a market economy, these lands will develop in ways that maximize their value. Therefore, those who can afford to purchase this land will undoubtedly engage in developing luxury homes or*

[114]AUBE Architectural and Urban Planning Design Company of France, AUBE Design Co., Ltd. In Shenzhen法國歐博建築與城市規劃設計公司,深圳市歐博設計有限公司(2010). Shenzhen Peninsula City State深圳半島城邦. *Architecture Technique*建築技藝. (02):62–71.

[115]China Resources Land (Shenzhen) Development Co., Ltd華潤置地(深圳)發展有限公司 (2017). China Resources Shenzhen Bay·Yuefu 華潤深圳灣·悅府. *China Real Estate* 中國房地產. 2017(26):79.

*apartments. The development of high-end residences is closely linked to the original value of the land....*[116]

Conversely, when local villagers, who have long resided in the area, describe the Shenzhen Bay sea area, the true significance for them lies in "grain" and "livelihood." There is no aesthetic expression of middle-class values, everything is simply about daily life.

*"During that time, reclaiming land from the sea and obtaining grain from the sea was truly challenging. We, along with the villagers, all dressed in undershirts, rolled up our pants, and poured mud into the sea using basins. The coastal area was all muddy... and we had to pile it up, basin by basin, waiting for it to dry before it could be used. It depended on the weather—work could only be done during low tide; it was impossible during high tide." Chen Xuliang said....*[117]

*"Previously, this entire area used to be farmland for the villagers of Nanguang Village (南光村)," Mai Huandi pointed to the location of the "Sea-Print Great Wall (海印長城)" on the map and emphasized, "It was the most suitable place for farming. With the 'Sea-Print Great Wall' as the boundary, one side was farmland, and the other side was the sea. Although occasionally seawater flooded into the farmland, Mai Huandi was never worried about her livelihood...".*[118]

Evidently, globalization encompasses not only political and economic aspects but also pertains to matters of aesthetics. This aesthetic functions as an intermediate, regulating the creation of physical space within the metropolis. Mike Featherstone categorizes this aesthetic as postmodern culture.[119] In his view, a globalized new investment model will reconstruct local social-space, and waterfront docks and other urban fringe areas will transition from industrial zones to new settlements for the emerging middle class. Simultaneously, they will become places for tourism and cultural consumption. The Shenzhen Bay and its coastal area, which were virtually inaccessible and unknown to citizens before 2000, have transformed into a developmental nature that attracts high-end industry talent and constructs an international elite community. This developmental nature is, in essence, a middle-class nature, with the key factor being humans rather than pure, objective nature—this is what Neil Smith considers the primacy of society over nature.[120] The symbolic environment created using Shenzhen Bay as a scenic resource also represents new cultural capital and symbolic experiences. They are adorned with a sacred cloak, borrowed

---

[116] According to the interview with Mr. Zhu Rongyuan, 2023-09-14.

[117] Gu, X. J.辜曉進 (Eds.) (2018). *In the Name of News III: Selected Works of News Graduation Design from Shenzhen University - Frontier Observations* 新聞的名義III:深圳大學新聞畢業設計作品選·前沿觀察. Beijing: Renmin University of China Press 北京: 中國人民大學出版社. (pp. 50).

[118] Gu, X. J. 辜曉進(Eds.) (2018). *In the Name of News III: Selected Works of News Graduation Design from Shenzhen University - Frontier Observations* 新聞的名義III:深圳大學新聞畢業設計作品選·前沿觀察. Beijing: Renmin University of China Press 北京:中國人民大學出版社. (pp. 53).

[119] Feather S. (1990). *Consumer Culture and Postmodernism*. Sage Publications. (pp. 228–230).

[120] Smith, N. (2008). *Uneven development nature, capital, and the production of space (third ed.)*. Athens: University of Georgia Press. (pp.45–91).

by design agencies, real estate companies, and even local governments. Shenzhen Bay Park, Talent Park, and even Happy Coast are public spaces where anyone and everyone can enter. However, distinctions do exist. Pierre Bourdieu[121] pointed out that aesthetic tastes contribute to the differentiation of people based on different economic statuses and also serve to mask inequalities. For the local villagers and oyster farmers who have left their homeland along the Shenzhen Bay coastline, this involves the complex and subtle local politics within the process of globalization. Additionally, scholars such as Manuel Castells,[122] Saskia Sassen,[123] Aljun Appadurai,[124] and others have observed, from various perspectives and to different extents, that a sense of belonging and identity rooted in nation-state is significantly impacted within this globalization context. This is also why urban parks with national forms, such as those found in the Shangbu-Luohu area, will not spring up along the Shenzhen Bay coastline. Here, globalization, the talents it demands, and the environmental aesthetics that go along with it are the main points of emphasis.

## 6.5 Summary

For centuries prior to the reform and opening-up period, the ancestors who inhabited the coast of Shenzhen Bay engaged in fishing, oyster farming, fruit cultivation, and agriculture on this coastal plain, achieving their livelihoods with relatively minimal interference to the natural environment. The blasting of Shekou after the reform and opening up marked the beginning of a large-scale topographic transformation along the coast of Shenzhen Bay. Simultaneously, the human–land relationship adhered to the topography underwent changes. From the period of "Three-Processing and One Compensation" to the period of regional integration, various urban landscapes such as Shekou Industrial Zone, Nantou Development Zone, Overseas Chinese Town, High-tech Park, Headquarters Base, and Western Corridor emerged along the coast of Shenzhen Bay, each representing the transformation of this human–land relationship at different times. As the most naturally characteristic urban landscape type along the coast of Shenzhen Bay, the Shenzhen Bay Park Group, established after the turn of the new millennium, vividly embodies this transformation. In 1980, for the four parks that constitute this park group, there was hardly any land, but after decades of land reclamation and urban development, by 2017, the Shenzhen Bay Park Group, stretching nearly 20 km, had been able to

---

[121] Bourdieu P. (1984). *Distinction: A Social Critique of the Judgment of Taste*. Massachusetts: Havard University Press. (pp.11–17).

[122] Castells, M. (2010). *The power of identity Volume II (second ed., with a new preface.)*. Malden, MA: Wiley-Blackwell. (pp.30–56).

[123] Sassen, S. (2018). *Cities in a world economy*. Sage Publications. (pp. 228–230).

[124] Appadurai, A. (1996). *Modernity at large: Cultural dimensions of globalization (Vol. 1)*. Univ of Minnesota Press. (pp. 158–177).

complement other urban landscapes built on reclaimed land, turning Nanshan District into a "hot spot" for popular parks.

The saying goes, "for every meter the reclamation advances, the sentinels advance one step." Along with land reclamation activities, there have been changes in the Guangdong–Hong Kong border defense line. Its formation can be traced back to the early Cold War period when the newly established People's Republic of China and the British Hong Kong government belonged to different camps. Consequently, the Shenzhen Bay-Shenzhen River-Dapeng Bay area was constructed as a political border with ideological opposition. The maritime territory of Shenzhen Bay, as a natural element, became part of the border. However, the border is not only a boundary of national sovereignty but also extends to the level of social practice, governing the joys and sorrows of the people in the Bao'an County and its surrounding area. Several large-scale waves of fleeing to Hong Kong during the Cold War were related to Shenzhen Bay, where many fleeing individuals drowned. The reason why Shenzhen Bay became a major route for fleeing to Hong Kong is closely related to the conditions provided by the natural environment. It stirred up political opposition between mainland China and British Hong Kong during the Cold War, becoming a kind of isulative nature. In the later stages of the Cold War, the antagonistic ideological opposition gradually eased, and "war and revolution" gradually transformed into "peace and development." The influx of Hong Kong capital to the SSEZ starting in the 1980s also marked a shift in the relationship between Guangdong and Hong Kong from geopolitics toward geoeconomics. Around the year 2000, the border defense lines of both regions gradually transformed from a harsh military image to a softer image closer to the citizens, and Shenzhen Bay also changed from being a heavily guarded border area to urban parks where citizens could see the sea up close. The spatial features of Shenzhen Bay Park, which emerged in this process, are also closely related to its historical context of the Cold War frontier.

In the late Cold War period, globalization tightened relationships between nations and regions, while the global rise of network society gave birth to metropolitan areas, a new spatial form. For the emerging PRD metropolitan area, transforming Shenzhen and Guangzhou into central cities with independent innovation capabilities through the "headquarters + base" model is essential for restructuring the regional industrial structure and enhancing its global competitiveness. For Shenzhen, enhancing independent innovation capabilities necessitates the attraction of high-end industry talent. Despite its lofty aspirations, Shenzhen has a weak talent base, prompting the government to clearly recognize and emphasize the relationship between urban industrial upgrading and talent development. Whether advocating for the construction of internationalized communities in 2010 or preparing for the establishment of international talent communities in 2021, Shenzhen's emphasis on talent development is evident in its consideration of urban space, serving as a crucial aspect of leading the industrial upgrading in the PRD and implementing regional cooperation policies. In this context, the Talent Park along the shores of Shenzhen Bay was opened in 2017. As a symbolic representation of Shenzhen's talent honor system, the planning and design of the Talent Park structurally reflect its political motivations: (1) adopting the abstract, international modernist garden style; (2)

emphasizing "named" individuals while downplaying the significance of "the Anonymous"; and (3) catering to the lifestyle and work patterns of the surrounding urban elites. As an innovation node in the global city network, the growth of high-income individuals along the shores of Shenzhen Bay has given rise to new consumption patterns and environmental aesthetics. High-end commercial complexes represented by Happy Coast and Shekou Sea World, as well as high-end residential areas represented by Shenzhen Bay No.1 and CITIC Mangrove Bay, all reflect these consumption patterns and environmental aesthetics. Moreover, whether high-end commercial or residential, they construct physical spaces and marketing discourses centered around the Shenzhen Bay and its surrounding park group. As Gandy understood, urban parks have become new commodities transformed from nature under the interaction of real estate and landscape aesthetics.[125] As a natural entity of a specific cultural form, the Shenzhen Bay is not only a developmental nature that attracts international talents and enhances milieux of innovation but also a gentrified nature that promotes capital reproduction and creates an image of middle-class environment. Nonetheless, this contrasts significantly with the descriptions of the Shenzhen Bay by the local villagers who long inhabited the area.

The coastal topography from Shenzhen Bay to the Pearl River Estuary, like the Shangbu-Luohu plain and the Meilin-Gangxia hilly areas mentioned previously, is also a typical topography within the SSEZ. Around 2010, this area concluded approximately 20 years of large-scale land reclamation, providing new space for Shenzhen to lead the PRD metropolitan area in industrial autonomous innovation while also changing the coastal human-land relationship. Consequently, on the shoreline of Shenzhen Bay, a new human–land relationship has formed between the international urban elites who work in headquarters bases, live in high-end residences, consume in upscale commercial complexes, and spend their leisure time in the Shenzhen Bay Park Group and the waters of Shenzhen Bay. The coastal terrain here has been constructed into an environment and spatial imagery that accommodate this relationship.

Consequently, along the shores of Shenzhen Bay, a new man–land relationship has formed among the international urban elites and the maritime area of Shenzhen Bay. These elites typically work at headquarters bases, live in high-end residences, shop in upscale commercial complexes, and spend their leisure time in the Shenzhen Bay Park Group. The coastal topographic features of this area have been developed into an environment and spatial imagery that accommodates and expresses this relationship, creating a setting that is in harmony with the lifestyles and preferences of these urban elites.

[125] Gandy, M. (2003). *Concrete and clay: reworking nature in New York City*. MIT Press. (pp.5).

# Chapter 7
# Conclusion

## Contents



**Abstract** As the concluding chapter, this chapter engages in theoretical discussion based on the empirical studies presented in the preceding chapters and distills the main conclusions: (1) Urban parks represent a cultural form of both space and nature; (2) There exists a bidirectional shaping between cities and urban parks; (3) Topography can integrate cities and urban parks on a conceptual level.

**Keywords** Theory · Nature · Cultural form · Topography



X. Liu, *Fabricating Nature*, The Urban Book Series,
https://doi.org/10.1007/978-981-95-8208-2_7

## 7.1 Discussions

### *7.1.1 The Role of the State and Government Behind the Production of Nature*

We can observe that specific social and historical contexts intertwine with the planning, design, construction, and management of urban parks through the case studies presented above. In the built environment, these case studies respond to viewpoints that objectify natural elements within cities. Moreover, they to some extent reintegrate the relationship between humans and nature, subject and object, and society and nature into a unified whole.

Theoretical responses should also consider the specificity contributed by empirical cases. In the late 1970s, Shenzhen became an experimental ground for China's reform and opening-up policy, gradually leading nationwide reforms across various sectors and levels toward a market economy. However, the driving force behind these reforms was not the free market but rather the state and government, distinguishing China's reform and opening-up from Western neoliberalism. The significance of studying Shenzhen lies in its close association with neoliberal globalization, while being distinct due to China's specific institutional environment.

Studying the production of nature through urban parks should also pay attention to this important distinction. In the case of the Luo Hu Park Group, the government employs ethnic-themed parks as explicit national symbols, directly organizes festivals, and engages in city marketing while also improving urban security through the park. In the Meilin-Gangxia Park Group, the government initiated a "Rectification Action" to improve the appearance of the city, colorize urban greening and the mountain vegetation. And the Shenzhen Bay Park Group has transformed the former Cold War border into an urban park, creating an elite environment that fosters independent innovation and industrial upgrading. Government actions mediate all these production processes of nature. In these three urban park groups, capital is indeed present. Whether it's the influx of Hong Kong capital to Shenzhen in the 1980s, the establishment of an international city in the 1990s, or the integration of the PRD region in recent decades, capital has been a significant force behind the production of nature. However, the empirical cases continue to demonstrate that the state and government, through permission and promotion, realize the capital demands. This is because, in China's economic and urban development processes, the government not only influences the distribution of the "cake" but also participates in its production. The government is deeply involved and cannot be separated from economic and city discussions. This is in stark contrast to classical economic models that advocate for minimizing government intervention in the market. The construction and development processes of the three urban park groups, as expressions of the production of nature, closely align with industrial development and capital accumulation. Numerous urban policies and government actions, common across the three cases, influence the demands for industrial development and capital accumulation.

The process of modernization in China differs from the fictitious notion of the laissez-faire state in the West; it relies on a strong state capacity as a necessary condition for development.[1] This has not changed even in the market-oriented SSEZ. Simultaneously, for developing countries in the Third World, the state, not capital, is the center of organizing society, space, and nature. Understanding the state and subsequently the actions of local governments mediated by state power should be one of the starting points for theoretical reflection in China's built environment field.[2]

### *7.1.2 The Complexity of Producing Nature in the Contemporary Urban Context*

In addition to their commonalities, the three cases also exhibit their respective uniqueness, which contributes to the complexity of nature production in the contemporary urban context. We can summarize this uniqueness in two dimensions: (1) variation over time and (2) variation based on topography.

In the 1980s, the construction of the Shangbu-Luohu Park Group emphasized national styles, representing an external manifestation of urban park construction as a form of production of nature which is deeply mediated by the state power. In terms of functionality, Lychee Park began to serve as a pivot for urban marketing and attracting foreign investment in the late 1980s, while Honghu Park also started to function as a propaganda medium to enhance urban security and resist the erosion of capitalist culture during the same period. This reflects the production of nature that was evident in the SSEZ in the 1980s, and indeed throughout China. It aligned with the initial stages of globalization, emphasizing its identity as an independent nation-state while simultaneously attempting to reap benefits and avoid harm in the process of globalization. When constructing the Meilin-Gangxia Park Group in the 1990s, emphasizing the identity of the nation-state was no longer a priority. Instead, establishing an "international city" to further integrate Shenzhen into the global market became urgent. Hence, the government conducted a "Rectification Action" to clear out shanties, tin shacks, and temporary residents from these parks, reshaping the originally sheltering nature into an aesthetic nature to enhance the image of Futian as the new city center. Moreover, the construction of "garden-style international city" and "ecological scenic forest" in the late 1990s also reshaped the aesthetic image of urban greening and mountainous areas. The primary motivation behind all these actions was to attract multinational corporations to settle in and further integrate Shenzhen into the global market. After 2000, the trend of regional

[1] Huang C. C. P.黄宗智(2022). *The Dualistic Unity of State and Society in China*國家與社會的二元合一. 桂林:廣西師範大學出版社Guilin: Guangxi Normal University Press. (pp.12;30–32).

[2] Hsia C.J.夏鑄九(2015). *A Glimpse of Devil Appearance: Planning and Design in Urban and Regional Processes*窺見魔鬼的容顏. Taipei: Tonsan Publications臺北:唐山出版社. (pp. 1–5).

integration in the PRD region intensified, representing a new spatial form emerging in the age of network information and an effort to compete for dominance and centrality in the global city network among cities in the PRD metropolitan area. In pursuit of this objective, Shenzhen transformed the former Cold War border into urban parks, enhancing its image as a waterfront city and attracting international talents to foster independent innovation. The "variation over time" in the three cases signifies that Shenzhen's specific developmental periods embed the process of production of nature within their historical and social contexts.

Plains, hilly areas, and reclaimed land serve as the foundation for the three cases, leading to specific and diverse ways in which "nature" intersects with urban social life. The plain topography in the Shangbu-Luohu Park Group allows for the construction of large-scale water systems intertwined with artificial landform, reflecting the aesthetic and design principles of "national forms" and expressing the power of the nation-state. The gentle slopes on the plains are suitable for planting lychee trees due to their water and fertilizer requirements; these lychee trees were deeply integrated into the social life of the SSEZ in the late 1980s. The transformation of the city–water relationship embodied in Honghu Park was also closely related to its location on the plain, forcing Honghu Park in its early establishment years to look up to the high ground of Buji Town, which is both a geographical and economic highland. Because of its hilly topography, the Meilin-Gangxia Park Group becomes the site for temporary settlement sites and self-formed habitations. The former allows the government to balance population resettlement and saving land resources, while the latter provides basic livelihoods and shelter for temporary residents. Furthermore, these mountainous hill parks, towering over the urban environment, serve as a medium for showcasing the city's visual image. In an international city, planting colorful tree species transforms them into natural spectacle and visual landmarks. The Shenzhen Bay Park Group is gradually reclaiming the sea area to increase urban construction land. Simultaneously, it transforms from a Cold War frontier separating two camps into a people-friendly urban park group. In this process, the meaning of the "sea area" to the city changes from "oyster fields" and "livelihoods" to a "seascape" that creates an elite environmental image, shaping the urban form of the Shenzhen Bay area and strengthening the "milieux of innovation" of the Shenzhen Bay area. These are each case's "variation based on topography." These are the unique characteristics that arise from "variation based on topography." Nature, or more specifically topography, is not merely an object to be reshaped at human will; it exerts its own agency in various forms, influencing urban social life by either supporting or constraining it. The significance of this study also lies in bringing nature into history through specific forms of writing, giving voice to their agency.

In the intersection and intermingling of such variations, it becomes evident that the distinction between second nature (agricultural, utilitarian nature) and third nature (garden, aesthetic nature) is inadequate for understanding the production of nature. For instance, during the Lychee Festival, is the lychee forest in Lychee Park an example of agricultural nature or garden nature? What about the construction of "ecological scenic forests" aimed at promoting foreign trade, finance, and real estate development? Are the Shenzhen Bay area and its park group, which aim to draw in high-end industry talent and foster urban consumption scenes, examples of utilitarian or aesthetic nature? This classification, which is based on the study of traditional gardens, largely fails to address the complexity of the production of nature in the contemporary urban context. This inadequacy reflects the need to reconsider our framework as we shift our focus from gardens to landscape architecture. Consequently, this study adopts "topography" rather than "nature" as its analytical entry point. Topography studies do not emphasize the hierarchical classification of nature, but rather acknowledge the inherent complexity and ambiguity in its production. They recognize that humans continuously "writing" on the land to reshape "place," rather than merely evolving from second nature to third nature.

## 7.2 Main Conclusions and Inspirations

### 7.2.1 *Urban Parks Represent a Cultural Form of Both Space and Nature*

Typically, urban parks are perceived as manifestations of nature within cities, known as urban nature. Due to their lush verdancy, they are regarded as permanent existing of romantic beauty, secluded from the urban clamor. However, research into the topography prior to the construction of urban parks reveals several insights: firstly, their existence is not a natural occurrence but the result of dual transformations in ownership and spatial form; secondly, these transformations in ownership and spatial form are intricately intertwined with changes at the urban, national, and even global levels; thirdly, the construction and management of these parks are clearly dominated by the structural macro objectives of local governments, exhibiting a pronounced political nature; fourthly, the dominance from above and the everyday life from below often engage in conflicts and compromises around the construction of the meaning of nature.

It is evident that urban parks are not merely natural spaces physically located within cities, but rather they are urbanized nature constructed through specific social and historical contexts. The production of space is a logical outcome of the production of nature. From this relational perspective, urban parks serve as a cultural form of space akin to streets, residential areas, and public buildings; simultaneously, they represent a cultural form of nature, akin to rivers, farmlands, and traditional gardens, embodying specific meanings and reflecting the dynamic social and historical contexts.

Historical research holds significance independent of practical planning and design, but within the built environment field, theoretical insights usually need to have practical implications. So, what is the practical significance of understanding urban parks as cultural forms of space and nature? Firstly, it tells us that planning and design practices for constructing space and nature should not stem from Platonic ideals of form but from the vibrant reality of social life; Secondly, this social life exists within a web of complex and subtle conflicts, competitions, and compromises, necessitating keen observation and awareness in planning and design practices. This also pertains to the social stance and role of planners and designers. Lastly, recognizing this helps us understand that the verdant nature in our world is not a spontaneous occurrence but a result of convoluted social processes.

### 7.2.2 *There Exists a Bidirectional Shaping Between Cities and Urban Parks*

In the generalized discourse, the development process of cities shapes the spatial form, functional layout, and attractions of urban parks; this expresses a unidirectional shaping of parks by the city. Conversely, the influence of parks on cities is primarily manifested in providing recreational spaces, beautifying urban environments, and promoting traditional culture. Such discourses are only a step away from deeply engaging with the political and economic contexts of cities, and it is this gap that diminishes the potential for planners and designers to gain a voice in practice. The generalized discourse structure lacks detailed discussion on how urban parks construct cities. The reason for this is that, although this structure recognizes that urban parks carry bottom-up demands, it fails to acknowledge a frequently obscured discourse: they also carry top-down governance practices.

In empirical case studies, whether it's the national gate image shaped by national-style parks in Shangbu-Luohu area, the "international city" identity fostered by parks planting colorful tree species in Meilin-Gangxia area, the international elite communities formed around modernism compositions and middle-class urban imagery in Shenzhen Bay area, and so forth, all demonstrate the agency of urban parks in facilitating the realization of top-down urban ambitions and their corresponding political and economic demands. Moreover, the case of Honghu Park indeed shows that nature was negatively affected by urban development at specific times, yet the construction of ecological scenic forests since the start of the new millennium suggests that nature has become more "natural" due to urban development. Thus, from an analytical perspective, nature in the process of urban development is not necessarily the victim; this depends on the specific practices and contexts, and one should not rigidly adhere to the binary opposition of city versus nature. However, it is undeniable that only by delving into the historical and social contexts of urban park construction can one comprehend the bidirectional shaping between cities and parks.

Understanding the bidirectional shaping between city and urban parks also provides relevant practical insights. First, this understanding helps decision-makers recognize the crucial role of urban parks in promoting the city's structural and macro objectives. Urban parks are not economically and politically valueless remnants in contemporary urban spaces; they require careful consideration; Second, it helps planners and designers recognize nature's agency. Urban nature can align with the economic and political logic of the city by appropriately modifying, editing, and guiding it through relevant technical means. This approach allows design to construct meaning for nature rather than having design serve only its own ends.

### 7.2.3 *Topography Can Integrate Cities and Urban Parks on a Conceptual Level*

The construction of buildings, parks, and even cities occur upon their respective physical topography, a fact that is both commonplace and unquestionable. However, if the understanding of terrain transcends its physical significance to encompass life practices, then topography has the potential to serve as a framework. This framework can conceptually integrate buildings, parks, and cities on a discursive and conceptual level, articulating their relationships with life practices. Thus, they become a form of "topographical arts."[3] In this sense, it is less about topography integrating these elements than it is about life practices doing so, with topography acting as a point of entry or an intermediary.

Whether from the phenomenological perspective of emotions, memories, myths, and bodily experiences, or from the political-economic perspective of agriculture, industry, urbanization, and globalization, all can be integrated into a discourse framework related to topography. The key lies in human subjectivity and life practices, rather than merely the physical significance of topography. Considering this understanding, this research tends toward a political-economic perspective, utilizing topography as a lens to conceptually integrate the city and urban parks. Whether it's "Nationalism-oriented Nature," "Sheltering Nature," or "Insulative nature," the seven empirical cases in this study, by utilizing topography to investigate the transitions of socio-natural relationships, demonstrate that topography can facilitate the integration of the city and urban parks, as well as the city and nature, at the conceptual level.

Based on this conceptual integration, the possibility of integration at the level of planning and design practice emerges. First, in generalized discussions, the technical and material aspects such as functional facilities, attractions, and transportation organization primarily reflect the integration of cities and urban parks. Topography, as a theoretical concept, focuses not only on the material level but also on the aspects of culture and life, or, more accurately, the fusion of both. Topography has the potential to integrate narrow perspectives fragmented across different disciplines in planning and design, much like its role in unifying landscape history and the history of landscape architecture in historical research. Second, the concept of topography's expansiveness and continuity implies that its attention to a site extends beyond the site itself. This also ensures that the planning and design of the physical space of urban parks can be deeply connected with the broader urban form and urban social life that encompass them.

---

[3] Leatherbarrow, D. (2015). *Topographical stories: studies in landscape and architecture*. University of Pennsylvania Press. (pp. 3).

## 7.3 Innovations

### *7.3.1 The Innovation in Theoretical Perspective and Research Methodology*

(1) This study approaches from a socio-natural theoretical perspective, which to some extent bridges the dichotomy between society and nature. For historical research, innovation in perspective equates to innovation in methodology. As a historical study in the field of the built environment, this research distances itself from the commonly adopted writing approach centered on form and style. While this approach undeniably holds its significance, it tends to offer a static and objectified description of the subject matter, largely overlooking the dynamic social and historical contexts. In the historical writing on urban parks, this issue manifests in the static description of natural elements within parks, neglecting their historicity. Nature element is merely treated as a decoration or container, not deeply engaged in the historical process, representing a significant shortfall in the knowledge production of the built environment field. Therefore, historical research in the built environment needs to "return to nature": not just to nature as a physical entity, but to a vibrant, integrated nature within everyday life at conceptual and social levels. Through such a theoretical perspective, this study meticulously examines the relationships between Litchi Park and urban marketing, Meilin Park and temporary residents, Shenzhen Bay and the context of the Cold War, etc., aiming to respond to the social–nature dichotomy derived from the modern knowledge production system in the built environment field.

(2) Through the analysis of social and historical contexts, this study reflects to some extent on the logic of formal autonomy. Furthermore, in response to the autonomy logic in the field of the built environment that posits a form of in architectural, garden and park designs, this study also makes an attempt to address. In empirical cases, whether it be the national-style parks in Shangbu-Luohu area or the modernism style parks in Shenzhen Bay, they all emerge within specific contexts: there is no ahistorical national style, nor an ahistorical modernism style. The discourse on the autonomy of architectural, garden and park forms may facilitate practice to a certain extent, but this is only possible through rigid demarcation of disciplinary boundaries. However, the discourse power gained in this manner is unlikely to proceed smoothly in today's knowledge production environment, which emphasizes integration and diversity.

(3) By constructing a research framework through topography, this research integrates landscape history and the history of landscape architecture to some extent. The commonality of social–nature dichotomy and the logic of autonomy lies in their discursive structures being ahistorical and decontextualized. This is also the real reason for the gulf between landscape history and the history of landscape architecture, although on the surface, this gulf seems to be caused by the differences in the spatial scales described by each. In fact, the ambiguity of the landscape concept can only be clarified within a historical context rather than through garden art.

Consequently, landscape history and the history of landscape architecture, as two branches of historiography that describe human–land relationships, have the potential to be deeply integrated. Under such theoretical understanding, this study begins with the landscape history of the PRD metropolitan area and whole Shenzhen city, then moves into the landscape history of various urban areas and from there delves into the history of corresponding urban parks, in an attempt to bridge landscape history and the history of landscape architecture.

### *7.3.2 Innovations in Empirical Materials and Research Subject*

(1) This study selects contemporary city following the Reform and Opening-up period for the historical study of urban parks. Presently, research on urban parks in China predominantly focuses on the modern parks constructed between the Opium War and the establishment of the People's Republic of China. However, contemporary urban parks developed after the period of reform and opening-up, despite their ubiquitous presence in urban spaces, are largely absent from historical research. This gap can be attributed to methodological issues within the field of historical research in built environment. The selection of Shenzhen as the research site for investigating the history of urban parks holds innovative value at the empirical level, given its status as a testing field for China's reform and opening-up policies.

(2) This study is based on the urban–rural interaction in Shenzhen, collecting materials that focus on its production, living, and environmental history before the Reform and Opening-up periods. Existing studies on the urban history and the history of urban construction of Shenzhen often depict the city as one constructed on a "blank slate," which overlooks many complex social relationships such as the nationalization of land ownership and changes in human–land interactions. Recognizing this, the present study meticulously examines the process of how urban park land transitions from collective land to state-owned land. Additionally, it pays attention to collecting materials on the production and livelihood of villages that once managed this land before the reform and opening-up period. This includes historical topographic maps, clan genealogies, agricultural output statistics before the reform and opening-up, as well as land acquisition compensation agreements after the reform and opening-up.

(3) This study intentionally presents materials that have been overlooked by generalized historical studies and official narratives. Existing studies on urban history and urban development often naturalize urban policies and government actions, resulting in the omission of "the Anonymous" emerged in Shenzhen's urbanization process from historical research. Residents of long-established villages who lost their land due to park construction, temporary residents forced to relocate, and Shenzhen Bay oyster farmers unable to continue their ancestral occupation, all of them need to be back into historical discourse through a micro-historical

perspective. Otherwise, there is a significant missed opportunity to examine urbanization from a socio-natural perspective. With this understanding, this study deliberately focuses on and presents the relationship between urban parks and "the Anonymous" during the collection of empirical materials.

(4) This study also intentionally explores the interplay between Shenzhen's urban parks and globalization, collecting relevant materials in the process. Shenzhen is not only a testing field for China's reform and opening-up but also a pioneering city leading China's integration into the global city network. Whether during the "Three-Processing and One Compensation" period, the Initial Phase of Establishing an International City, or later during the integration of the PRD metropolitan area, Shenzhen's development has been closely intertwined with globalization since the 1970s. Therefore, in collecting empirical materials, this study expands its perspective to focus on local practices within the context of globalization. For example, attention is paid to the relationship between the Lychee Festival and neoliberalism, the relationship between mountain park vegetation beautification and the construction of an international city, and the relationship between the Shenzhen Bay Park Group and international elite communities.

## 7.4 Shortcomings

### *7.4.1 Insufficient Depth in Micro-Level Research*

Although the research intends to present "the Anonymous" from a micro-historical perspective, due to various practical constraints, it was not possible to contact relevant individuals for interviews. Consequently, the study largely relies on secondary sources such as newspapers, books, etc., for relevant information, thus limiting the depth of micro-level investigation.

### *7.4.2 Ownership of Park Land Was Not Thoroughly Explored*

For the city of Shenzhen, the transformation of rural collective land into urban state-owned land is a highly complex issue. Specifically concerning urban parks, the land areas of many urban parks have undergone multiple changes over time, that even the management personnel who worked long years in these parks may not be fully aware of specific data. While this study has made efforts to search for land acquisition agreements for each park and related personnel and materials to provide supporting evidence, it still cannot guarantee absolute accuracy.

# Acknowledge and Afterwords

This book is based on my doctoral dissertation completed through a joint PHD program between Southeast University in mainland China and the City University of Hong Kong. No matter how passionate one might be about academic research, writing a doctoral dissertation inevitably involves periods of tedium and frustration. Sustaining the momentum to write until the final draft requires a certain illusion: the belief that this research holds extraordinary significance. Now, it is time to dispel this illusion. To do so, I must reflect on the many sparks that ignited this humble work.

In June 2011, Mr. Ma Xiaowei invited me and nine other graduate students to visit and learn from his landscape design company, namely Ager, in Shanghai for two months. "Why can professionals who coordinate human relationships, like lawyers, become mayors or even presidents, while those of us who coordinate human-nature relationships can't ever become mayors?" This question, posed by Mr. Ma, has lingered in my mind throughout my subsequent studies. At the end of August, after the internship at Ager concluded, I visited Shenzhen for the first time. My new internship was located on the east side of Central Park in Tianmian Design City. My projects were primarily real estate landscapes, often requiring me to calculate the 35% green space ratio and strive to make the designs appear "high-end." Ironically, I lived in Gangxia Village, East Second Lane, on the west side of Central Park. Even on the eighth floor, my window never received a single second of sunlight each day, let alone any green space. The cheap rent in the urban village attracted many migrant workers and recent graduates, including me. My memories of Gangxia include not narrow building gaps, but the absence of building gaps. The streets were bustling, yet perpetually muddy in Shenzhen's southern weather. The steamy rice noodle restaurants, the coming and going of migrant workers, the telephone booths with various dialects, the inexpensive grocery stores, and the dimly lit, ambiguous places with red lights created a surreal cityscape that often made me question if this was truly Shenzhen. How could a city be like this? It was worlds apart from the South Xiangyang Road in Shanghai's Xuhui District, where I had been just months before.



X. Liu, *Fabricating Nature*, The Urban Book Series,
https://doi.org/10.1007/978-981-95-8208-2

However, as I stepped out of this area, within a hundred meters, the gleaming International Finance Center and the lush Central Park repeatedly confirmed that I was indeed in this international metropolis. My interest in Shenzhen may have started with such contradictory experiences. In the following decade, I repeatedly came and went from Shenzhen, entangled with it in an unbreakable bond. However, considering the massive influx of young human resources into Pearl River Delta cities like Shenzhen around the turn of the millennium, this phenomenon is not difficult to understand. Fascinating personal experiences are like glistening but scattered pearls, difficult to transform into academic thought. A doctoral dissertation, however, should be as orderly and coherent as a necklace. Therefore, its completion relies on the guidance and connections provided by many supervisors and friends.

First and foremost, I must express my gratitude to my two supervisors—Professor Charlie Xue and Professor Wang Xiaojun. Professor Xue, known for his expertise in historical research within the field of architecture, provided me with an invaluable opportunity to pursue joint training at the City University of Hong Kong. He emphasized the importance of historical materials in research, and his patient guidance helped me shift from abstract theoretical debates to the collection and organization of empirical materials about Shenzhen. Professor Wang is one of the earliest scholars in China to conduct systematic research on modern American landscape design. Initially, I also hoped to delve deeper into modern landscape research, following his early work. "Those are introductory materials. Your generation of young scholars should focus on contemporary Chinese issues." Professor Wang's far-reaching insight changed my initial, immature ideas and became a guiding light throughout my doctoral studies. Both supervisors are gentle and modest gentlemen. They have shown great kindness and tolerance for my occasional lapses in conduct, allowing me to complete my studies without any inner conflict.

Secondly, I would like to express my gratitude to the numerous scholars and practitioner I have encountered over the past decade. Professor Liu Chunqing from Jiangxi Agricultural University, for instance, was the first to broaden my understanding of landscape architecture beyond small-scale spatial design to the larger scales of scenic spots and urban green systems. In life, she consistently offered me sound advice and sincere care at every crossroads. Then there is Mr. Ma from Ager in Shanghai, who not only sparked my interest in the relationship between landscape architecture and politics but also strove to provide opportunities and knowledge to the younger generation. Professor Liu Bingyi from Tongji University challenged my preconceived notions about aesthetics with his question, "Why must we see the big in the small? Isn't it also acceptable to see the big in the big? "The aesthetics of seeing the big in the small only emerged after ancient Chinese cities began to densify." His words prompted me to consider the relationship between aesthetics and non-formal factors. Professor Hendrik Tieben from the Chinese University of Hong Kong was the first to introduce me to the connections between society and space, as well as between cities and entrepreneurialism. He often critiqued my design work as "Olmstedian," which I initially took as a compliment, unaware of the historical and theoretical issues it entailed. It took years of study for me to appreciate his persistent guidance. Peter Smulders of Attractions International, a former vice president of Belt Collins, also deserves mention. Over the past few decades, he has deeply witnessed and participated in urban development in China and Asia, and his extraordinary design, modeling, and hand-drawing skills have always been a source of admiration for me. Additionally, Professor Hsia Chu-joe from Southeast University often provided sharp and thought-provoking insights, such as, "There is no architectural theory, only social theory." His history writing course enlightened me on the basics of historical research methodology. Southeast University is a stronghold of historical and theoretical research in the built environment field, and I have had the pleasure of learning from Professor Chen Wei, Professor Li Baihao, Professor Ge Ming, Professor Li Hua, Professor Wu Xiao, Professor Lai Delin, Professor Zhu Jianfei, Professor Cheng Yuning, Associate Professor Gu Kai, and Zhu Haohao. Their wisdom has been as refreshing and enriching as a spring breeze. Beyond Southeast University, I have also received guidance and assistance in various forms from Professor Yang Rui and Professor Zhu Yufan of Tsinghua University, Professor Yao Peng and Professor Feng Xiao of Beijing Forestry University, Professor Xu Hao of Nanjing Forestry University, Professor Zhang Yichi of Chongqing University, Professor He Fang of Shenzhen University, and Associate Professor Lu Yi of City University of Hong Kong. I consider myself extremely fortunate to have had the opportunity to seek advice and learn from these esteemed scholars, whose brilliance is evident throughout this humble work.

Then, during my research in Shenzhen, I received selfless help from many seniors, leaders, and friends. They are Chen Shouya, Ye Guidong, Zhang Xinsi, Fu Keqin, Shen Baojiu, Guo Binghao, Huang Fuquan, Luo Xiaobing, Zou Songfa, Wang Xiaoming, Yang Yibiao, Huang Chongyuan, Wu Qingxiang, Chen Yixin, Xie Chaoping, Zhu Rongyuan, Zhu Jie, and He Fang follows the order of interviews. Additionally, some seniors, leaders, and friends, although not specifically interviewed, provided invaluable assistance in material collection and contact facilitation. They include Xie Liangsheng, Fu Weimin, Ouyang Dimei, Long Dandan, Zhuang Meimei, Shen Xiaolan, Xie Zuogui, He Henghui, Zheng Haiyan, Zeng Xiangyi, Zhu Ruichong, Wen Jianpeng, Liang Gang, and many others. Despite their different stages in life—some are now octogenarians, some are in their sixties, and some are in the prime of their careers—they share a common trait: a willingness to dedicate their youth to the vigorous construction of the Special Economic Zone. Several generations have marked these decades with their relay efforts, even though forty years have passed swiftly. The fortunate support my research received is partly due to their helpful, open, and generous nature. Additionally, I believe it is also because the forty-plus years of Special Economic Zone construction have instilled in many participants a sense of history and an eagerness to document, reflect on, and understand this period. Whether it is Mr. Guo Binghao, who edited the oral history of Vice Mayor Luo Changren, Mr. Chen Yixin, who summarized and organized the history of Shenzhen's urban planning and the construction history of the Futian Central District, or Mr. Xie Liangsheng, who encouraged me to approach both historical materials and experiential insights—these interactions made me acutely aware of their profound sense of mission in recording history. I am extremely grateful for the opportunity to engage deeply with them and benefit from their wisdom and experience.

I am also grateful to many of my peers who have walked this journey with me. Among them are my colleagues from Southeast University: Qian Yun, Zhang Guanting, Zou Hao, Sun Tianzhi, Xu Ruofei, Wei Xiao, Chen Xin, Hu Mengjun, Shen Mengying, Zhuang Xuntong, and Wu Zhigao. From the City University of Hong Kong, I extend my gratitude to Xiao Yingbo, Sun Cong, Zhang Lujia, Chen Yingting, Qiu Yue, and Ma Kaiyue. Additionally, I am thankful for the support and camaraderie of Mu Xiaodong, Zhang Xianchun, Chen Yongming, Zhong Le, Zhang Biao, Xia Geyang, Wang Zongqi, Cui Kai, Zhu Siyu, He Shan, Tang Jia, Wu Binsheng, and Dai Kunli. Many of these individuals are young scholars of considerable intellectual depth, and it has been my privilege to engage in deep and meaningful conversations with them. Their insights and support have been invaluable, and I am honored to have had the opportunity to collaborate and learn alongside such a distinguished group of peers.

Finally, I must express my indescribable gratitude for the unwavering support and dedication of my family.

With this, I hope to put an end to the illusion that has driven the writing of this doctoral dissertation: it is, in fact, a mindset of overestimating one's own efforts. My abilities as an author are limited, and my level is modest. Having traversed numerous paths, interacted with numerous individuals, and received copious amounts of help and inspiration, I have fortunately completed this work. The sincere help and inspiration I received have already made the points this research aims to express shine like starlight, and I am merely the agent who has sought out these stars and fortunately compiled them into a written form. Thus, this research is nothing extraordinary.

The term "Shenzhen," in my view, embodies a variegated aesthetic. This aesthetic is similar to its coastal monsoon climate: for idealists who strive against all odds, Shenzhen is like brilliant sunshine after a typhoon, with clear skies stretching for miles and shimmering waters; yet for transient and grassroots individuals, it is the drizzling, sticky weather of a southern breeze. They combine and give rise to this variation with different concentrations, intensities, methods, strategies, and policies.

# Bibliography

## Periodical

Alon-Mozes, T., & Gilad-Ilsar, S. (2020). Modern park for a modern city: Planning Tel Aviv's Yarkon Park during the 1960s–1970s. *Studies in the History of Gardens & Designed Landscapes, 40*(1), 80–94.

Appadurai, A. (2000). Grassroots globalization and the research imagination. *Public Culture, 12*(1), 1–19.

Appleton, J. (1986). Some thoughts on the geology of the picturesque. *The Journal of Garden History, 6*(3), 270–291.

Attorre, F., Francesconi, F., Pepponi, L., Provantini, R., & Bruno, F. (2003). Spatia-temporal analyses of parks and gardens of Rome. *Studies in the History of Gardens & Designed Landscapes, 23*(3), 293–306.

AUBE Architectural and Urban Planning Design Company of France, AUBE Design Co., Ltd. In Shenzhen法國歐博建築與城市規劃設計公司,深圳市歐博設計有限公司 (2010). Shenzhen Peninsula City State深圳半島城邦. *Architecture Technique*建築技藝. *02*, 62–71.

Badiu, D. L., Onose, D. A., Niță, M. R., & Lafortezza, R. (2019). From "red" to green? A look into the evolution of green spaces in a post-socialist city. *Landscape and Urban Planning, 187*, 156–164.

Bao, M. H.包茂宏 (2004). Martin Melosi and his Study of the U.S. Urban Environmental History 馬丁.麥樂西與美國城市環境史研究. *Journal of Chinese Historical Geography*中國歷史地理論叢, *04*, 114–126+160.

Castree, N. (1995). The nature of produced nature: Materiality and knowledge construction in Marxism. *Antipode: A Radical Journal of Geography, 27*(1), 12–48.

Chen, T. 陳濤 (2000). Reflections on building Shenzhen as an "International Garden City." 建設深圳"國際花園城市"的思考. *Ecological Science*生態科學. (pp. 91–93).

Chen, Y. Q. 陳蘊茜 (2004a). Changes in tourism and entertainment spaces from the Late Qing Dynasty to the Republic of China: An examination centered on parks論清末民國旅遊娛樂空間的變化——以公園為中心的考察. *Historical Review*史林, *2004*(05), 93–100+124.

Chen, Y. Q. 陳蘊茜 (2005). On the conflict between colonialism and nationalism in everyday life: From a perspective of the Parks in Modern China日常生活中殖民主義與民族主義的衝突——以中國近代公園為中心的考察. *Journal of Nanjing University (Philosophy, Humanities and Social Sciences)* 南京大學學報(哲學.人文科學.社會科學版), (05).



X. Liu, *Fabricating Nature*, The Urban Book Series,
https://doi.org/10.1007/978-981-95-8208-2

Chen, Y. Q. 陳蘊茜 (2006a). Reconstruction of Space and the Worship towards Sun Yat-sen: An Investigation with Yat-sen Park as its Center in Republic of China空間重組與孫中山崇拜——以民國時期中山公園為中心的考察. *Historical Review*史林, *2006*(01), 1–18+123.

Chen, T., Li, G., & Liang, Z.陳濤,李剛,梁正陽 (1999). Tree species selection and demonstration forest construction in Shenzhen's ecological landscape forests深圳生態風景林樹種選擇與示範林營造. *Chinese Landscape Architecture*中國園林. (04), 44–46.

Chen, H. S., Li, Z. G., Xiao, Y., et al. 陳宏勝,李志剛,肖揚等 (2022). Research on the development of new cross-border districts from the perspective of regional function: A comparison between Qianhai and Hengqin Cooperation Zones基於區域功能視角的邊界新區發展研究——對深圳前海合作區與珠海橫琴合作區的比較. *City Planning Review*城市規劃, *46*(04), 21–29.

Chen, C., & Wang, K. 陳晨,王柯 (2019). The Path of Architectural Creation about the National Form by the Architects of The Republic of China in Nanjing民國建築師在南京的民族形式建築創作軌跡(1927—1937). *Architect*建築師, *200*, 107–114.

Chen, K. S., Yang, R., & Liu, B. B. 陳可石,楊瑞,劉冰冰 (2013). Study on the evolution and development of Shenzhen's cluster spatial structure深圳組團式空間結構演變與發展研究. *Urban Development Studies*城市發展研究, *20*(11), 22–26.

Chen, X., & Zhao, J. J. 陳曦,趙紀軍 (2020). A Comprehensive Study of Huagangguanyu Park: A Discussion on the Research of Modern Chinese Garden History花港觀魚公園研究面面觀——兼論中國現代園林史研究. In *Proceedings of the 2020 annual conference of the Chinese society of landscape architecture*中國風景園林學會2020年會論文集(下冊) (Vol. 2, pp. 95–101).

Chi, W. X., & Lin, G. S. 茌文秀,林廣思 (2021). Study on the safeguard mechanism of large-scale landscape planning projects—A case study of the Greenway Network in Pearl River Delta 大尺度景觀規劃專案的實施保障機制研究——以珠三角綠道網為例. *Chinese Landscape Architecture*中國園林, *37*(09), 25–30.

China Resources Land (Shenzhen) Development Co., Ltd華潤置地(深圳)發展有限公司 (2017). China Resources Shenzhen Bay·Yuefu 華潤深圳灣·悅府. *China Real Estate*中國房地產. *2017*(26), 79.

Crompton, J. L. (2001). The impact of parks on property values: A review of the empirical evidence. *Journal of Leisure Research, 33*(1), 1–31.

Crompton, J. L. (2013). The health rationale for urban parks in the nineteenth century in the USA. *World Leisure Journal, 55*(4), 333–346.

Cui, G. H. 崔功豪 (1989). Review and Prospect of Research on China's Urbanization in Last Decade近十年來中國城市化研究的進展. *Areal Research and Development*地域研究與開發, (01), 1–5+62.

Cui, X. S.崔孝松 (2016). An explanatory study on the historical evolution of administrative divisions in Shenzhen深圳行政區劃歷史沿革考釋. *Practice and Theory of SEZS*特區實踐與理論, (04), 71–74.

Cui, G. H., & Ma, R. C. 崔功豪,馬潤潮 (1999). Urbanization from Bellow in China: Its development and mechanisms中國自下而上城市化的發展及其機制. *Acta Geographica Sinica*地理學報, *1999*(02), 12–21.

Cui, H., Pan, M., & Wu, S. 崔宏軼,潘夢啟,吳帥 (2020). Changes and countermeasure suggestions of science and technology talents policies in China's economic special zones: A case study of Shenzhen我國經濟特區科技人才政策變遷及對策建議——以深圳為例. *Jiang-huai Tribune*江淮論壇, *303*(05), 30–36.

Dai, H. B. 戴海斌 (2005). The Central Park and Beijing Society in Early Republic of China中央公園與民初北京社會. *Social Sciences of Beijing*北京社會科學, (02), 45–53+121.

Dai, Y. F.戴一峰 (2011). A multianalysis on the space of city in Modern China—Focus on Park of City多元視角與多重解讀:中國近代城市公共空間——以近代城市公園為中心. *Journal of Social Sciences*社會科學, (06), 134–141.

Dai, B., & Huang, H. 戴斌,黃璜 (2016). Theories and strategic for the tourism integration: The case study of the Beijing-Tianjin-Hebei區域旅遊一體化的理論建構與戰略設計——以京津冀為例. *Human Geography*人文地理, *31*(03), 128–135.

David, T., Zhang, S. Y., & Henry, D., et al. 大衛·湯普森,張樹儀,亨利·道爾頓,等 (2012). North Lake Wetland Park in Happy Coast of Shenzhen Oct深圳華僑城歡樂海岸北湖濕地公園. *Landscape Architecture*風景園林, (02), 86–91.

David, T., Zhang, S. Y., & Henry, D., et al. 大衛·湯普森,張樹儀,亨利·道爾頓,等 大衛·湯普森,張樹儀,亨利·道爾頓,等 (2013). Shenzhen Oct Bay深圳華僑城歡樂海岸. *Landscape Architecture*風景園林, *2013*(03), 68–76.

Deng, H. B., Zhang, Z. H., & Deng, Y. Z. 鄧煥彬,張智慧,鄧遠征 (2011). Study on countermeasures of regional transportation integration in the Pearl River Delta Region珠三角區域交通一體化對策研究. *Transpo World*交通世界(運輸.車輛), *2011*(11), 128–130.

Du, J. (2008). Don't underestimate the rice fields. *Urban Transformation*, 218–225.

Duan, J. Q., & Zhang, H. 段建強,張樺 (2021). Interactions between urban space and landscape: Research on Transformation mechanism of Urban Public Parks in Shanghai's urbanization from early modern to nowadays城園互動:上海近現代城市化中公共園林轉型機制研究. *Landscape Architecture*風景園林, 28(06), 38–43.

Düempelmann, S. (2014). Herrington S. Plotting time in landscape architecture. *Studies in the History of Gardens & Designed Landscapes, 34*, 1–14.

Dümpelmann, S. (2005). La battaglia del fiore. Gardens, Parks and the City in Fascist Italy. *Studies in the History of Gardens & Designed Landscapes, 25*(1), 40–70.

Fan, Y.T., Cui, Q.Y., & Xiao, J. 范雅婷,崔秦毓,肖靖 (2021). Shenzhen bay waterfront in the agenda of leap-forward urbanism: Historical and reflective studies on its formation and publicity躍進式城市化視角下的深圳灣濱水帶沿革及其公共性演變研究. *Chinese & Overseas Architecture*中外建築, *11*, 26–32.

Fan, S., Liu, Z., & Xu, X. 范思敏,劉志成,許曉明 (2020). Development and evolution of the Chicago's Lakefront Park System芝加哥湖濱公園系統發展演變歷程. *Landscape Architecture*風景園林, *27*(02), 66–71.

Friedmann, J. (2006). Four theses in the study of China's urbanization. *International Journal of Urban and Regional Research, 30*(2), 440–451.

Friedmann, J., & Wolff, G. (1982). World city formation: An agenda for research and action. *International Journal of Urban and Regional Research, 6*(3), 309–344.

Fu, K.Q., Lai, S.W., & Wang, F. 傅克勤,賴樹翁,王鋒 (2006). Brief notes on the planning of Shenzhen Garden Expo Park深圳園博園規劃小記. Chinese Landscape Architecture中國園林, 2006(10), 60–63.

Fung, S. 馮仕達 (2017). Longing and belonging in Chinese Garden History中國園林史的期待與指歸. *Heritage Architecture*建築遺產, (006), 39–47.

Gabriel, N. (2011). The work that parks do: Towards an urban environmentality. *Social & Cultural Geography, 12*(02), 123–141.

Gao, D. 高岱 (2017). William Hoskins and the studies of landscape history威廉·霍斯金斯与景观史研究. *Academic Research*学术研究, *2017*(12), 103–107, 178.

Gao, L. L.郜亮亮 (2023). Reform practice, logic of change, and future evolution direction of China's Rural Land Property Rights System中國農地產權制度的改革實踐、變遷邏輯及未來演進方向. *China Review of Political Economy*政治經濟學評論, *14*(01), 48–76.

Gao, W., Lu, Y., Qu, X., et al. 高雯琪,陆颖,屈霄,等 (2021). River habitat assessment under urbanization: A case study in Shenzhen城镇化背景下河流生境评价——以深圳市为例. *Acta Ecologica Sinica*生态学报, *41*(22), 8783–8793.

Gosden, C., & Marshall, Y. (1999). The cultural biography of objects. *World Archaeology, 31*(2), 169–178.

Gu, C. L., Chen, G., Huang, C.Y., et al. 顧朝林,陳果,黃朝永等 (2001). On the development of new industrial space in Shenzhen: The result of economic globalization, industrial restructuring, and relocation論深圳新工業空間開拓——經濟全球化、產業結構重建與轉移的結果. Economic Geography.經濟地理, (03), 261–265.

He, F. 何昉 (2010). Shenzhen landscape architecture: From the inheritance of landscape architecture mechanism of China to the creation of regional style深圳園林:從繼承中國風景園林體系到地域風格的創建. *Landscape Architecture*風景園林, *2010*(05),70–72.

He, F. 何昉 (2020). The 40 years of prosperity and development of landscape heritage and innovation風景園林傳承創新,大鵬展翅騰飛卌年. *Architectural Practice*建築實踐, (11), 168–175.

He, F., & Ye, F. 何昉,葉楓 (2003). Creating a "living green natural museum" in the urban center of Shenzhen: Overall planning ideas for Lianhua Mountain Park創建深圳城市中心區"活的綠色自然博物館"——蓮花山公園總體規劃思想. *Chinese Landscape Architecture*中國園林, (06), 11–15, 6–7, 10, 23.

Henderson, J. C. (2013). Urban parks and green spaces in Singapore. *Managing Leisure, 18*(3), 213–225.

Hou, S. (2013). Cities without borders: From American urban history to urban environmental history. *Journal of Renmin University of China, 27*(03), 20–29.

Hu, Q. F. 胡其舫 (1999). Overview, significance, and influence of Modern Parks in Jiangsu Province江蘇近代公園概貌及其意義和影響. *Chinese Landscape Architecture*中國園林, *1999*(06), 68–71.

Huang, H. H.黃煥華 (1990a). Pine tree devastating epidemic in Shenzhen—Pine Wilt Nematode Disease深圳市松樹毀滅性疫病—松材線蟲萎蔫病. *Guangdong Forestry Science and Technology (Forestry and Environmental Science)*廣東林業科技, (06), 34–36.

Huang, D. H. 黃東漢 (2013). Everlasting memories of the Mangrove Forest永志難忘紅樹林. *Remembrance*記憶, *93*(02):35–38.

Jackson, P. (2000). Rematerializing social and cultural geography. *Social & Cultural Geography, 1*(1), 9–14.

Jiang, Y., Zhang, Q. Y., Li, F. D., et al. 蒋钰,张秋英,李发东,等 (2023). Impact of rapid urbanization on Shenzhen water environment快速城市化对深圳水环境的影响分析. *Pearl River*人民珠江, *44*(08), 8–17, 24.

Kohn Pedersen Fox Associates KPF建築設計事務所 (2021). Shenzhen Bay No. 1深圳灣1號. *Architectural Practice*建築實踐, (03), 176–183.

Laine, J. P. (2017). Borders and borderscapes under contemporary globalization. *Progress in Geography, 36*(12), 1531–1539.

Latham, A., & McCormack, D. P. (2004). Moving cities: Rethinking the materialities of urban geographies. *Progress in Human Geography, 28*(6), 701–724.

Lei, Y.雷頤 (2008). Parks: Past and present公園古今事. *Yanhuang Chunqiu*炎黃春秋, *2008*(04), 65–68.

Lei, S., & Li, J. 雷聲,李峻 (1998). Design and practice of sediment discharge channel in Sungang Detention District筍崗滯洪區排沙道工程設計實踐. *Guangdong Water Resources and Hydropower*廣東水利水電, *1998*(S1), 22–24.

Li, Z. S. 李錚生 (1994). Planning of Shenzhen's Green Space System深圳市綠地系統規劃. *City Planning Review*城市規劃, *1994*(01),53–55.

Li, D. Y.李德英 (2003). Social conflicts in Parks: A case study of Parks in Modern Chengdu公園裡的社會衝突——以近代成都城市公園為例. *Historical Review*史林, *2003*(01), 1–11+123.

Li, J. Z. 李建中 (2013). From "illegal crossing" to "setting sail." 從"督卒"到"起錨". *Remembrance*記憶 *93*(02), 3–10.

Li, L., Li, W., & Wang, L. 李淩月,李雯,王蘭 (2021). Urban entrepreneurialism and green renewal of industrial heritage: Mechanisms and impacts—Implications and reflections on Highline Park都市企業主義視角下工業遺產綠色更新路徑及其影響——廢棄鐵路蛻變高線公園. *Landscape Architecture*風景園林, *28*(01), 87–92.

Li, Y. C., & Meng, X. Y. 李穎春,孟心宇 (2023). Detailed study on the spatial changes of the bund riverside Grassland in Modern Shanghai近代上海外灘濱江草地空間變遷詳考. In Ma, X. Q., Tsukada, T., & Tang, H. R.馬學強,塚田孝,唐惠榮 (Eds.), *Proceedings of the international academic symposium on spatial morphology and historical cultural evolution of waterfront cities*濱水城市空間形態與歷史文化演進國際學術研討會論文集. Commercial Press 北京: 商務印書館 (pp. 62–82).

Li, B. Y., Wu, C. S., Liu, M., et al. 李寶勇,吳春生,劉牧等 (2020). Research and Enlightenment on Planning and Construction of Modern Nanchang Park and Green Space in the Republic of China 民國南昌公園綠地規劃建設研究及啟示(1926–1949). *Chinese Landscape Architecture*中國園林, *36*(08), 101–106.

Liang, C. W., Liu, W., & Zhang, Y. Y. 梁成文,劉威,張玉一 (2010). Rapid transit planning of Shenyang under the background of regional integration區域一體化背景下瀋陽市快速軌道交通線網規劃. *City Planning Revies*城市規劃, *34*(03), 71–75.

Liao, H. L. 廖虹雷 (2016). Historical evolution of Shenzhen Markets深圳墟歷史沿革. *Shenzhen Historical Records*深圳史志, (52), 63–65.

Lin, Z. 林峥 (2019). "Homeland Nostalgia" of the "Worldly Man": On the Park Narratives in Kang Youwei's Overseas Travelogues. "世界人"的鄉愁——論康有為海外遊記中的公園書寫. *Collection of Studies on Modern Chinese Literature*中國現代文學研究叢刊, (01), 151–166.

Lin, L., & Yin, K. 林淩,殷克勝 (1993). Economic development strategy and outline planning of Nanyou Development Zone (Excerpt) 南油開發區經濟發展戰略與規劃綱要(摘編). *China Opening Journal*開放導報. *1993*(04), 41–44.

Ling, D. X. 淩大燮 (1983). Changes in China's forest resources我國森林資源的變遷. *Agricultural History of China*中國農史, *1983*(02), 26–36.

Liu, H. Y. 劉懷玉 (2003). Henri Lefebvre and several trends of critique of everyday life in the 20th century west. 列斐伏爾與20世紀西方的幾種日常生活批判傾向. *Seeking Truth*求實, (05), 44–50.

Liu, J. Q. 刘家麒 (2010). The innovative development of landscape construction program in Shenzhen深圳园林绿化建设的发展创新之路. *Landscape Architecture*风景园林, (05), 58–59.

Liu, D. Y. 劉東洋 (2011). Urban notes VII: A view of topography from my rear window城市筆記之七:從我家後窗望見的地形學. *The Architect*建築師, *2011*(03), 105–111.

Liu, W. W. 劉偉偉 (2015). Multiple streams analysis of policy termination: Empirical study of the custody and repatriation system 政策終結的多源流分析——基於收容遣送制度的經驗研究. *Journal of Public Management*公共管理學報, *12*(04), 21-38+154.

Liu, Y.S.劉亦師 (2016). A preliminary study on the rise and popularization of roadside tree planting in the Republic China民國時期行道樹種植觀念之形成與普及探述. *Landscape Architecture*風景園林, (04), 80–87.

Liu, Y. S. 劉亦師 (2017). Leisure and recreation under socialism: A preliminary study on New Urban Parks in Beijing in the early 1950s社會主義城市休閒空間——建國初北京新公園建設探論. *New Architecture*新建築, *2017*(06), 35–41.

Liu, Y. S. 劉亦師 (2018a). "People's Fairyland": A study on Taoranting Park "人民的樂園"——陶然亭公園營建記 (1952–1966). *Archicreation*建築創作, *2018*(05), 162–169.

Liu, Y., & Liu, L. D. 劉淼,劉利東 (2010). Research on Tianjin Binhai New Area under Beijing-Tianjin-Hebei Region Frame京津冀區域框架下的濱海新區發展研究. *Urban Development Studies*城市發展研究, *17*(05), 18–21.

Liu, Y. J., Luo, X. L., Tian, D., et al. 劉永敬,羅小龍,田冬等 (2014). Preliminary study on China's cross-border new district in forming mechanisms, spatial organization, and governance mode 中國跨界新區的形成機制、空間組織和管治模式初探. *Economic Geography*經濟地理, *34*(12), 41–47.

Liu, Q. Q., & Wang, X. 劉琪琪, 王欣 (2018). Research on the Development of Moonlake Public Garden in Ancient City of Ningbo寧波月湖古代城市公共園林流變研究. *Landscape Architecture*風景園林, *2018*(01).

Liu, X. Y., & Wang, X. 劉新宇,王曉俊 (2021). A study of F. L. Olmsted and his designing ideas from the perspective of American History in 19th century 19世紀美國史視角下的奧姆斯特德及其設計思想. *Chinese Landscape Architecture*中國園林, *37*(02), 106–111.

Liu, X. Y., Wang, X. J., & Xue, Q. L.劉新宇,王曉俊,薛求理 (2022). The Practice and representation of Urban Parks in Shenzhen special economic zone in the 1980s 20世紀80年代深圳經濟特區城市公園的實踐與表現. *Landscape Architecture Frontiers*景觀設計, *10*(04), 70–83.

Liu, X. Y., Xue, Q., & Wang, X. 劉新宇,薛求理,王曉俊 (2022). Morphological transformation of Boston Park System during the 19th century and its reasons十九世紀波士頓公園體系的形態演變歷程及其動因分析. *Chinese Landscape Architecture*中國園林, *38*(09), 134–139.

Liu, F. X., & Zhao, J. L.劉方馨,趙紀軍 (2021). Cultural export and representation in the making of overseas Chinese gardens in the 1980s20世紀80年代境外中國園營造中的文化輸出與再現. *Chinese & Overseas Architecture*中外建築, (01), 86–90.

Liu, F., Zhao, J., & Han, Y. 劉方馨,趙紀軍,韓依紋 (2021). Research on the characteristics of urban green space and practice in China from the perspective of "ecological view" in the initial stage of economic reform and opening-up改革開放初期"生態觀"視角下中國城市綠地發展與實踐特點研究. *Landscape Architecture Academic Journal*園林, *38*(12), 68–74.

Liu, X., Zhao, H., Zheng, D., et al. 劉杏玲,趙煥庭,鄭德延等 (1988). Evolution and Development of Shenzhen Bay 深圳灣的演變與開發利用. *Journal of Tropical Oceanography*熱帶海洋, 48–56.

Loughran, K. (2020). Urban parks and urban problems: An historical perspective on green space development as a cultural fix. *Urban Studies, 57*(11), 2321–2338.

Luo, X. L., & Shen, J. F. 羅小龍,沈建法 (2010). From "front shop, back factory" to Hong Kong-Shenzhen Metropolis: The changing relationship between Hong Kong and Shenzhen in the Past Three Decades從"前店後廠"到港深都會:三十年港深關係之演變. *Economic Geography*經濟地理, *30*(05), 711–715.

Luo, Y. C., Xiao, L., Lv, B. W., & Lin, G. S. 羅雨晨,蕭蕾,呂博文,林廣思 (2023). Historical facts and characteristics of the construction of Guangzhou Parks in the 1980s and 1990s1980~1990年代廣州公園的建設史實與特點. *South Architecture*南方建築, (06), 42–51.

Luo, Z. X., Zhou, S. H., & He, J. M. 羅子昕,周素紅,何嘉明 (2019). Cross-border social networks: A case study of Guang-Fo Metro, China區域一體化下城市居民跨邊界社會網路構建——以廣佛地鐵為例. *Urban Development Studies*城市發展研究, *26*(06), 70–78.

M'Closkey, K., & VanDerSys, K. (2023). Topography as the groundwork for landscape design—Interview with Karen M'Closkey and Keith VanDerSys. *Landscape Architecture Frontiers, 11*(4).

Ma, L. J. (2002). Urban transformation in China, 1949–2000: A review and research agenda. *Environment and Planning A, 34*(9), 1545–1569.

Maass, J. (1969). Where architectural historians fear to tread. *Journal of the Society of Architectural Historians, 28*(1), 3–8.

Mang, J.莽疾 (1998). Resettlement areas: An Important form of providing affordable housing for low-income families by the government 安置區: 國家為低收入家庭提供廉租房的一種重要形式. *China and Foreign Real Estate Guide*中外房地產導報, (10), 38–39.

Mao, H. S.毛華松 (2014). Formation analysis of Chinese Ancient Parks——Analysis of Urban Parks Development in Song Dynasty論中國古代公園的形成——兼論宋代城市公園發展. *Chinese Landscape Architecture*中國園林, (01), 116–121.

Mao, H. S., Wang, X. C., & Wu, Y. H. X. 毛華松,王雪純,吳映華夏 (2021). Park-oriented development of modern urban landscape: Cognition, construction, and achievements近代城市山水風景公園化:認知、營建與事功. *Landscape Architecture*風景園林, *28*(06), 25–31.

Meilin 1st Village: A beautiful "urban business card" in Shenzhen 梅林一村: 深圳一張靚麗的"城市名片". (2009). *China Property Management*中國物業管理, *70*(01), 41–42.

Meng, Z. Z. 孟兆禎 (2010). The united endeavour for the construction programs in Shenzhen in Honor of the 30th Anniversary of Shenzhen Special Economic Zone眾志成城的深圳建設 紀念深圳特區30周年. *Landscape Architecture*風景園林, *2010*(05), 56–57.

Meng, Z. Z. 孟兆禎 (2020). Chinese Traditional Garden Flourishing in Shenzhen: In Honor of the 40th Anniversary of Shenzhen Special Economic Zone大鵬風舞中國傳統園林——紀念深圳特區成立40周年. *Landscape Architecture*風景園林, *27*(10), 10–12.

Olwig, K. R. (1980). Historical geography and the society/nature "problematic": The perspective of JF Schouw, GP Marsh and E. Reclus. *Journal of Historical Geography*, *6*(1), 29–45.

Park, R. E. (1915). The city: Suggestions for the investigation of human behavior in the city environment. *American Journal of Sociology, 20*(5), 577–612.

Peng, C. X. 彭長歆 (2014). The beginning of Public Park in China: The American and English Gardens of Thirteen Factories of Canton中國近代公園之始——廣州十三行美國花園和英國花園. *Chinese Landscape Architecture*中國園林, *30*(05), 108–114.

Peng, C. X., & Wang, Y. T. 彭長歆,王豔婷 (2021). Landscaping outside the City Wall: Canton's Gardens in Late Qing Dynasty and the Moder Transformation of Lingnan Gardens城外造景:清末廣州景園營造與嶺南園林的近代轉型. *Chinese Landscape Architecture*中國園林, *37*(11), 127–132.

Qian, Q., Wang, T., Xiao, J., et al. 千茜,王濤,肖潔舒等 (2011). Shenzhen's Bay Era: Landscape design for Shenzhen Bay Park 深圳進入"灣"時代 深圳灣公園景觀設計解析. *Landscape Architecture* 風景園林, (04), 32–37.

Qu, X., Liu, H., Yang, M., et al. 屈霄,劉晗,陽敏,等 (2022). Characteristics of fish communities and driving factors under urbanization in typical river basin in Shenzhen City, China城鎮化背景下深圳典型流域魚類群落特徵及驅動因數. *Acta Ecologica Sinica*生態學報, *42*(24), 10029–10040.

Rose, J. (1938). Freedom in the Garden-A contemporary approach. *Pencil Points, 19*, 639–644.

Rosen, C. M., & Tarr, J. A. (1994). The importance of an urban perspective in environmental history. *Journal of Urban History, 20*(3), 299–310.

Shatkin, G. (2007). Global cities of the South: Emerging perspectives on growth and inequality. *Cities, 24*(1), 1–15.

Shen, S., Hu, W., Yang, F., et al. 沈姍姍, 胡汪涵,楊凡等 (2022). Historical changes of Plant Landscape in Zhongshan Park of Hangzhou杭州中山公園植物景觀歷史變遷探析 (1912–2021). *Chinese Landscape Architecture*中國園林, *38*(09), 99–104.

Shen, R. S., & Sun, M. X. 沈汝生,孫敏賢 (1947). A study on urban geography of Chengdu成都都市地理之研究. *Acta Geographica Sinica*地理學報, *1947*(Z1), 14–30.

Shenzhen Happy Coast Shenzhen, China 深圳歡樂海岸中國深圳. (2020). *World Architecture Review* 世界建築導報, *35*(02):74–79.

Shenzhen Innovation and Development Institute. (2018). Luo Jinxing: Crab eater who won the "first auction" of land use rights in China. *Special Zone Economy, 355*(8), 10–11.

Shi, M. (1998). From imperial gardens to public parks: The transformation of urban space in early twentieth-century Beijing. *Modern China, 24*(3), 219–254.

Sini, R. (2023). Singapore's postcolonial ethnic revival in parks design. *Studies in the History of Gardens & Designed Landscapes, 43*(3), 225–250.

Smith, A. (2014). From green park to theme park? Evolving legacy visions for London's Olympic Park. *Arq: Architectural Research Quarterly, 18*(4), 315–323.

Solecki, W. D., & Rosenzweig, C. (2004). Biodiversity, biosphere reserves, and the Big Apple: A study of the New York Metropolitan Region. *Annals of the New York Academy of Sciences, 1023*(1), 105–124.

Song, Y. H., & Liu, L. J. 宋穎慧,劉麗君 (2005). Zhongxin Hongshuwan: Precision Ensures Victory中信紅樹灣 精准奠定勝局. *Chinese and Foreign Real Estate Times*中外房地產導報. *2005*(09), 32–37.

Sun, S. X. 孫紹先 (1984). Port construction in Shekou Industrial Zone蛇口工業區的港口建設. *Port & Waterway Engineering*水運工程, *Z1*, 78–81.

Sun, X., & Hu, X. 孫筱祥,胡緒渭 (1959). Planning and design of Huagangguanyu Park in Hangzhou杭州花港觀魚公園規劃設計. *Architectural Journal*建築學報, (05), 19–24.

Tan, Z., & Xiao, Y. B.譚崢,肖映博 (2020). Bridge and door: The evolutionary public space of the Luohu Port in the perspective of "borderscape" 橋及閘——"關境"視角下的羅湖口岸公共空間演化史.*New Architecture*新建築, *2020*(04), 11–17.

Taylor, H. A. (1995). Urban public parks, 1840–1900: Design and meaning. *Garden History*, 201–221.

Taylor, D. E. (1999). Central Park as a model for social control: Urban parks, social class and leisure behavior in nineteenth-century America. *Journal of Leisure Research, 31*(4), 420–477.

Thompson, C. W. (2002). Urban open space in the 21st century. *Landscape and Urban Planning, 60*(2), 59–72.

Thompson, D., Zhang, S., Dalton, H., et al. 大衛·湯普森,張樹儀,亨利·道爾頓等 (2013). Shenzhen OCT Bay 深圳華僑城歡樂海岸. *Landscape Architecture*風景園林 *2013*(03), 68–76.

Thongtha, S., Teamkao, P., Boonapatcharoen, N., Tripetchkul, S., Techkarnjararuk, S., & Thiravetyan, P. (2014). Phosphorus removal from domestic wastewater by Nelumbo nucifera Gaertn. and Cyperus alternifolius L. *Journal of Environmental Management, 137*, 54–60.

Von Hoffman, A. (1988). "Of greater lasting consequence": Frederick Law Olmsted and the Fate of Franklin Park, Boston. *The Journal of the Society of Architectural Historians, 47*(4), 339–350.

Waitt, G. (2008). Urban festivals: Geographies of hype, helplessness, and hope. *Geography Compass, 2*(2), 513–537.

Wang, Y. Y. 王益厓 (1935). A study on urban geography of Wuxi無錫都市地理之研究. *Acta Geographica Sinica*地理學報, *1935*(03), 23–63.

Wang, J. Y.汪菊淵 (1992). Retrospect and prospect of urban greening and landscape construction in China我國城市綠化、園林建設的回顧與展望. *Chinese Landscape Architecture*中國園林, *1992*(01), 17–25.

Wang, S. Y.王淑英 (1993). Achievements in the research on controlling pine wilt disease in Shenzhen 深圳控制松材線蟲病研究取得階段性成果. *Forest Pest and Disease Communication (Forest Pest and Disease)*森林病蟲通訊, (02), 31.

Wang, B. Z.王炳真 (2001a). Ecological Landscape Forest: The Guarantee of a Garden City 生態風景林:花園城市的保障. *Chinese Forestry Science and Technology*中國林業, *24*, 14–15.

Wang, J. T. 王勁韜 (2011a). On the public nature of Chinese ancient gardens and its impact on urban life: Taking gardens of song dynasty as an example中國古代園林的公共性特徵及其對城市生活的影響——以宋代園林為例. *Chinese Landscape Architecture*中國園林, *27*(05), 68–72.

Wang, M.王敏 (2011b). Open of Shanghai's Parks in Modern Times論近代上海公園開放. *Historical Review*史林, *2011*(01), 1–13+188.

Wang, D. D.王丹丹 (2012). Study of the Opening Process of Public Gardens in Beijing in the early Republic of China (1914–1929) 民國初期(1914–1929年)北京公共園林開放初探. *Landscape Architecture*風景園林, *2012*(06), 101–103.

Wang, Y. S. 王元舜 (2014a). Exploration on the origin and evolution of Soviet architectural theory "National in Form, Socialist in Content"蘇聯"民族形式、社會主義內容"建築理論來源與演變之初探. *New Architecture*新建築, (04), 91–93.

Wang, N., Deng, G., & Zhao, W. T.王娜,鄧舸,趙偉韜 (2019). Research on the historical transition of space utilization of Zhongshan Park in Shenyang City瀋陽中山公園空間利用的歷史變遷研究. *Chinese Landscape Architecture*中國園林, *35*(02), 139–144.

Wang, T., & Dong, X. 王濤,董心瑩 (2018). Enhance the ecological value of urban green space and develop public green space in a common effort: Taking the west section design of Shenzhen Bay Park as an example提升生態價值、協同共建公共綠地——以深圳灣濱海休閒帶西段為例. *Chinese Landscape Architecture* 中國園林, *34*(S2), 75–79.

Wang, Y., Man, C. X., Xu, A. L., et al. 王雨,滿彩霞,徐瑷琳等 (2022). Spatial transformation and environmental governance innovation in China from the perspective of Urban political ecology: An analysis based on the dynamic evolution of the urban-water relationship in Shenzhen 城市政治生態學視角下的空間轉型與中國環境治理創新——基於深圳城——水關係動態演變的分析. *Progress in Geography*地理科學進展, *41*(09), 1755–1769.

Wang, F. Y., & Wang, K. Y. 王甫園,王開泳 (2019). Connectivity and distribution pattern of regional greenways and ecological recreation spaces in the Pearl River Delta Urban Agglomeration珠江三角洲城市群區域綠道與生態遊憩空間的連接度與分佈模式. *Progress in Geography*地理科學進展, *38*(03), 428–440.

White, R. (1985). American environmental history: The development of a new historical field. *Pacific Historical Review, 54*(3), 297–335.

Worster, D. (1984). History as natural history: An essay on theory and method. *Pacific Historical Review, 53*(1), 1–19.

Wu, Y. R. 吳友仁 (1979). On the issue of socialist urbanization in China關於我國社會主義城市化問題. *City Planning Review*城市規劃, (05), 13–25.

Wu, R. W.吳人韋 (1998). Development history of urban green spaces in foreign cities國外城市綠地的發展歷程. *City Planning Review*城市規劃, (06), 39–43.

Wu, Q. Y. 吳啟焰 (2011). The critic of the neoliberalism urban studies in China新自由主義城市空間重構的批判視角研究. *Scientia Geographica Sinica*地理科學, *31*(07), 769–774.

Wu, Q. G., & Yang, K. Z. 吳群剛,楊開忠 (2010). Thinking on the integrated development of Beijing-Tianjin-Hebei Region關於京津冀區域一體化發展的思考. *Urban Problems*城市問題, *2010*(01), 11–16.

Wu, L., & Zhang, Q. 武力,張強 (2015). Four transformations of the rural social governance structure鄉村社會治理結構的四次變革. *National Governance*國家治理, *38*(14), 14–25.

Xiao, J., & Cao, K. 肖競,曹珂 (2014). Nowhere to place our field: Study on the invasion about cultivated land in China during the globalization of space production era我們無處安放的農田——空間生產全球化格局下我國耕地資源侵佔問題研究. *Urban Planning International*國際城市規劃, *29*(02), 59–67.

Xiong, Y. Z. 熊月之 (1996). A study of Zhang Garden: A public space in late Qing Dynasty Shanghai 張園晚清上海一個公共空間研究. *Memories and Archives*檔案與史學, *1996*(06), 31–42.

Xiong, Y. Z. 熊月之 (1998). Opening of private gardens and expansion of public spaces in late Qing Shanghai晚清上海私園開放與公共空間的拓展. *Academic Monthly*學術月刊, *1998*(08), 73–81.

Xiong, Y. Z.熊月之 (2013). The parks and daily life in modern Shanghai近代上海公園與社會生活. *Journal of Social Sciences*社會科學, *2013*(05), 129–139.

Xu, G. D.徐國弟 (1994). Conceptualization of development strategy positioning for the Beijing-Tianjin-Hebei Urban agglomeration京津冀城市群體發展戰略定位構想. *Urban Development Research*城市發展研究, *1994*(02), 18–22.

Xu, H. 許浩 (2002). A study on history of Japanese modern urban green space and system對日本近代城市公園綠地歷史發展的探討. *Chinese Landscape Architecture*中國園林, (03), 62–65.

Yang, L. L. 杨赉丽 (2010). A Memory of Shenzhen记忆深圳. *Landscape Architecture*风景园林, (05), 60–62.

Yao, S. M.姚士謀 (1992). Characteristics, types, and spatial distribution of urban agglomerations in China我國城市群的特徵、類型與空間佈局. *Urban Problems*城市問題, *1992*(01), 10–15+66.

Ye, L. Y. 葉林宜 (1995). The flood control effect of the Shenzhen reservoir combined with Sungang Floodgate on the Shenzhen river深圳水庫結合筍崗洩洪閘對深圳河的防洪作用. *Pearl River*人民珠江, (06), 6–8+14.

Yeung, Y. M. (2011). Rethinking Asian cities and urbanization: Four transformations in four decades. *Asian Geographer, 28*(1), 65–83.

Yeung, Y. M., Lee, J., & Kee, G. (2009). China's special economic zones at 30. *Eurasian Geography and Economics, 50*(2), 222–240.

Young, T. (1995). Modern urban parks. *Geographical Review*, 535–551.

Yu, X. X. 於希賢 (1995). Preliminary study on the historical natural environment changes in Beijing北京市歷史自然環境變遷的初步研究. *Journal of Chinese Historical Geography*中國歷史地理論叢, (01), 45–58.

Yu, Y., & Zhang, Q. Y. 余洋,張琦瑀 (2021). Product of cultural change: Analysis of evolution process and mechanism of modern parks in Northeast China文化變遷的產物:東北近代公園演化進程與機制解析. *Landscape Architecture*風景園林, *28*(06), 32–37.

Yuan, S. N., & He, B. J. 苑思楠, 何蓓潔 (2020). The transformation of architectural theories and education after modernism in between autonomy and heteronomy: Speculation and orientation現代主義之後自律與他律價值演變中的建築理論與教學——思辨與定位. *The Architect*建築師, *2020*(02),40–47.

Zhang, Y. X. 張宇星 (1995). Spatial fractal characteristics of urban and urban agglomeration morphology城市和城市群形態的空間分形特性. *New Architecture*新建築, *1995*(03), 42–46.

Zhang, Z. E.張志恩 (1996). Historical evolution and historical evaluation of Lu Xun Park魯迅公園的歷史沿革與歷史評價. *Chinese Landscape Architecture*中國園林, *1996*(04), 22–24.

Zhang, J. H. 章俊華 (2001). City Green Belt Area in Japan (A)日本城市綠地空間(上). *Chinese Landscape Architecture*中國園林, (05), 38–42.

Zhang, T. W. 張庭偉 (2003). The misunderstood globalization and the misled selling place in China 對全球化的誤解以及經營城市的誤區. *City Planning Review*城市規劃, *2003*(08), 6–14.

Zhang, T. W. 張庭偉 (2004a). Neo-liberalism, city management, city governance and city competitiveness in China新自由主義·城市經營·城市管治·城市競爭力. *City Planning Review*城市規劃, *2004*(05), 43–50.

Zhang, J. X. 張京祥 (2004b). Transformation of regional development and planning in the globalization background全球化背景中的區域發展與規劃轉變. *Urban Planning International*國外城市規劃, *2004*(03), 1–4.

Zhang, X. S. 张信思 (2010). The landscape architecture profession in Shenzhen will have a brighter future a celebration of the 30th anniversary of the founding of Shenzhen special economic zone 深圳园林明天会更好 纪念特区成立30周年的回顾与展望. *Landscape Architecture*风景园林, (05), 63.

Zhang, X. S. 張信思 (2017a). Editorial Board of Guangdong Landscape Journal廣東園林期刊編輯部. *Guangdong Landscape Architecture*廣東園林, *39*(05), 4–9.

Zhang, K.張亢 (2020a). Reconstructing nature and politics: On Bruno Latour's political ecology 重構自然與政治:論拉圖爾的政治生態學. *Journal of Dialectics of Nature*自然辯證法通訊, *42*(01), 19–26.

Zhang, Y. C. 張亦弛 (2020b). Transforming Urban Parks in the international settlement of Shanghai at the turn of the 19th and 20th centuries十九、二十世紀之交的上海公共租界城市公園功能變遷研究. *Chinese Landscape Architecture*中國園林, *36*(06), 125–129.

Zhang, L. P. 张利萍 (2023). Long-term characteristics of drinking water quality in rapidly urbanization area: A case study in Shenzhen reservoir (1991–2018) 快速城市化地区饮用水源水质长期变化特征分析——以深圳水库(1991–2018年)为例. Guangdong Water Resources and Hydropower广东水利水电, (08), 31–35.

Zhang, W., & Chen, Y. 章文,陳勇 (2001). Repatriation: When will it emerge from sorrow? .收容遣送,何時走出悲辛? *South Reviews*南風窗, *14*, 36–38.

Zhang, T. J., & Chen, B. Q張天潔,程秉鈴 (2021). City cultural capital and local rebuilding: Analysis of Singapore Park construction and renewal process城市文化資本與地方重塑:新加坡公園建設與更新歷程解析(1968–2020). *Landscape Architecture*風景園林. (06), 56–63.

Zhang, T. J., & Cheng, B. L.張天潔,程秉鈴 (2021). City cultural capital and local rebuilding: Analysis of Singapore Park construction and renewal process城市文化資本與地方重塑:新加坡公園建設與更新歷程解析 (1968–2020). *Landscape Architecture*風景園林, *28*(06), 56–63.

Zhang, Q., & Feng, L. 張清勇,豐雷 (2020). Who are the owners of China's Coastal Wetland? The institutional change and controversy of the ownership of Coastal Wetland. 誰是中國沿海灘塗的所有者?——灘塗所有權的制度變遷與爭議. *China Land Science*中國土地科學, *34*(09), :24–31.

Zhang, T. 張庭偉, & LeGates, R. (2009). Changing planning paradigm in China in the post-neoliberal era後新自由主義時代中國規劃理論的範式轉變. *Urban Planning Forum*城市規劃學刊, *05*, 1–13.

Zhang, X. C., Luan, X. F., Yue, W. Z., et al. 張衔春,欒曉帆,岳文澤等 (2023). Spatial restructuring of land development right under the regional governance framework區域治理框架下的土地發展權空間重構研究. *City Planning Review*城市規劃, *47*(08), 21–28.

Zhang, Y., & Wang, J. 張亦弛,王珺哲 (2019). Jean-Charles Adolphe Alphand and the Promenades in the 19th century Paris阿爾方與19世紀巴黎城市公園建設. *Landscape Architecture*風景園林, *26*(11), 108–112.

Zhao, Y. J. 趙燕菁 (2004). Rapid development and spatial extension: A case study by Shenzhen 高速發展與空間演進——深圳城市結構的選擇及其評價. *City Planning Review*城市規劃, 06, 32–42.

Zhao, J. J. 趙紀軍 (2009a). 60 years of landscape policies and development in China (2): Soviet experience新中國園林政策與建設60年回眸(二)蘇聯經驗. *Landscape Architecture*風景園林, (02), 98–102.

Zhao, J. J. 趙紀軍 (2009b). 60 years of landscape policies and development in China (4): Garden revolution新中國園林政策與建設60年回眸(四) 園林革命. *Landscape Architecture*風景園林, (05), 75–79.

Zhao, J. J. 趙紀軍 (2011). A historical study on the evolution of Shouyi Park, Wuchang武昌首義公園歷史變遷研究. *Chinese Landscape Architecture*中國園林, *27*(09), 70–73.

Zhao, S. Q., & Bai, X. Z. 趙松喬,白秀珍 (1950). Preliminary study on urban geography of Nanjing 南京都市地理初步研究. *Acta Geographica Sinica*地理學報, *1950*(02), 39–71.

Zheng, Q. X., & Liu, Y. L. 鄭瓊現,劉鳶淩 (2013). The restrictive function of parks: A case study of the power struggle between Modern Shanghai Parks and State Power公園的限權功能——以近代上海公園與國家權力的博弈為例. *Social Science Front*社會科學戰線, *2013*(11), 166–170.

Zhou, Y. Y. 周運源 (2010). Several thoughts on the construction of Nan Station New City in the regional economic integration development of the Pearl River Delta南站新城建設在珠三角區域經濟一體化發展中的若干思考. *Urban Insight*城市觀察, (S1), 122–125.

Zhu, S. G. 朱士光 (2000). The relationship between the ecological environment characteristics of the Guanzhong Region in the Western Han Dynasty and the Capital City Chang'an西漢關中地區生態環境特徵與都城長安相互影響之關係. *Journal of Shaanxi Normal University: Philosophy and Social Sciences Edition*陝西師範大學學報:哲學社會科學版, *29*(3), 8.

Zhu, Y. M. 朱英明 (2001). Trends in the regional development of urban agglomerations in China 我國城市群區域聯繫發展趨勢. *Urban Problems*城市問題, *2001*(06), 22–24.

Zhu, W. H. 朱伟华 (2010). A dream of the city of parks the 30 years' development of parks in Shenzhen为了"公园之城"的梦想 深圳公园30年. *Landscape Architecture*风景园林, (05), 64–66.

Zhu, H. B. 朱惠斌 (2013). Urban administrative center location selection and migration type: A case study of Shenzhen-Dongguan-Huizhou Metropolitan Area城市行政中心區位選擇與遷移類型——以深莞惠都市區為例. *Tropical Geography*熱帶地理, *33*(05), 527–532.

Zhu, N.朱楠 (2017). Observations overseas in Late Qing Dynasty and the introduction of park concepts晚清海外見聞與公園觀念的引進. *Cultural Studies*文化研究, (01), 274–286.

Zhu, J. F. 朱劍飛 (2021). What is "the Everyday"? Reading across multiple spheres for an architectural agenda什麼是"日常生活"? 建築學的遠望與策劃. *Time + Architecture*時代建築, *2021*(05), 14–17.

Zhu, Z. J., Yan, C. Y., Wang, X., et al. 朱震峻,嚴晨怡,王欣等 (2017). On the Historical Status of Wuxi Modern Gardens論無錫近代園林的歷史地位. *Chinese Landscape Architecture*中國園林, *33*(10), 47–51.

Zou, B. 鄒兵 (2017). The evolution of urban spatial structure in Shenzhen and the utility evaluation of its planning深圳城市空間結構的演進歷程及其中的規劃效用評價. *Urban & Rural Planning*城鄉規劃, *06*, 69–79.

## Newspaper

Chen, L. F.陳來複 (1988, June 29). Opening of the first lychee festival in Shenzhen: Hong Kong guests attend lychee product exhibition深圳首屆荔枝節開幕 香港來賓出席品荔會. *Ta Kung Pao*大公報, *04*.

Dang, G. Y. 党國英 (2009, September 14). Shenzhen's "rectification action": Struggle between the city and outsiders深圳"梳理行動"、城市與外來者的角力. *Southern Metropolis Daily* 南方都市報.

Feng, L., & Ya, N. 豐雷,亞牛 (2004, May 21). Have you found the house? 你找到房子沒?. *Southern Metropolis Daily* 南方都市報, (A39).

Gong, J. N. 龔江南 (2002, March 1). Recommended Plan for Meilin Park Born Last Night梅林公園推薦方案昨晚誕生. *Shenzhen Evening News*深圳晚報. (A10).

Guan J. 關鍵 (2006, January 23). Overseas Chinese Town Invests 1 Billion Yuan in Building "Happy Coast"華僑城10億元建"歡樂海岸". *Shenzhen Economic Daily*深圳商報. (A02).

Guan, G., Lin, R., & Li, B.關國林 林若飛 李博. (2001, December 20). Buji: Building Guangdong's Premier Town This Way布吉:這樣打造廣東第一鎮. *Shenzhen Ecnomic Daily*深圳商報, A1.

He, J. 何俊 (2022, February 25). Thirty years: The changes and rise of Bao'an Central Area三十年，寶安中心區的變遷與躍升. *Southern Metropolis Daily*南方日報. (BD02).

Huang, Y. L., & Li, H. 黃楊略,李紅 (2001, September 27). Futian: Making Shenzhen's Face More Beautiful 福田: 讓深圳的臉面更靚麗. *Shenzhen Economic Daily*深圳商報. (A4).

Jiang, D. N.姜東南 (1998, September 11). No "Nail" that can't be pulled out – Lessons from the demolition of illegal buildings in Donggualing 沒有拔不掉的"釘子"——冬瓜嶺違章建築群拆除的啟示. *Shenzhen Economic Daily* 深圳商報. (1).

Jin, C., & Chen, S. Z. 金城,陳善哲 (2004, August 9). Shenzhen's "Rectification Action": An exemplar of rapid urbanization in China 深圳"梳理行動":急速城市化的中國標本. *21st Century Business Herald* 21世紀經濟報導. (001).

Li, H. 李灝 (1988, June 29). Opening Speech for the First Lychee Festival in Shenzhen深圳市首屆荔枝節開幕詞. *Shenzhen Special Zone Daily*深圳特區報, 01.

Li, Z. Q. 李振岐 (2019a, January 4). Donggualing, Have You Ever Grown Winter Melons 冬瓜嶺，種過冬瓜嗎? *Daily Sunshine*晶報 (16).

Li, H. & Guan, Y. D.李紅，管亞東 (2004a, May 29). Yesterday's Demolition of Huangmugang Resettlement Area 黃木崗安置區昨日開拆. *Shenzhen Business Daily* 深圳商報. (A3).

Li, H., & Guan, Y. 李紅,管亞東 (2004b, May 29). How did the smooth progress of the citywide "cleanup" proceed successfully: Owners and residents consider the overall situation 全市"梳理"之最何以進展順利:業主居民顧全大局. Shenzhen Economic Daily 深圳商報. (A3).

Liang, E. P. 梁二平 (2022, July 16). Rarely seen "Shenzhen Fair" on ancient maps古地圖上難得一見的"深圳墟". *Shenzhen Evening News*深圳晚報, A07.

Lin, T. H. 林天宏 (2010, December 8). People will vote with their feet人民會用腳投票. *China Youth Daily* 中國青年報.

Liu, X. D. 劉顯東 (2004, September 15). Sorting Action: Who is Struggling with Whom梳理行動:誰與誰角力. *Daily Sunshine*晶報.

Liu, J., Cai, L.H. 劉鍵,蔡良煥 (2008, May 29). Comprehensive launch of the main project of overseas Chinese town happy coast 華僑城歡樂海岸專案主體工程全面啟動. *Shenzhen Special Zone Daily* 深圳特區報. (A01).

Long, L.玲瓏 (2010, March 1st). Lianhuashan Hill: The Settlement Site is my first home 蓮花山:安置區有我第一個家. *Shenzhen Evening News* 深圳晚報.

Ma, X., Zhang, L., & Bai, X. J.馬璿,張淩,白景雪 (2009, December 20). A blank sheet of paper lets dreamers freely express themselves 一張白紙任理想者揮灑筆墨. *Shenzhen Special Zone Daily*深圳特區報 (A03).

Mi, P. M. 米鵬民 (2004, January 19). High-tech Industry: The primary driving force for constructing an international city 高新產業是建設國際化城市第一動力. *Shenzhen Business Daily*深圳商報. (T00).

Peng, S. 彭森 (May 21, 2004). 800-meter Green Belt to be Fully Displayed 800米綠化帶將完整展現. *Shenzhen Evening News* 深圳晚報. (A7).

Peng, Y. 彭琰 (2021, June 24). Encouraging the development of headquarters enterprises: Shenzhen Introduces supportive policies鼓勵總部企業發展 深圳出臺政策支援. *Shenzhen Business Daily*深圳商報.

People's Daily Commentary Department人民日報評論部 (2020, October 26). Embarking on Splendor on a Blank Sheet of Paper在一張白紙上演繹精彩. *People's Daily*人民日報(005).

Qiu, Y. 秋盈 (1988, June 28). Visiting ding bo before the Litchi festival荔枝節前訪丁波. *Shenzhen Special Zone Daily*深圳特區報. (01).

Qiu, H. 邱禾 (2010, August 2). Lianhua mountain and donggualing resettlement areas: Carrying the passionate years of entrepreneurship 蓮花山、冬瓜嶺安置區:承載創業時的激情歲月. *Shenzhen Evening News* 深圳晚報. (24).

Recollection: Unforgettable experience of Shenzhen competing for the international “Garden City” title 回憶:深圳角逐國際“花園城市”的難忘經歷. *Shenzhen Special Zone Daily* 深圳特區報, 2010–12-06.

Shen, Z.沈祖 (2019, August 10). Comrade Zhou Ding’s Life周鼎同志生平. *Shenzhen Special Zone Daily*深圳特區報, A04.

Shenzhen Economic Special Zone Developed Land Road Map深圳經濟特區已開發土地路名圖. *Shenzhen Special Zone Daily*深圳特區報, 1983-02-28(05).

Shenzhen Special Zone Daily深圳特區報. (1981, September 8). Vice Premier Gu Mu and Wan Li Discuss Strengthening Urban Construction萬裡、穀牧副總理談加強城市建設 [News article]. *Shenzhen Special Zone Daily*深圳特區報, 01.

Shenzhen Urban Daily Comprehensive Report深圳都市報綜合報導 (2018, January 16). Historical Overview of Second-Line二線關歷史. *Shenzhen Urban Daily*深圳都市報. (A02).

Sun L. (May 21, 2004). Temporary settlement sites in Huangmugang completely demolished on the 28th 黃木崗臨時安置區28日全拆除. *Southern Daily*南方日報. (001).

Sun, Z. H. & Lou, L. 孫中海,婁荔 (2002, January 13). Meilin back mountain in devastating condition 梅林後山滿目瘡痍. *Shenzhen Economic Daily* 深圳商報. (01).

Sun, R. F., & Zhao, J. 孫榮飛,趙傑 (2005, August 12). “Calling for the effectiveness of the ‘Four Difficulties’ in Shenzhen.” “四個難以為繼”呼喚效益深圳. *First Financial Daily*第一財經日報. (A06).

Wang, Z. Y., Li, C., & Sun, X. W. 王智勇,李純,孫新旺 (2017). Interpretation, evaluation, and enlightenment of the park system planning in Nanjing Capital Plan民國南京《首都計畫》中公園系統規劃的解讀、評價及啟示. *Chinese Landscape Architecture*中國園林, *33*(11), 81–86.

Xinhua News Agency新華社 (1988, June 29). Lychee festival held in Shenzhen深圳舉辦荔枝節. *Overseas Chinese Daily*華僑日報, 2.

Xu, S.徐升 (2001, August 1). The finalized greenery plan for the demolition site of the Lotus Mountain Park resettlement area 蓮花山公園安置區拆除場地綠化方案敲定. *Shenzhen Economic Daily*深圳商報. (A5).

Xu, H. 許嘽 (2020a, May 6). The sea belongs to the state; there is no so-called “ancestral sea owned collectively.”海域屬於國家所有不存在所謂“集體所有的祖宗海”. *Zhuhai Special Zone News*珠海特區報. (02).

Yi, X. H. 伊宵鴻 (2017, April 15). Shenzhen creates a new business card as the “world famous flower city.”深圳打造“世界著名花城”新名片. *Shenzhen Evening News*深圳晚報. (B08).

Yuchun雨純 (1981, June 12). Shenzhen Specialties深圳特產. *Shenzhen Special Zone Daily*深圳特區報, 04.

Zeng, J. Y.曾紀允 (1988, July 7). Opening of the Shenzhen Folk Technology Enterprise Product Exhibition深圳民間科技企業產品展覽開幕. *Shenzhen Special Zone Daily*深圳特區報, 01.

Zhang, Y. G. 張玉閣 (2017b). Deepening Cooperation between Shenzhen and Hong Kong: The Key to the Construction of the Guangdong-Hong Kong-Macao Greater Bay Area 深港合作:粵港澳大灣區建設的關鍵. *China Opening Journal* 開放導報, *2017*(04), 3.

Zhang, X. Y., Guan, G. L., & Chen, S. 張興衍 關國林 陳姝 (2006, October 25). Qiaocheng Wetland: The lake water is expected to become clear next year 僑城濕地湖水明年可望變清. *Shenzhen Business Daily*深圳商報. (A04).

Zhang, X. W., & Rao, J. 張興文,饒潔 (2002, January 14). Ye Minhui, Director of the Municipal Foreign Trade and Economic Cooperation Bureau, Discusses the Train of Thought for Foreign Trade and Economic Work This Year 市外經貿局局長葉民輝談今年外經貿工作思路. *Shenzhen Special Zone Daily*. (01).

## Electronic Literature


Administration of Ocean and Fisheries of Guangdong Province 廣東省海洋漁業局 (2014, May 1). Announcement on the establishment of fishing prohibited areas in parts of the sea in Shenzhen Bay 廣東省海洋與漁業局關於在深圳灣部分海域設置禁漁區的通告. Retrieved from http://ocean.nbt.edu.cn/show/showdetail.php?id=241586

Beijing Zai Ming Law Firm 北京在明律師事務所 (2023, March 3). *How to define collective ownership of coastal tidal flats? Experts say relevant laws are missing* 沿海農民灘塗集體產權如何界定?專家:相關法律缺失亟待完善. Retrieved from https://baijiahao.baidu.com/s?id=1761876830377141777&wfr=spider&for=pc

Chen, Y. 陳玉 (2004b, December 22). *Demolition of the Oyster Fields in Qianhai* 清拆前海蠔田. Retrieved from https://news.sina.com.cn/c/2004-12-22/16234599655s.shtml

Chen, H. X. 陳海先 (2013, April 24). "*Establishment of coastal defense in Shenzhen throughout the ages.*" 歷代深圳海防的設立 Retrieved from https://www.szzx.gov.cn/content/2013-04/24/content_8987085.htm

Ding, B.丁波. *Recollections of Days at the China National Publications Import& Export CO., LTD.*回憶在中圖的日子. Retrieved from http://cnpeak.com/page/ztdw01.html

Du, Y. 杜黯 (2018, July 12). *Ten years of construction: The story of Shenzhen Talent Park*深圳人才公園十年煉成記. Retrieved from https://www.sohu.com/a/240668591_161794

Feng, Q.馮慶 (2022, September 1). *Cai Hongliang: From washing feet in the fields to brainwashing in the city, Caiwuwei is a microcosm of rural urbanization in Shenzhen*蔡洪亮:從洗腳上田到洗腦進城，蔡屋圍是深圳農村城市化的縮影.[EB/OL]. Retrieved from https://mp.weixin.qq.com/s?__biz=MzAxMjA1NDk1MA==&mid=2650016078&idx=1&sn=89e72fb24a77d4571ccba81fde7026e6&chksm=83b71187b4c09891bacab946cb286519b15e959dbfdd3a55c4df65dcc3ef269e9cb21a5e71f5&scene=27

Globalization and World Cities Research Network (GaWC). (2018, November 13). *The world according to GaWC 2018*. Retrieved from https://www.lboro.ac.uk/microsites/geography/gawc/world2018t.html

Guangdong Provincial People's Government廣東省人民政府 (2014, December 11). *Notice of the General Office of the People's Government of Guangdong Province on Issuing the Integrated Planning of Ecological Security System in the Pearl River Delta Region* 廣東省人民政府辦公廳關於印發珠江三角洲地區生態安全體系一體化規劃 (2014–2020). Retrieved from http://www.gd.gov.cn/gkmlpt/content/0/143/post_143525.html#7

Guangdong Provincial People's Government廣東省人民政府 (2020, March 24). *Notice on Issuing the "Shenzhen Bao'an and Longgang Urbanization Land Management Measures" by the Shenzhen Municipal People's Government*深圳市人民政府關於印發《深圳市寶安龍崗兩區城市化土地管理辦法》的通知. Retrieved from http://www.gd.gov.cn/zwgk/wjk/zcfgk/content/post_2939390.html

He, Y. H. 何有華 (2012, August 29). *Former temptations: Crossing the water to the right* 從前的誘惑，涉水向右. Retrieved from http://www.szdag.gov.cn/dawh/szzg/content/post_99164.html

He, H. Y. 何煌友 (2019, March 23). *LuoHu Village Area (from south to north)* 羅湖村區域(由南往北), 1980. Retrieved from http://www.szlh.gov.cn/ssznzt/jy/lszpdb/content/post_5027271.html

Li, H. Y. 李海元 (2008a, November 3). *Massive Transformation Along the Guangdong-Hong Kong Border: 4 Billion Investment in Building a New Border*粵港一線大變革4億建新邊界. Retrieved from https://news.sohu.com/20081103/n260399819.shtml

Li, C. H. 李昌鴻 (2019b, February 28). *Shenzhen's GDP Surpasses Hong Kong for the first time, becoming the economic leader of the Greater Bay Area*深圳GDP首次超香港 成大灣區經濟龍頭. Retrieved from http://news.wenweipo.com/2019/02/28/IN1902280011.htm

Li, G. L., Guo, X., & Feng, L. F. 李國良,國喜,馮立峰 (2008, October 24). *My experience: Border Guard Posts Have Become Scenic Spots* 我的親歷:邊防哨所成了風景線. Retrieved from https://news.sohu.com/20081024/n260234835.shtml

Li, G. L., Hao, X. T., & Feng, L. F. 李國良,郝雪濤,馮立峰 (2008, October 30). *My Experience: The Extended Border Defense Line*我的親歷: 延長的邊防線. Retrieved from https://news.sohu.com/20081030/n260343578.shtml

Lin, T.林濤 (2018, November 14). *20 Years of China Hi-Tech Fair: Witnessing Shenzhen's Industrial Transformation and Innovation Development Dreams*高交會20年:見證深圳產業轉型與創新發展夢想. Southern News Network南方新聞網. Retrieved from https://www.sohu.com/a/275211650_222493.

Liu, J. S., & Song, J. T. 劉家順 宋京濤. (2007, August 17). *Border Defense Sentiments along the Shenzhen Bay*深圳灣畔戍邊情.Retrieved from https://news.cctv.com/military/20070817/109034.shtml

Nanshan District People's Government of Shenzhen深圳市南山區人民政府 (2021, February 18). *Nanshan becomes a gathering place for 'Internet Celebrity Parks,' truly living up to the name of 'The District of a Hundred Gardens'* 南山成"網紅公園"聚集地 "百園之區"名副其實. Retrieved from http://www.szns.gov.cn/xxgk/qzfxxgkml/bmdt/content/post_8555918.html

Overseas Chinese Town Group Corporation華僑城集團公司. (2011, August 17). *Overseas Chinese Town Happy Coast Holds Trial Operation Ceremony, Phase I Projects Successively Unveiled*華僑城歡樂海岸舉行試業盛典 一期項目相繼亮相. Retrieved from http://ysp.net.sasac.gov.cn/n2588025/n2588124/c3983961/content.html

People's Government of Guangdong Province廣東省人民政府. (March 10, 2022). *Regulations on the management of Shenzhen's basic ecological control line*深圳市基本生態控制線管理規定. Retrieved from: http://www.gd.gov.cn/zwgk/wjk/zcfgk/content/post_2532255.html?eqid=a032e0d200011b5a00000006643f56a0

Shenzhen Bay Super Headquarters Base Development and Construction Command Office深圳灣超級總部基地開發建設指揮部辦公室 (2019, June). *International consultation for overall planning, development and operation of shenzhen bay super headquarters base (consultation requirements)* 深圳灣超級總部基地片區統籌開發建設與運營國際諮詢(諮詢要求). Retrieved from https://www.renrendoc.com/paper/217209080.html

Shenzhen Government Online深圳政府線上. (2023, January 24). *Shenzhen Lychee Festival*深圳荔枝節. Retrieved from http://www.sz.gov.cn/szstoryfr/202301/content/post_10391219.html

Shenzhen Institute of Standards and Technology深圳市標準技術研究院 (2021, November 16). *"Yi Shenzhen" reports on the Municipal Standards Institute: Shenzhen releases the first national Evaluation Guidelines for International Talent Community* 《壹深圳》報導市標準院:深圳發佈全國首部《國際人才街區評價指南》. Retrieved from https://www.sist.org.cn/xwzx/mtbd/202112/t20211206_2364350.html

Shenzhen Municipal People's Government深圳市人民政府 (1990, September 30). *Notice on issuing the trial implementation plan for the construction and management of settlement areas for temporary residents in Shenzhen special economic zone*. 深圳市人民政府關於印發《深圳特區暫住人員安置區建設管理試行方案》的通知. Retrieved from www.sz.gov.cn/zfgb/1900/gb9/content/post_10036113.html

Shenzhen Municipal People's Government深圳市人民政府 (1994, March 30). *Notice of the Shenzhen Municipal People's Government on Suspending the Approval of "Three Processing and One Compensation" Projects and Others within the Special Economic Zone*深圳市人民政府關於在特區內停止審批三來一補等專案的通知. Retrieved from http://www.sz.gov.cn/zfgb/1994/gb51/content/post_10090800.html

Shenzhen Municipal People's Government深圳市人民政府 (1998, June 5). *Notice on further implementation plan for cleaning up and dismantling various illegal buildings*关于进一步清理和拆除各类违法违章建筑实施方案的通知. Retrieved from https://www.sz.gov.cn/zfgb/1998/gb143/content/post_10123163.html

Shenzhen Municipal People's Government深圳市人民政府 (2007, May 9). *Comprehensive remediation plan for the marine environment in the Shenzhen Bay Shekou Area*深圳灣蛇口片區海洋環境綜合整治工作方案. Retrieved from http://www.sz.gov.cn/zfgb/2007/gb546/content/post_4988477.html

Shenzhen Municipal Urban Management and Comprehensive Law Enforcement Bureau深圳市城市管理與綜合執法局. (2021). *Compilation explanation of the five-year action plan (2021–2025) for building Shenzhen as a "World-Famous Flower City."* 深圳市打造"世界著名花城"五年行動計畫. Retrieved from http://cgj.sz.gov.cn/attachment/0/766/766058/8645699.pdf

Shenzhen Nanshan District Government深圳市南山區政府 (2021, February 18). *Nanshan becomes a gathering place for "Internet-Famous Parks," living up to its reputation as the "District of a Hundred Parks."* 南山成"網紅公園"聚集地 "百園之區"名副其實. Retrieved from http://www.szns.gov.cn/xxgk/qzfxxgkml/bmdt/content/post_8555918.html

Shenzhen News Website深圳新聞網 (2021, February 1). *Why are there so few stairs on the seaside boardwalk of Shenzhen Bay Park? Latest response!* 深圳灣公園濱海不到上下階梯為什麼這麼少?最新回應! Retrieved from https://wxd.sznews.com/BaiDuBaiJia/20210201/content_460695.html

Shenzhen Talent Work Website深圳人才工作網. (2021, November 15). *The first local standard "Evaluation Guidelines for International Talent Community" released in Shenzhen*全國首部《國際人才街區評價指南》地方標準在深發佈. Retrieved from http://www.sztalent.org/content/2021-11/15/content_24739626.htm

Shenzhen Urban Planning and Land Resources Commission深圳市規劃和國土資源委員會. (2013, August). *Controlling detailed plan for the Shenzhen Bay Super Headquarters Base*深圳灣超級總部基地控制性詳細規劃. Retrieved from https://www.360docs.net/doc/4a18825764.html

Shenzhen Urban Planning and Natural Resources Bureau深圳市規劃和自然資源局 (2014, September 19). *Bulletin on the main data results of the land change survey in Shenzhen for the years 2010–2012* 深圳市*2010–2012*年度土地變更調查主要資料成果的公報. Retrieved from http://pnr.sz.gov.cn/xxgk/sjfb/tjsj/content/post_5837064.html

Shenzhen Urban Planning and Natural Resources Bureau深圳市規劃和自然資源局 (2022, April 26). *Major Data Bulletin (Charts) of the Third National Land Survey in Shenzhen*深圳市第三次全國國土調查主要資料公報(圖表). Retrieved from https://pnr.sz.gov.cn/gkmlpt/content/9/9744/post_9744274.html#4290

Tang, Y. 唐毅 (2007, May 30). *Shenzhen bay: The disappearing indigenous oyster farmers* 深圳灣漸遠的原住蠔民. Retrieved from https://news.sina.com.cn/o/2007-05-30/095211923480s.shtml

Wang, J.王靜 (2018a, December 13). *Zhou Ding: It's unimaginable to achieve today's success without good policies*周鼎:如果沒有好的政策，是無法想像能有今天. Retrieved from http://www.china.com.cn/cppcc/2018-12/13/content_74258098.htm

Wang, Y., Zhang, J., Wang, Z., et al. 王雨,張京祥,王梓懿等 (2022). *Rescaling and cross-border governance transformation of the northern metropolis of Hong Kong under the background of "City-Regions"* 城市—區域"背景下香港北部都會區的尺度重構和跨界治理轉型. Urban Planning International國際城市規劃. https://doi.org/10.19830/j.upi.2022.054

Wu, Y. Y. 武媛媛 (2017, November 16). *How was Shenzhen's first talent-themed park created? Chief designer explains…* 深圳首個人才主題公園是怎麼煉成的?總設計師說…. Retrieved from https://ilonggang.sznews.com/lgnews/content/2017-11/16/content_17764288.htm

Wu, Z.吳臻 (2023, February 28). *What a Predicament! In a famous park in Shenzhen, tourists have to "Climb" in to take photos!* 好狼狽!深圳一網紅公園，遊客要"翻"進去打卡!Retrieved from https://new.qq.com/rain/a/20230228A01L4Q00

## Dissertation

Luan, F. 欒峰 (2004). *Study on the causal mechanisms of rapid urban spatial form evolution since the reform and opening-up*改革開放以來快速城市空間形態演變的成因機制研究 (PHD thesis, Tongji University同濟大學).

Luo, P. 羅佩 (2007a). *Research on the evolution of urban form in Shenzhen*深圳城市形態演進研究 (PHD thesis, Sun Yat-sen University中山大學).
Luo, H. L. 羅華莉 (2012). *The research on the narrative of ancient Chinese Public Garden*中國古代公共園林故事性研究 (Doctoral thesis, Beijing Forestry University北京林業大學).
Peng, H. X. 彭皓炘 (2010). *Ambiguous parks—A political economy analysis of the Riverside Park Development in Taipei*曖昧的公園—臺北市公園綠地向堤外移轉的政治經濟學分析 (Master's thesis, Graduate Institute of Building and Planning, College of Engineering, National Taiwan University,臺灣大學建築與城鄉研究所).
Wang, F. H.王富海 (2003). *Research on the evolution of urban space in Shenzhen*深圳城市空間演進研究 (PHD thesis). Peking University北京大學.
Zhang, Y. Q. 張勇強 (2004). *Research on self-organization of urban space development: A case study of Shenzhen*城市空間發展自組織研究——深圳為例 (Doctoral thesis, Southeast University東南大學).
Zheng, H. X. 鄭浩欽 (2020). *Characteristics and mechanisms of spatiotemporal evolution of Shenzhen reclamation from 1975 to 2020* 1975–2020年深圳填海空間時空演變的特徵及其機制研究. (Master's thesis, Shenzhen University深圳大學).

## Codes and Standards

Ministry of Housing and Urban-Rural Development of the PRC中華人民共和國住房和城鄉建設部, General Administration of Quality Supervision, Inspection and Quarantine of the PRC 中華人民共和國國家品質監督檢驗檢疫總局. (2016). *Park Design Standards*公園設計規範: GB51192-2016. China Architecture & Building Press北京: 中國建築工業出版社.
Ministry of Housing and Urban-Rural Development of the PRC中華人民共和國住房和城鄉建設部. (2017). *Basic terminology standard for landscape architecture*風景園林基本術語標準: CJJ/T91–2017. China Architecture & Building Press北京: 中國建築工業出版社.

## Monographs and Collected Papers

Akinci, G., & Crittle, J. (2008). *Special economic zone: Performance, lessons learned, and implication for zone development (English)*. World Bank Group.
Andersson, T., Conan, M., & Whangheng, C. (2008). Swedish mid-century utopia: Park design as a tool for societal improvements. In M. Conan & C. Wangheng (Eds.), *Gardens, city life and culture: A world tour*. Dumbarton Oaks Research Library and Collection.
Angelo, H. (2021). *How green became good: Urbanized nature and the making of cities and citizens*. University of Chicago Press.
Appadurai, A. (1996). *Modernity at large: Cultural dimensions of globalization*. University of Minnesota Press.
Appadurai, A. (2001). Grassroots globalization and the research imagination. In A. Appadurai (Ed.), *Globalization* (Vol. 2). Duke University Press.
Australia PT Design Consultants Limited & Chaoli Sifang Architectural Design Firm, Shenzhen 澳大利亞柏濤設計諮詢有限公司, 深圳市朝立四方建築設計事務所 (2012). *Architectural Detail Design: Shenzhen Zhongxin Hongshuwan Volume*建築細部設計 深圳中信紅樹灣篇. Dalian University of Technology Press大連: 大連理工大學出版社.
Bach, J. (2017). Shenzhen: From exception to rule. In O'Donnell et al. (Eds.), *Learning from Shenzhen: China's post-Mao experiment from special zone to model city*. The University of Chicago Press.

Bacon, F. 法蘭西斯. 培根 (2012). *The New Atlantis*新大西島. Beijing: The Commercial Press北京: 商務印書館.

Bear, C. (2017). Socio-Nature. In D. Richardson, N. Castree, M. F. Goodchild, & A. Kobayashi (Eds.), *International encyclopedia of geography: People, the earth, environment and technology*. Wiley.

Benton-Short, L., & Short, J. R. (2013). *Cities and nature*. Routledge.

Berrizbeita, A. (1999). The Amsterdam Bos: The modern public park and the construction of collective experience. In J. Corner (Ed.), *Recovering landscape: Essays in contemporary landscape architecture*. Princeton Architectural Press.

Birnbaum, C. A., & Karson, R. S. (2000). *Pioneers of American landscape design*. McGraw Hill.

Blackbourn, D. (2007). *The conquest of nature: Water, landscape and the making of modern Germany*. Random House.

Bloch, M. (1992). *The historian's craft*. Manchester University Press.

Bourdieu, P. (1984). *Distinction: A social critique of the judgment of taste*. Havard University Press.

Brenner, N. 尼爾·博任納 (2019). *City, territory planet: Essays in critical urban theory*城市、地域、星球 批判城市理論 (Trans. by Li Z.G.李志剛 et al.). Commercial Press北京:商務印書館.

Burke, P. (2005). *History and social theory*. Polity Press.

Cannon, T., & Jenkins, A. (Eds.). (2002). *The geography of contemporary China: The impact of Deng Xiaoping's decade*. Routledge.

Cao, M. 曹牧 (2022). *Resources, environment, and rights: Modern pollution transformation and impact of Wangzi River in Tianjin*資源、環境與權益——天津牆子河的近代排汙轉型與影響. In Gao, B., & Hu, H. 高波, 胡恒 (Eds.), How much space is there in "Space"? Revisiting multidimensional history"空間"還有多少空間?重放多維度歷史. China Social Sciences Press北京:中國社會科學出版社.

Cao, R. X., & Yan, S. X. 曹榮新,嚴書翔 (Eds.). (1993). *Shenzhen Talent Market* 深圳人才市場. Hai Tian Press深圳:海天出版社.

Capital Design and Technology Commissioner's Office國都設計技術專員辦事處 (2006). *Capital Planning*首都計畫. Nanjing Publishing House南京:南京出版社.

Carson, R. (2002). *Silent spring*. Houghton Mifflin.

Castells, M. (1983). *The city and the grassroots: A cross-cultural theory of urban social movements*. E. Arnold.

Castells, M. (2010). *The power of identity Volume II (2nd ed., with a new preface.)*. Wiley-Blackwell.

Castells, M. (2011). *The rise of the network society: The information age: Economy, society, and culture*, Volume I (2nd ed., Vol. 12). Wiley.

Castells, M., & Hall, P. (2014). *Technopoles of the world: The making of 21st century industrial complexes*. Routledge.

Central Compilation and Translation Bureau of the Communist Party of China中共中央文獻研究室 (2004). *Deng Xiaoping Chronology (1975–1997)* 鄧小平年譜 (1975–1997). Central Literature Publishing House北京:中央文獻出版社.

Chen, Z.陳植 (1930). *On urbanism and parks*都市與公園論. The Commercial Press北京: 商務印書館.

Chen, J. P. 陳潔萍 (2011a). *Place-writing: The Study on Topography in the expanded field in contemporary architecture, urbanism, and landscape design*書寫場地:當代建築、城市、景觀設計中擴展領域的地形學研究. Southeast University Press南京:東南大學出版社.

Chen, Y. X.陳一新 (2015a). *A thirty-year historical study on urban planning and construction of Shenzhen Futian Central Business District (CBD)* 深圳福田中心區 (CBD)城市規劃建設三十年歷史研究. Southeast University Press南京:東南大學出版社.

Chen, Y. X. 陳一新 (2015b). *Shenzhen Futian CBD: A thirty-year history of planning and construction*深圳福田中心區 (CBD)城市規劃建設三十年歷史研究. Southeast University Press南京:東南大學出版社.

Chen, Y. X. 陳一新 (2022). *The history of urban planning in Shenzhen*深圳城市規劃簡史. China Social Sciences Press北京: 中國社會科學出版社.

Cheng, S. Y. 程遂營 (2002). *Research on the ecological environment of Kaifeng during the Tang and Song Dynasties*唐宋開封生態環境研究. China Social Sciences Press中國社會科學出版社.
China Development Research Foundation中國發展研究基金會 (2022). *Research on the integration of city clusters in China*中國城市群一體化研究. China Development Press北京:中國發展出版社.
Chinese Academy of Sciences, Editorial Committee of Flora of China中國科學院中國植物志編輯委員會 (1979). *Flora of China (Volume 27)*中國植物志(第27卷). Science Press北京:科學出版社.
Cloke, P., Crang, P., & Goodwin, M. (2005). *Introducing human geographies* (2nd ed.). Routledge.
Cohen, P. A. (2010). *Discovering history in China: American historical writing on the recent Chinese past*. Columbia University Press.
Collingwood, R. G. (1960). *The idea of nature*. Oxford University Press.
Conan, M. (1999a). Introduction. In M. Conan (Ed.), *Perspectives on garden history*. Dumbarton Oaks.
Conan, M. (1999b). *From vernacular gardens to a social anthropology of gardening*. Perspectives on Garden Histories. Dumbarton Oaks.
Corner, J. (2014). *The landscape imagination: Collected essays of James Corner, 1990–2010*. Princeton Architectural Press.
Cosgrove, D. E. (1998). *Social formation and symbolic landscape*. Univ of Wisconsin Press.
Cosgrove, D., & Daniels, S. (Eds.). (1988). *The iconography of landscape: Essays on the symbolic representation, design and use of past environments* (Vol. 9). Cambridge University Press.
Crang, M. (2013). *Cultural geography*. Routledge.
Cranz, G. (1982). *The politics of park design: A history of urban parks in America*. MIT press.
Cronon, W. (1992). *Nature's metropolis: Chicago and the Great West*. W.W. Norton.
Darby, W. J. (2020). *Landscape and identity: Geographies of nation and class in England*. Routledge.
Dong, G., & Zhang, Y董觀志 張穎 (2008). *Tourism + Real Estate: Business Model of Overseas Chinese Town* 旅遊+地產:華僑城的商業模式. Sun Yat-sen University Press中山大學出版社.
Douglas, I. (2013). *Cities: An environmental history*. Bloomsbury Publishing.
Du, J. (2020). *The Shenzhen experiment: The story of China's instant city*. Harvard University Press.
Duan, Y. B.段亞兵 (2018). *Pioneering Chronicles of Shenzhen*深圳拓荒紀實. People's Publishing House北京:人民出版社.
Editorial Committee of "Guangdong Forest."《廣東森林》編委會 (1990). *Guangdong Forest* 廣東森林. Guangdong Science and Technology Press 廣州:廣東科技出版社, China Forestry Publishing House中國林業出版社.
Emma, X. M., & Blackwell, A. (2017). The political architecture of the first and second lines. In O'Donnell et al. (Eds.), *Learning from Shenzhen: China's post-Mao experiment from special zone to model city*. The University of Chicago Press.
Ewen, S. (2016). *What is urban history*. Polity Press.
Farole, T., & Akinci, G. (2011). *Special economic zones: Progress, emerging challenges, and future directions (English)*. World Bank Group.
Feather, S. (1990). *Consumer culture and postmodernism*. Sage.
Febvre, L. (2013). *A geographical introduction to history*. Routledge.
Fei, X. T. 費孝通 (1986). *On small towns and others*論小城鎮及其他. Tianjin People's Publishing House天津: 天津人民出版社.
Feng, B. Y. 馮邦彥 (2002). *A study on the industrial structure of Hong Kong*香港產業結構研究. Management Publishing House北京:經營管理出版社.
Fraser, D., & Sutcliffe, A. (1983). *The Pursuit of urban history*. E. Arnold.
Gandy, M. (2003). *Concrete and clay: Reworking nature in New York City*. MIT Press.
Grantham, A. (2011). *Via Ports: From Hong Kong to Hong Kong* (Vol. 1). Hong Kong University Press.
Gu, M. 穀牧 (2009). *Memoirs of Gu Mu*穀牧回憶錄. Central Literature Publishing House北京:中央文獻出版社.

Gu, X. J. 辜曉進 (Eds.). (2018). *In the Name of News III: Selected works of news graduation design from Shenzhen University – Frontier Observations* 新聞的名義III:深圳大學新聞畢業設計作品選·前沿觀察. Renmin University of China Press 北京:中國人民大學出版社.

Gu, C. L., 顧朝林et al. (1999). *Economic globalization and urban development in China: A study on the development strategy of Chinese Cities across the Century*經濟全球化與中國城市發展——跨世紀中國城市發展戰略研究. Commercial Press北京: 商務印書館.

Guangdong Institute of Botany廣東省植物研究所 (1976). *Vegetation of Guangdong* 廣東植被. Science Press 北京: 科學出版社.

Harvey, D. (1989). *The condition of postmodernity: An enquiry into the origins of cultural change*. Blackwell.

Harvey, D. (2005a). *A brief history of neoliberalism*. Oxford University Press.

Harvey, D. (2005b). *Spaces of neoliberalization: Towards a theory of uneven geographical development* (Vol. 8). Franz Steiner Verlag.

Harvey, D.哈威 (2010). *Justice, nature and the geography of difference*正義、自然和差異地理學. (Trans. by Hu D. P. 胡大平). Shanghai People's Publishing House上海: 上海人民出版社.

Harvey, D. (2016). *The ways of the world*. Profile Books.

Hisa, C. J. 夏鑄九 (2016). *On Heterotopias: Selected essays*異質地方之營造:理論與歷史(I). Tonsan Publications臺北: 唐山出版社.

Hobsbawm, E. J. From social history to the history of society從社會史到社會的歷史. In Cai, S. Q. (Eds.), *Representing the past: The theoretical perspective of social history*再現過去:社會史的理論視野. Zhejiang People's Publishing House杭州:浙江人民出版社.

Hoskins, W. G. (1955). The making of the English landscape. Hodder and Stoughton.

Hsia, C. J. 夏鑄九 (Ed. & Trans.) (1989). *Readings in social theories and the cultural form of spaces*空間的文化形式和社會理論讀本. Ming Wen Book, Ltd. 臺北:明文書局.

Hsia, C. J.夏鑄九 (2015). *A glimpse of devil appearance: Planning and design in urban and regional processes*窺見魔鬼的容顏. Tonsan Publications臺北:唐山出版社.(pp. 1–5).

Hsia, C. J. 夏铸九 (2020). *Representation of space: Rupture and repair*空间再现: 断裂与修复. Tongji University Press上海: 同济大学出版社.

Huang, P. C. (1990b). *The peasant family and rural development in the Yangzi Delta, 1350–1988*. Stanford University Press.

Huang, P. C. 黃宗智 (2022). *The dual unity of state and society*國家與社會的二元合一. Guangxi Normal University Press桂林:廣西師範大學出版社. (pp. 238–275).

Hughes, J. D. (2016). *What is environmental history?* (pp. 1–7). Wiley.

Hunt, J. D. (1999). *Approaches (new and old) to garden history*. Perspectives on Garden Histories. Dumbarton Oaks.

Hunt, J. D. (2000). *Greater perfections: The practice of garden history*. Thames & Hudson.

Hunt, J. D. (2013). *A cultural history of gardens in the modern age* (Vol. 6). Bloomsbury Publishing.

Hunt, J. D. (2014). *A world of gardens*. Reaktion Books.

Hunter, J. M. (1985). *Land into landscape*. George Godwin.

Ishikawa, M., 石川幹子 (2014). *City and green space: Striving to create a new urban environment*城市與綠地——為創造新型城市環境而努力. (Trans. by Lei, Y., et al.雷芸 等). China Architecture & Building Press北京:中國建築工業出版社.

Jackson, J. B. (1984). *Discovering the vernacular landscape*. Yale University Press.

Ji, C. (Author) 計成 原著, & Wang, S. Z.王紹增 (Eds.). (2013). *The craft of gardens reader*園冶讀本. China Architecture & Building Press北京:建築工業出版社.

Jiang, Y. X.蔣有緒 (1999). Urban forestry in the new century: On the construction of ecological landscape forests and the demonstrative significance of Shenzhen City新世紀的城市林業方向:試論生態風景林建設兼論深圳市的示範意義. In: Wei X. Q. 魏小琴 (Eds.), *A covenant of the century: Collected works on the construction of ecological landscape forests in Shenzhen*世紀之約:深圳市生態風景林建設文集. China Forestry Publishing House.

Jiang, R., & Nan, Z. 蔣榮耀,南兆旭 (2019). *Time, mountains, and rivers: A century of Shenzhen historical images* 歲月山河:百年深圳歷史影像紀. Haitian Publishing House 深圳:海天出版社.

Joas, H., & Knöbl, W. (2009). *Social theory: Twenty introductory lectures*. Cambridge University Press.

Jones, D. (2015). *Crime, protest, community, and police in nineteenth-century Britain*. Routledge.

Kirkby, R. J. R. (1985). *Urbanisation in China: Town and country in a developing economy, 1949–2000 AD*. Croom Helm.

Konijnendijk, C. C., Annerstedt, M., Nielsen, A. B., & Maruthaveeran, S. (2013). *Benefits of urban parks. A systematic review*. A Report for IFPRA, Copenhagen & Alnarp.

Lan, X. H. 蘭小歡 (2021). *Immersed in affairs: The Chinese Government and Economic Development*置身事內 中國政府與經濟發展. Shanghai People's Publishing House上海:上海人民出版社.

Le Goff, J.雅克·勒高夫. (1988). The New History新史學. In Cai S. Q. (Eds.), *Representing the past: The theoretical perspective of social history*再現過去:社會史的理論視野. Zhejiang People's Publishing House杭州:浙江人民出版社.

Le Goff, J. (2015). *Must we divide history into periods?* Columbia University Press.

Leatherbarrow, D. (2015a). *Topographical stories: Studies in landscape and architecture*. University of Pennsylvania Press.

Leatherbarrow, D. (2015b). Is landscape architecture? In G. Doherty & C. Waldheim (Eds.), *Is landscape...?: Essays on the identity of landscape*. Routledge.

Leatherbarrow, D. (2016). Not anywhere, not only here. In C. Girot & D. Imhof (Eds.), *Thinking the contemporary landscape* (1st ed.). Princeton Architectural Press.

Lefebvre, H. (1991). *Critique of everyday life, volume I, Introduction* (Trans. by John Moore). Verso.

Lefebvre, H.亨利·列斐伏爾 (2015). *Space and Politics*空間與政治 (Trans. by Li C. 李春). Shanghai People's Publishing House上海: 上海人民出版社.

Levenson, J. (1970). *Liang Chi-chao and the mind of modern China* (2nd rev. ed.). University of California Press.

Li, M.李敏 (1987). *Chinese modern parks: Development and evaluation*中國現代公園——發展與評價. Beijing Science and Technology Press北京:北京科學技術出版社出版.

Li, H.李灝 (1993). Taking practical steps and actively and steadily developing Shenzhen into an international city. In Lin, Z.林祖基 (Eds.). *Essays on Shenzhen – An International City*深圳-國際性城市論文集. China Economic Publishing House 北京:中國經濟出版社.

Li, R. K. 黎瑞坤 (1997). Discussion on Modern Urban Greening Construction and Management現代化城市園林建設與管理淺談. In Shenzhen Municipal Foreign Training Leadership Group 深圳市涉外培訓領導小組 (Eds.), *Research series on international successful urban management experiences: Hong Kong urban management studies*國際性城市成功管理經驗研究叢書:香港城市管理研究.

Li, D. 李定 (2008b). How Shenzhen Explored the Path of Market Economy深圳是如何探索市場經濟之路的. In *Shenzhen municipal committee of the Chinese People's political consultative conference, cultural and historical committee, and learning committee*深圳市政協文史和學習委員會 (Eds.). A Miracle of a City一個城市的奇跡. China Social Sciences Press北京:中國文史出版社.

Li, Y. T. 李詠濤 (2010). *Da Dao 30: National Memories on Shennan Avenue*大道30 深南大道上的國家記憶. Shenzhen Press Group Publishing House深圳:深圳報業集團出版社.

Li, Z. 李子彬. (2020). *My Time as the Mayor of Shenzhen*我在深圳當市長. CITIC Publishing Group中信出版社.

Li, H. J., Huang, H. H., & Ye, Y. R. 李華傑,黃涵蓀,葉煜榮 (1982). The Rise of Shekou Industrial Zone – An investigation into the establishment of Shekou Industrial Zone by China Merchants Group 蛇口工業區的崛起——對招商局創辦蛇口工業區的調查. *Academic Research*學術研究, (01), 13–18.

Liang, S. C., & Lin, Z. 梁思成 著 林洙 (Eds.) (2007). *Grand simplicity and beauty: Liang Sicheng's most beautiful architectural writings*大拙至美: 梁思成最美的文字建築. China Youth Publishing House北京:中國青年出版社.

Lin, Z. J. 林祖基 (1993). Actively and steadily building Shenzhen into an international city based on reality 從實際出發積極穩步地把深圳建設成為國際性城市. In Lin, Z. J. 林祖基 (Eds.). *Essays on Shenzhen – An international city* 深圳-國際性城市論文集. China Economic Publishing House 北京:中國經濟出版社. (pp. 4–11).

Lin, Z.林峥 (2022). *Park Beijing: Cultural production and literary imagination*公園北京 文化生產與文學想像. Peking University Press北京: 北京大學出版社.

Lin, Z. J. 林祖基 (Eds.). *Essays on Shenzhen – An International City*深圳-國際性城市論文集. China Economic Publishing House北京:中國經濟出版社. (pp. 8).

Liu, S. H. 柳尚華 (Eds.) (1999). *Contemporary 50 years of Chinese landscape architecture*中國風景園林當代50年. China Architecture & Building Press北京:中國建築工業出版社.

Liu, D. Y. 劉東洋 (2012). Seeing topography from the window behind my house從我家後窗看見地形學. In Cousins, M., Chen, W., Li, H., Ge, M. 卡森斯, 陳薇, 李華, 葛明 (Eds.). *Architectural Research 02: Topography & mental space*建築研究02:地形學和心理空間. China Architecture & Building Press北京:中國建築工業出版社. (pp. 73–106).

Liu, S. Y. 劉守英 (2018b). *Land system and China's development*土地制度與中國發展. Renmin University of China Press北京:中國人民大學出版社.

Lu, D. (2006). *Remaking Chinese urban form: Modernity, scarcity and space, 1949–2005*. Routledge.

Lu, S. C., & Lu, D. Q. 盧受采 盧冬青 著 (2002). *A history of Hong Kong economy* 香港經濟史. Joint Publishing (Hong Kong) Co., Ltd 香港:三聯書店(香港)有限公司.

Luo, C. R. 羅昌仁 (1987). Improving urban planning in Shenzhen把深圳市的規劃工作做得更好. In Hai Tian Publishing House海天出版社. *Comprehensive Review of the General Plan of Shenzhen Special Economic Zone* 深圳經濟特區總體規劃評論集. Hai Tian Publishing House深圳:海天出版社. (pp. 31–35).

Luo, Q. H.羅清和 (2007b). *Economic development and industrial growth: A study on the industrial growth and development trends in Shenzhen*經濟發展與產業成長：深圳產業成長與發展趨勢研究. Shanghai Sanlian Bookstore上海: 上海三聯書店.

Luo, X. 羅新 (2020). *Rebels who refuse to conform: Criticism, doubt, and imagination*有所不為的反叛者 批判、懷疑與想像力. Shanghai Joint Publishing Company上海: 上海三聯書店.

Marks, R. (1998). *Tigers, rice, silk, and silt: Environment and economy in late imperial South China*. Cambridge University Press.

Marx, K., & Engels, F. 卡爾·馬克思, 弗裡德里希·恩格斯 (2001). *Marx-Engels collected works (Volume 44)* 馬克思恩格斯全集(第44卷). (Trans. by Central Compilation and Translation Bureau of Marxist, Leninist, Stalinist Works of the Central Committee of the Communist Party of China中共中央馬克思恩格斯列寧史達林著作編譯局). People's Publishing House北京: 人民出版社.

Mayhew, S. (2015). *A dictionary of geography*. Oxford University Press.

McGee, T., Lin, G. C., Wang, M., Marton, A., & Wu, J. (2007). *China's urban space: development under market socialism* (Vol. 29). Routledge.

Melosi, M. V. (2000). *Effluent America: Cities, industry, energy, and the environment*. University of Pittsburgh Press.

Meng, X. W., et al. 孟祥偉等 (Eds.) (2021). *In the City Center: Operation and management of overseas Chinese Town Wetland and the model of nature education* 在城中央——華僑城濕地運營管理及自然教育模式. China Forestry Publishing House 北京: 中國林業出版社.

Moran, J. (2010). *Interdisciplinarity*. Routledge.

Nash, R. F. (2014). *Wilderness and the American mind*. Yale University Press.

National Archives of Singapore新加坡國家檔案館 編. (2015). *Lee Kuan Yew's Governance Strategies*李光耀執政方略. People's Publishing House北京:人民出版社.

National Committee of the Chinese People's Political Consultative Conference, Committee on Culture, History, and Learning全國政協文史和學習委員會 (2009). *Selected compilation of historical and literary materials, Issue 153*文史資料選輯. 第153輯. China Social Sciences Press北京:中國文史出版社.

Nie, Z. G. 聶政光 (1994). *Research on Talent Faultage in Shenzhen*深圳人才斷層研究. Hai Tian Press深圳:海天出版社.

O'Donnell, M. A. (2017). Laying Siege to the Villages: The Vernacular Geography of Shenzhen. In O'Donnell et al. (Eds.), *Learning from Shenzhen: China's post-Mao experiment from special zone to model city*. The University of Chicago Press.

O'Donnell, M. A., Wong, W., & Bach, J. (Eds.). (2019). *Learning from Shenzhen: China's post-Mao experiment from special zone to model city*. University of Chicago Press.

Olmsted, F. L., Schuyler, D., Kaliss, G., & Schlossberg, J. (2015). *The Papers of Frederick Law Olmsted (Volumn III)*. Johns Hopkins University Press.

Ou, L. T. 歐力同 (1987). *Kong De and His Positivism*孔德及其實證主義. Shanghai Academy of Social Sciences Press上海:上海社會科學院出版社.

Palley, T. I. (2005). From keynesianism to neoliberalism: Shifting paradigms in economics. In A. Saad Filho & D. Johnston (Eds.), *Neoliberalism: A critical reader*. Pluto Press.

Pastoureau, M. (2014). *Green: The history of a color*. Princeton University Press.

Peng, Q.彭全民 (2015). *Shenzhen anecdotes: A casual discussion*深圳掌故漫談. Shenzhen Press Group Publishing House深圳:深圳報業集團出版社.

Porphyrios, D. (1981). *On the methodology of architectural history*. Academy Editions.

Qiu, D. 邱大昭 (1990). *On the Issue of Talent Fault in Professional and Technical Fields* 試論專業技術人才斷層問題. Chinese Talents中國人才.

Quest-Ritson, C. (2003). *The English garden: A social history*. David R. Godine Publisher.

Reed, M. (1990). *The Landscape of Britain: From the Beginnings to 1914*. Routledge.

Robbins, P. (2012). *Lawn people: How grasses, weeds, and chemicals make us who we are*. Temple University Press.

Robbins, P. (2020). *Political ecology: A critical introduction*. Wiley.

Rosenzweig, R., & Blackmar, E. (1992). *The park and the people: A history of Central Park*. Cornell University Press.

Rowe, W. T. (1992). *Hankow: Conflict and community in a Chinese city, 1796–1895*. Stanford University Press.

Rowe, C., & Slutzky, R.. 柯林·罗, 罗伯特·斯拉茨基 (2007). *Transparency* (Trans. by Jin, Q. Y., & Wang, Y. J. 金秋野 王又佳). China Architecture & Building Press北京: 中國建築工業出版社.

Ruan, Z. 阮志 (2014). *Entering Forbidden Zones: A History of Hong Kong's Border Restricted Areas*入境問禁 香港邊境禁區史. Joint Publishing (Hong Kong) Co., Ltd 香港:三聯書店(香港)有限公司.

Said, E. W. (2003). *Orientalism*. Penguin.

Said, E. W. (2012). *Representations of the intellectual* (p. xviii). Vintage Books.

Sassen, S. (1991). *The global city* (2nd ed.). Princeton University Press.

Sassen, S. (2014). *Expulsions: Brutality and complexity in the global economy*. Harvard University Press.

Sassen, S. (2018). *Cities in a world economy*. Sage.

Schmidt, A. (2014). *The concept of nature in Marx* (Vol. 8). Verso Books.

Scott, A. J. (2012). *A world in emergence: Cities and regions in the 21st century*. Edward Elgar Publishing.

Shen, B. J., Shen, T. X., & Shen, T. P. 沈葆久,沈天翔,沈天鵬 (1994). *Shenzhen new gardens: Abstract gardens*深圳新園林——抽象式園林. Haitian Publishing House深圳:海天出版社.

Shenzhen Institute of Building Design Co., Ltd. 深圳市建築設計研究總院有限公司(Eds.). (2019). *Shenzhen forty years: Industry and city*深圳四十年:產業與城市. China Architecture & Building Press北京:中國建築工業出版社.

Shenzhen Municipal Planning and Land Resources Commission深圳市規劃和國土委員會 (2010). *Urban planning practice in Shenzhen's Fifteen Years of Reform and opening-up: 1980–1995*深圳改革開放十五年的城市規劃實踐: 1980–1995年. Haitian Publishing House 深圳: 海天出版社.

Shenzhen Nanshan District Shekou Sub-district Office Compilation深圳市南山區蛇口街道辦 (2022). *Strolling in Shekou*蛇口漫步. Shenzhen Press Group Publishing House 深圳: 深圳報業集團出版社.

Shi, P., et al. 史培軍 等 (2012). *Landscape urbanization and ecological infrastructure construction: Shenzhen case*景觀城市化與生態基礎設施建設:以深圳為例. Science Press北京:科學出版社.

Shirahata, Y.白幡洋三郎. (2014). *A History of Modern Urban Parks: Origins of Westernization* 近代都市公園史:歐化的源流. New Star Press 北京: 新星出版社.

Skinner, G. W. (1977). *The City in late imperial China*. Stanford University Press.

Smith, N. (2010). *Uneven development: Nature, capital, and the production of space*. University of Georgia Press.

Spirn, A. W. (1996). Constructing nature: The legacy of Frederick Law Olmsted. In W. Cronon (Ed.), *Uncommon ground: Rethinking the human place in nature*. WW Norton & Company.

Spirn, A. W. (2000). *Ian McHarg, landscape architecture, and environmentalism: Ideas and methods in context*. Environmentalism in landscape architecture, Dumbarton Oaks.

State Council of the PRC issued "Regulations on Foreign Trade and the Increase of Foreign Exchange Earnings" 國務院頒發《關於大力發展對外貿易增加外匯收入若干問題的規定》. In Wang, Z. C. 王振川 (Eds.), *Yearbook of China's reform and opening-up new era*中國改革開放新時期年鑒. Press of Chinese Democratic Legal System北京: 中國民主法制出版社.

Steinberg, T. (2014). *Gotham unbound: The ecological history of greater New York*. Simon and Schuster.

Sun, X. X. 孫筱祥 (2011). *Garden Art and Garden Design*園林藝術及園林設計. China Architecture & Building Press北京:中國建築工業出版社.

Sze, J. (2015). *Fantasy islands: Chinese dreams and ecological fears in an age of climate crisis*. University of California Press.

Tafuri, M. (1980). *Theories and history of architecture*. Granada.

Teng, S., & Fairbank, J. K. (1982). *China's response to the West: A documentary survey, 1839–1923*. Harvard University Press.

Thernstrom, S. (1964). *Poverty and progress: Social mobility in a nineteenth century city*. Harvard University Press.

Timothy, J. G. (2006). *Millennium Park: Creating a Chicago Landmark*. University of Chicago Press.

Tournikiotis, P. (2001). *The historiography of modern architecture*. The MIT Press.

Trieb, M. (1999). Nature recalled. In J. Corner (Ed.), *Recovering landscape: Essays in contemporary landscape theory*. Princeton Architectural Press.

Trouillot, M. R. (2015). *Silencing the past: Power and the production of history*. Beacon Press.

Tu, Q. 塗俏 (2016). *Biography of Yuan Geng: Scenes of Reform 1978–1984* 袁庚傳: 改革現場 1978–1984. Hai Tian Press深圳: 海天出版社.

Vogel, E. F. (1989). *One step ahead in China: Guangdong under reform* (Vol. 2). Harvard University Press.

Vogel, E. F. (2017). Foreword. In O'Donnell et al. (Eds.), *Learning from Shenzhen: China's post-Mao experiment from special zone to model city*. The University of Chicago Press.

Wang, Z. C.王振川 (Eds.) (1979). *Yearbook of the New Period of Reform and opening-up in China*中國改革開放新時期年鑒. China Democratic Legal System Press北京:中國民主法制出版社.

Wang, R. B.王若兵. (1990). *Water resources chronicle of Shenzhen*深圳市水利志. Guangdong Science and Technology Press廣州:廣東科技出版社.

Wang, R. B.王若兵 (2001b). *Shenzhen water affairs annals*深圳水務志. Haitian Publishing House 深圳:海天出版社.

Wang, D. Y.王定躍 (Eds.) (2011). *Investigation and Planning of the Scenic Forest in Wutong Mountain Scenic Area*梧桐山風景區風景林調查與規劃. Shenzhen Press Group Publishing House深圳:深圳報業集團出版社.

Wang, S. W.王世巍 (2014b). *Population change in Shenzhen, 1979–2010*深圳市人口變遷研究 1979–2010.深圳:海天出版社 Haitian Press.

Wang, D. J. 王地久 (2018b). Analysis of the Agricultural Situation in Shenzhen in 1979 1979年深圳農民狀況分析. In Shenzhen Municipal Chronicles Office (Eds.) 深圳市史志辦公室. *Chronicles of Shenzhen's Reform and opening-up, Volume II* 深圳改革開放實錄. 第二輯. Shenzhen Press Group Publishing House 深圳: 深圳報業出版社. (pp. 1–10).

Wang, J. B. 王江波 (2022). *Research on the transformation of Shenzhen's land system*深圳土地制度變遷研究 (1979–2020). China Financial and Economic Publishing House北京: 中國財政經濟出版社.

Weber, M. (2008). *Confucianism and Taoism*儒教與道教 (Tran. by Hong, T. F. 洪天富). Jiangsu People's Publishing House南京:江蘇人民出版社.

Whatmore, S. (2007). Culture-Nature 文化-自然. In P. Cloke et al. (Eds.), *An introduction to human geography*人文地理概論. (Trans. by Wang Zhihong王志弘 et al.). Chuliu Book, LTD. 巨流圖書股份有限公司.

Wheatley, P. (1971). *The pivot of the four quarters: A preliminary enquiry into the origins and character of the ancient Chinese city*. Edinburgh University Press.

Williams, R. (1985). *Keywords: A vocabulary of culture and society*. Oxford University Press.

World Bank. (2003). *The East Asian miracle: Economic growth and public policy*. Oxford University Press.

World Bank Group (2015). *East Asia's changing urban landscape: Measuring a decade of spatial growth. Urban Development.* World Bank. https://openknowledge.worldbank.org/handle/10986/21159 License: CC BY 3.0 IGO

Wu, J. 吳家驊 (1999). *Landscape morphology: A comparative study of landscape aesthetics*景觀形態學 景觀美學比較研究. China Architecture & Building Press北京: 中國建築工業出版社.

Wu, J. C.吳景超 (2008). *The fourth types of nations*第四種國家的出路. Commercial Press北京:商務印書館.

Wu, R. S. 巫仁恕 (2010). From Sightseeing to Tourism: Changes in Suzhou's Tourism Activities and Spaces from the 16th to the Early 20th Centur y從遊觀到旅遊:16至20世紀初蘇州旅遊活動與空間的變遷. In R. S. Wu, B. Kang, & M. L. Lin巫仁恕 康豹 林美莉 (Eds.), *Viewing modernity in China from urban perspective*從城市看中國的現代性. Institute of Modern History, Academia Sinica臺北:中央研究院近代史研究所.

Wu, D., & Qin, Q. L. 吳鏑,秦群力 (1993). Vigorously Developing Trade and Tourism, Accelerating the Pace of Shenzhen's Construction as an International City 大力發展貿易、旅遊業，加快深圳建設國際性城市步伐. In Lin, Z. J.林祖基 (Eds.), *Shenzhen – Collection of Essays on an International City* 深圳-國際性城市論文集. China Economic Publishing House 北京:中國經濟出版社.(pp. 120–126).

Xia, Z. Z. 夏柱智 (2022). *Between the City and the Village Rural Revitalization in the Process of Urbanization*亦城亦鄉 城市化進程中的鄉村突圍. Guangxi Normal University Press桂林: 廣西師範大學出版社.

Xiao, X. D. 肖曉丹 (2018). *A historiography of European urban environmental history*歐洲城市環境史學研究. Sichuan University Press成都:四川大學出版社.

Xu, Y. C. 許淵沖 (Trans.) (2011). *300 Tang Poems: A Chinese-English Bilingual Edition* 唐詩三百首:漢英對照. Wuzhou Communication Press北京:五洲傳播出版社.

Xu, H.徐航 (Eds.) (2020b). *Shenzhen Bay No. 1: Standing Tall with Belief*深圳灣1號 因信念而屹立. Haitian Publishing House 深圳: 海天出版社.

Xu, Y. C. (Trans.) 許淵沖 譯. (2021). *Book of Poetry Book of Songs: Feng (Chinese-English Bilingual Edition)* 許淵沖譯詩經.風:漢英對照. China Translation and Publishing Corporation 北京:中譯出版社.

Xu, S. L.許仕廉 (2022). *Population issues in China*中國人口問題. Commercial Press北京:商務印書館.

Xu, Y. C., & Xu, M. (Trans.) 許淵沖 許明 譯. (2021). *Goden Treasure of Quatrains and Octaves (Chinese-English Bilingual Edition)* 許淵沖譯千家詩:漢英對照. China Translation and Publishing Corporation北京:中譯出版社.

Xue, C. (2006). *Building a revolution: Chinese architecture since 1980*. Hong Kong University Press.

Yeung, Y. M. 楊汝萬 (2004). *Asia-Pacific cities in the context of globalization*全球化背景下的亞太城市. Science Press北京: 科學出版社出版.

Yozaburo, S. 白幡洋三郎 (2014). *A History of Modern Urban Parks: The origin of Europeanization*近代都市公園史:歐化的源流. (Trans. by Li Wei and Nan Cheng李偉 南誠). New Star Press北京: 新星出版社.

Yu, P. 於沛 (2007). *Globalization and global history*全球化和全球史. Social Sciences Academic Press北京:社會科學出版社.

Zaitzevsky, C. (1982). *Frederick Law Olmsted and the Boston Park System*. Belknap Press.

Zaizhi, Z., & Chokkalingam, U. Success and sustainability: Lessons from Guangdong Province. In U. Chokkalingam et al. (Eds.), *Learning lessons from China's forest rehabilitation efforts: National level review and special focus on Guangdong Province*. CIFOR.

Zhang, Z. X.張振聲 (1999a). Unforgettable one year and three months 難忘的一年零三個月. In Witness Shekou compilation committee見證蛇口編委會 (Eds.) *Witnessing Shekou* 見證蛇口. Huacheng Publishing House廣州: 花城出版社.

Zhang, Y. B. (Eds.) 張一兵 校 (2006). *Three editions of old chronicles of Shenzhen*深圳舊志三種. Haitian Publishing House深圳:海天出版社.

Zhang, K. K. 張克科 (2010b). Witnessing the Advancement of Shenzhen's High-Tech Industries 見證深圳高新技術產業的前行步履. In Shenzhen Municipal Political Consultative Conference History and Learning Committee深圳市政協文史和學習委員會(Eds.). *The Rise of Shenzhen's Four Pillar Industries: High-Tech*深圳四大支柱產業的崛起 高新技術. Chinese history and literature Publishing House.北京:中國文史出版社. (pp. 143).

Zhang, M. J. 張明傑 (2010c). *The initial development of Jiang Zemin and China's economic special zones*江澤民與我國經濟特區的初創. Literature of Chinese Communist Party黨的文獻, (137), 23–28+120.

Zhang, Y. F. 張燕方. (2014). *Spatiotemporal images: Works from Zheng Zhongjian Photography Exhibition*影像時空:鄭中健攝影展作品集. Haitian Publishing House深圳:海天出版社.

Zhang, W., & Le, Y. C. 張偉,樂宜春 (2019). Shenzhen Port City: Spatial relationship transformation process and coordinated development strategy深圳港城空間關係變遷歷程與協調發展策略. In Urban transportation planning academic committee of China urban planning society 中國城市規劃學會城市交通規劃學術委員會 (Eds.). *Proceedings of the 2019 China urban transportation planning annual conference 2019*年中國城市交通規劃年會論文集. (pp. 10).

Zhang, Y., Wen, Y., Guo, W., et al 張玉閣, 文雅靖,郭萬達 (2022). *Construction of the Greater Bay Area and Forty Years of Shenzhen-Hong Kong Cooperation* 大灣區建設與深港合作四十年. Social Sciences Academic Press 社會科學文獻出版社.

Zhao, C. 趙川 (2002). *Ten major areas of high-tech industries in Shenzhen* 深圳高新技術產業的十大片區. Haitian Publishing House 深圳:海天出版社.

Zhao, C. 趙辰 (2007). Nationalism and classicism: The contradictions and tragedies in Liang Sicheng's architectural theoretical system民族主義與古典主義——梁思成建築理論體系的矛盾性與悲劇性. In *Misunderstanding of the "Facade": Architecture, theory, history*"立面"的誤會:建築·理論·歷史. Joint Publishing北京:生活·讀書·新知三聯書店.(pp. 39–40).

Zhong, J., & Guo, M. J. 鐘堅 郭茂佳 (2010a). *Documentation and Materials on Chinese Special Economic Zones* (Vol. 2) 中國經濟特區文獻資料(第2輯). Social Sciences Academic Press北京:社會科學文獻出版社.

Zhong, J., & Guo, M. J. 鐘堅 郭茂佳 (2010b). *Documentation and Materials on Chinese Special Economic Zones* (Vol. 1) 中國經濟特區文獻資料(第1輯). Social Sciences Academic Press北京:社會科學文獻出版社.

Zhou, D. 周鼎 (1987). A retrospective on urban planning and construction in Shenzhen深圳城市規劃和建設的回顧. In Hai Tian Publishing House 海天出版社. *Overall review of the general plan of Shenzhen Special Economic Zone*深圳經濟特區總體規劃評論集. Hai Tian Publishing House深圳:海天出版社.(pp. 19).

Zhu, S. X. 朱士秀 (1999). The Arrival of the Golden Phoenix: A Documentary on the Site Selection of Shekou Industrial Zone 金鳳飛來兮——蛇口工業區選址紀實. In Witness Shekou Compilation Committee (Eds.) 見證蛇口編委會 編, *Witnessing Shekou*見證蛇口. Huacheng Publishing House 廣州: 花城出版社.

Zou, H. 鄒暉 (2012). *Fragments and mirroring: The twofold discourse of comparative architecture*碎片與比照 比較建築學的雙重話語. The Commercial Press北京: 商務印書館.
Zukin, S. (1995). *The cultures of cities*. Blackwell.

## Yearbook, Local Chronicles and Local Literature

Bao'an County Local Chronicles Compilation Committee寶安縣地方誌編纂委員會 (1997). *Chronicles of Bao'an County*寶安縣誌. Guangdong People's Publishing House廣州: 廣東人民出版社.
Bujizhen Chronicles Compilation Committee布吉鎮志編纂委員會 (2009). *Buji Town Chronicle*布吉鎮志. Bujizhen Chronicles Compilation Committee深圳:布吉鎮志編纂委員會.
Caiwuwei Genealogy Compilation Committee蔡屋圍族譜編纂委員會 (2005). *Caiwuwei Genealogy of Shenzhen*深圳蔡屋圍族譜. Internal document.
Chen, H. 陳宏 (2006b). *1979–2000: Significant decisions and civil observations of Shenzhen* 1979–2000 深圳重大決策和時間民間觀察. Changjiang Literature and Art Publishing House 武漢: 長江文藝出版社.
Chen, B. A.陳秉安 (2010). *The Great Escape to Hong Kong*大逃港. Guangdong People's Publishing House 廣州: 廣東人民出版社.
Chen, H. 陳宏 (2011b). *Genealogy of Shenzhen indigenous people*深圳原住民家譜Shenzhen Press Group Publishing House深圳:深圳報業集團出版社.
Compilation Committee of Local Records of Futian District 深圳市福田區地方誌編纂委員會. Shenzhen City. *Futian District Annals 1979–2003*深圳市福田區志1979–2003. Fangzhi Publishing House北京:方志出版社.
Compilation Committee of Shangmeilin Village Chronicle《上梅林村志》編纂委員會 (2007). *Shangmeilin Village Chronicle*上梅林村志. Southern Publishing House海口: 南方出版社.
Editorial Committee of "Shenzhen Real Estate Yearbook."《深圳房地產年鑒》編委會 (2001). *Shenzhen real estat yearbook 2001*深圳房地產年鑒2001. Haitian Publishing House深圳:海天出版社.
Editorial Committee of the "Plan for the Coordinated Development of the PRD metropolitan area."《珠江三角洲城鎮群協調發展規劃》編委會 (2007). *Pearl river delta urban agglomeration coordinated development plan*珠江三角洲城鎮群協調發展規劃: 2004–2012. China Architecture & Building Press北京:建築工業出版社.
Guangdong Provincial Local Chronicles Compilation Committee 廣東省地方市志編纂委員會 (1998). *Guangdong annals: Forestry annals* 廣東省志·林業志. Guangdong People's Publishing House 廣州: 廣東人民出版社.
Guangdong Provincial Local Chronicles Compilation Committee 廣東省地方市志編纂委員會 (2000a). *Chronicles of Guangdong province: Public security* 廣東省志·公安志. Guangdong People's Publishing House 廣州: 廣東人民出版社.
Guangdong Provincial Local Chronicles Compilation Committee 廣東省地方市志編纂委員會 (2000b). *Chronicles of Guangdong province: Maritime and Island chronicles* 廣東省志·海洋與海島志. Guangdong People's Publishing House 廣州: 廣東人民出版社.
Guangdong Provincial Local Chronicles Compilation Committee 廣東省地方市志編纂委員會 (2000c). *Chronicles of Guangdong province: Resources and environment volume* 廣東省志資源.環境卷. Guangzhou: Guangdong People's Publishing House 廣州: 廣東人民出版社.
Guangzhou Institute of Geography廣州地理研究所 (1983). *Shenzhen landforms*深圳地貌. Guangdong Science and Technology Press廣州:廣東科技出版社.
Guangzhou Institute of Geography廣州地理研究所 (1985). *Atlas of natural resources and economic development in Shenzhen*深圳市自然資源與經濟開發圖集. Science Press北京: 科學出版社.

He, F. 何發. Counting the trees on which hill even with closed eyes閉著眼睛都能數出哪個山頭種了什麼樹. In Shenzhen Urban Planning and Natural Resources Bureau, Shenzhen Forestry Bureau深圳市規劃和自然資源局 深圳市林業局 (Eds.), *Memory forest: Oral history of Shenzhen forestry*記憶森林 深圳林業口述史 (1980–2020). n.d. Unpublished internal document.

Huang, Z. G. 黄鎮國 (1984). *Shenzhen's natural resources and economic development*深圳自然資源與經濟開發. Guangzhou Institute of Geography廣州:廣州地理研究所.

Huang, L. M. (Eds.) 黄麗滿 編 (1997). *Establishment and Development of the Special Economic Zone in Shenzhen (Shenzhen Volume)* 深圳市中國經濟特區的建立與發展(深圳卷). Party History Press of the Communist Party of China北京:中共黨史出版社.

Peng, Q. M.彭全民 (2019). *Shenzhen Huangbeiling Village Zhang Family Genealogy*深圳黄貝嶺村張氏族譜. Haitian Publishing House深圳:海天出版社.

Publicity Department of the Shenzhen Municipal Committee of the Communist Party of China, Shenzhen Press Group中共深圳市委宣傳部,深圳報業集團 (2001). *Clear and smooth—The record of the initial battle in 2003* 淨暢寧——2003初戰實錄. Haitian Publishing House 深圳:海天出版社.

Shenzhen Construction Bureau深圳市建設局 (1989). *Urban development strategy of Shenzhen*深圳市城市發展策略.

Shenzhen Development and Reform Commission 深圳市發展和改革委員會 (2017). *Compilation of Shenzhen's five-year plans*深圳市歷次五年規劃彙編. Haitian Publishing House深圳:海天出版社.

Shenzhen Futian District Historical Records Office深圳市福田區史志辦公室 (2012). *Futian tracing the roots* 福田溯源. Haitian Publishing House 深圳:海天出版社.

Shenzhen Futian District Local Records Compilation Committee. 深圳市福田區地方誌編纂委員會 (2012). *Futian district annals, Shenzhen 1979–2003*深圳市福田區志1979–2003. Fangzhi Publishing House. 北京:方志出版社.

Shenzhen Honghu Park Management Office深圳市洪湖公園管理處 (2009). *Lotus*荷花. China Science and Education Publishing House北京:中國科教出版社.

Shenzhen Local Chronicles Compilation Committee深圳市地方誌編纂委員會 (2006). *Chronicles of Shenzhen: Volume on legal system and administrative affairs*深圳市志·法制政務卷. Fang Zhi Publishing House北京: 方志出版社.

Shenzhen Local Chronicles Compilation Committee 深圳市地方誌編纂委員會 (2014). *Shenzhen Chronicles: Volume on reform and opening-up*深圳市志·改革開放卷. Fangzhi Publishing House北京:方志出版社.

Shenzhen Luohu District Chronicle Editorial Committee Office深圳市羅湖區志編委會辦公室 (2012). *Nanshan district chronicle of Shenzhen*深圳市羅湖區志. Fangzhi Publishing House 北京:方志出版社.

Shenzhen Municipal Chronicles Compilation Committee 深圳市地方誌編纂委員會. (2009a). *Chronicles of Shenzhen City: Volume on political parties and political power*深圳市志·政黨政權卷. Fangzhi Publishing House 北京: 方志出版社.

Shenzhen Municipal Chronicles Compilation Committee深圳市地方誌編纂委員會. (2009b). *Chronicles of Shenzhen: Volume on infrastructure construction*深圳市志·基礎建設卷. Fangzhi Publishing House北京: 方志出版社.

Shenzhen Municipal Chronicles Compilation Committee深圳市地方誌編纂委員會. (2014). *Chronicles of Shenzhen: Volume on the tertiary industry*深圳市志·第三產業卷. Fangzhi Publishing House北京: 方志出版社.

Shenzhen Municipal Chronicles Office深圳市史志辦公室 (Eds.). (1997). *Establishment and Development of China's special economic zones: Shenzhen volume*中國經濟特區的建立和發展:深圳卷. Communist Party History Publishing House北京:中共黨史出版社.

Shenzhen Municipal Chronicles Office深圳市史志辦公室 (2020). *Overview of Shenzhen Villages, Volume II: Futian Nanshan Volume*深圳村落概覽·第二輯: 福田南山卷. South China University of Technology Press廣州:華南理工大學出版社.

Shenzhen Municipal History and Chronicles Office深圳市史志辦公室 (1996). *Brief Records of the Nineteen Towns in Shenzhen*深圳市十九鎮簡志. Haitian Publishing House深圳:海天出版社.
Shenzhen Municipal Local Chronicles Compilation Committee 深圳市地方誌編纂委員 (2014a). *Chronicles of Shenzhen: Volume on legal system and administrative affairs*深圳市志·法制政務卷. Fangzhi Publishing House 北京: 方志出版社.
Shenzhen Municipal Local Chronicles Compilation Committee深圳市地方誌編纂委員會 (2014b). *Chronicles of Shenzhen: Volume on infrastructure construction*深圳市志·基礎建設卷. Fangzhi Publishing House北京: 方志出版社.
Shenzhen Municipal Political Consultative Conference Cultural深圳市政協文化文史和學習委員會. Historical, and Learning Committee (Eds.) (2010a). *The Rise of Shenzhen's four main industries: High-Tech*深圳四大支柱產業的崛起·物流. China WenShi Publishing House北京:中國文史出版社.
Shenzhen Municipal Political Consultative Conference Cultural深圳市政協文化文史和學習委員會. Historical, and Learning Committee (2010b). *The Rise of Shenzhen's four pillar industries: High-Tech*深圳四大支柱產業的崛起·高新技術. China WenShi Publishing House北京:中國文史出版社.
Shenzhen Museum深圳博物館 (2019). *Historical changes in the administration boundary (second line) of Shenzhen special economic zone*深圳經濟特區管理線(二線)歷史變遷研究. Haitian Publishing House深圳:海天出版社.
Shenzhen Nanshan District Chronicle Editorial Committee Office深圳市南山區區志編委會辦公室 (2012). *Nanshan District Chronicle of Shenzhen*深圳市南山區志. Fangzhi Publishing House北京:方志出版社.
Shenzhen Real Estate Yearbook Editorial Committee《深圳房地產年鑒》編輯委員會 (1991). *Shenzhen real estate yearbook 1991*深圳房地產年鑒1991. Haitian Publishing House深圳:海天出版社.
Shenzhen Special Economic Zone Research Association深圳經濟特區研究會 (2010). *A special history on the reform and opening-up of Shenzhen special economic zone*深圳經濟特區改革開放專題史. Haitian Publishing House深圳:海天出版社.
Shenzhen Special Economic Zone Yearbook Editorial Committee 深圳經濟特區年鑒編輯委員會 (1986). *Shenzhen special economic zone yearbook 1985*深圳經濟特區年鑒1985. Shenzhen Special Zone Yearbook Press深圳:深圳特區年鑒社.
Shenzhen Special Economic Zone Yearbook Editorial Committee 深圳經濟特區年鑒編輯委員會 (1996). *Shenzhen special economic zone yearbook 1996*深圳經濟特區年鑒1996. Shenzhen Special Zone Yearbook Press深圳:深圳特區年鑒社.
Shenzhen Statistical Bureau深圳市統計局 (2021). *Shenzhen statistical yearbook 2021*深圳統計年鑒2021. China Statistics Press北京:中國統計出版社.
Shenzhen Statistical Information Bureau 深圳市統計資訊局 (1999). *Shenzhen statistical yearbook 1999*深圳統計資訊年鑒1999. China Statistical Publishing House 北京:中國統計出版社.
Shenzhen Statistical Information Bureau深圳市統計局 (2001). *Shenzhen statistical yearbook 2001*深圳統計年鑒2001. China Statistical Publishing House北京:中國統計出版社.
Shenzhen Statistical Information Bureau深圳市統計局 (2003). *Shenzhen statistical yearbook 2003*深圳統計年鑒2003. China Statistical Publishing House北京:中國統計出版社.
Shenzhen Statistical Information Bureau深圳市統計局 (2005). *Shenzhen statistical yearbook 2005*深圳統計年鑒2005. China Statistical Publishing House北京:中國統計出版社.
Shenzhen Statistical Information Bureau深圳市統計局 (2011). *Shenzhen statistical yearbook 2011*深圳統計年鑒2011. China Statistical Publishing House北京:中國統計出版社.
Shenzhen Urban Management and Comprehensive Law Enforcement Bureau深圳市城市管理和綜合執法局. (2022). *Stories of Shenzhen Parks*深圳公園的故事. Publishing House of Shenzhen Press Group深圳:深圳報業集團出版社.
Shenzhen Urban Planning and Land Development Research Center深圳市規劃國土發展研究中心. (2019). *Shenzhen land resources*深圳市土地資源. Science Press北京:科學出版社.

Shenzhen Urban Planning and Land Resources Commission深圳市規劃和國土資源委員會. (2011). *Guiding urban transformation through transformation planning – Shenzhen urban master plan*轉型規劃引領城市轉型——深圳市城市總體規劃 (2010–2020). China Architecture & Building Press北京:中國建築工業出版社.

Shenzhen Urban Planning Bureau 深圳市規劃局 (2007). *Shenzhen 2030: Toward a pioneer metropolis based on sustainable development*建設可持續發展的全球先鋒城市 深圳2030城市發展策略. China Architecture & Building Press北京:中國建築工業出版社.

Shenzhen Yearbook Compilation Committee深圳年鑒編纂委員會 (1997). *Shenzhen yearbook 1997*深圳年鑒社1997. Shenzhen Yearbook Press深圳: 深圳年鑒社.

Shenzhen Yearbook Compilation Committee深圳年鑒編纂委員會 (2005). *Shenzhen yearbook 2005*深圳年鑒2005. Shenzhen Yearbook Press深圳: 深圳年鑒社.

Tao, Y. T., & Lu, Z. G. 陶一桃 魯志國 (2008). *Historical discourse on China's economic special zones*中國經濟特區史論. Social Sciences Literature Press北京:社會科學文獻出版社.

Urban Planning and Natural Resources Bureau of Shenzhen深圳市規劃和自然資源局*Memories of Land System Reform in Shenzhen*深圳土地制度改革記憶. Unpublished. n.d.

Urban Planning, Land and Resources Commission of Shenzhen Municipality, Shenzhen Center for Design 深圳市規劃和國土資源委員會 深圳市城市設計促進中心 (2014). *Collection of Shenzhen bay cloud city international design competition entries* 深圳灣雲城市國際競賽作品集. Lingnan Fine Arts Publishing House 廣州: 嶺南美術出版社.

Wang, X. M.王曉明 (1997). Hong Kong Urban Garden Greening Funds and Utilization香港城市園林綠化經費及使用. In Shenzhen Municipal Foreign Training Leadership Group深圳市涉外培訓領導小組 (Eds.), *Research series on international successful urban management experiences: Hong Kong Urban management studies*國際性城市成功管理經驗研究叢書:香港城市管理研究.

Wen, W.文維 (2021). Building Shenzhen into a National and Global Maritime Center City and Hub City將深圳打造成國家乃至全球海洋中心城市和樞紐城市 In. Shenzhen Urban Planning and Natural Resources Bureau深圳市規劃和自然資源局 (Eds.). *Forty years of Shenzhen special economic zone: Oral history interviews on planning and land use*深圳經濟特區四十年規劃國土口述歷史訪談錄. Unpublished.

Wu, Z. J. 吳子俊 (Eds.). (2005). *Shenzhen city sorting action chronicle: Adjusting the direction of urbanization with a scientific development perspective* 深圳市梳理行動紀實:用科學發展觀調校城市化發展方向. Haitian Publishing House 深圳:海天出版社.

Wu, J. X.吳錦雄 (Eds.). (2016). *Village histroy and intangible cultural heritage of Futian District*深圳市福田區村史與非遺文化. Shenzhen Huifengtang Cultural Communication Co., Ltd 深圳: 深圳市惠風堂文化傳播有限公司.

Yan, F. Y. 顏鳳迎 (2018). Special zone window: Three lives and three worlds of Shekou Port特區視窗:蛇口港的三生三世. In Y. H. Xu & W. Liu許永軍 劉偉 (Eds.), *Shekou: The place where dreams begin – paying tribute to 40 years of reform and opening-up*蛇口 夢開始的地方——致敬改革開放40年. People's Daily Press北京:人民日報出版社.

Zan, Q. J., Tan, F. Y., et al. 昝啟傑,譚鳳儀 等著 (2016). *Demonstration and evaluation on ecological restoration of Wetland in Overseas Chinese Town*. 華僑城濕地生態修復示範與評估. Ocean Press北京: 海洋出版社.

Zeng, J. T. 曾錦棠 (2009). *Old news of Shenzhen*深圳舊聞. Shenzhen Press Group Publishing House深圳:深圳報業集團出版社.

Zhang, J. 張軍 (1999b). Grand projects bringing benefits to descendants: Reflections on the Shenzhen ecological landscape forest construction project 功在當代福蔭子孫的巨集偉工程——深圳市生態風景林建設工程隨想. In: Wei, X. Q. 魏小琴 (Eds.), *A Covenant of the century: Collected works on the construction of ecological landscape forests in Shenzhen* 世紀之約:深圳市生態風景林建設文集. China Forestry Publishing House 北京: 中國林業出版社.

Zhou, G. M.周光明 (1999). Preface. 前言 In Wei X. Q. 魏小琴 (Eds.). *A covenant of the century: Collected works on the construction of ecological landscape forests in Shenzhen*世紀之約:深圳市生態風景林建設文集China Forestry Publishing House北京:中國林業出版社.